Using the American Community Survey

Benefits and Challenges

Panel on the Functionality and Usability of Data
from the American Community Survey

Constance F. Citro and Graham Kalton, *Editors*

Committee on National Statistics

Division of Behavioral and Social Sciences and Education

NATIONAL RESEARCH COUNCIL
OF THE NATIONAL ACADEMIES

THE NATIONAL ACADEMIES PRESS
Washington, D.C.
www.nap.edu

THE NATIONAL ACADEMIES PRESS 500 Fifth Street, N.W. Washington, DC 20001

NOTICE: The project that is the subject of this report was approved by the Governing Board of the National Research Council, whose members are drawn from the councils of the National Academy of Sciences, the National Academy of Engineering, and the Institute of Medicine. The members of the committee responsible for the report were chosen for their special competences and with regard for appropriate balance.

The project that is the subject of this report was supported by contract number YA132304CN0006 between the National Academy of Sciences and the U.S. Census Bureau. Support of the work of the Committee on National Statistics is provided by a consortium of federal agencies through a grant from the National Science Foundation (award number SBR-0453930). Any opinions, findings, conclusions, or recommendations expressed in this publication are those of the author(s) and do not necessarily reflect the views of the organizations or agencies that provided support for the project.

Library of Congress Cataloging-in-Publication Data

Using the American community survey : benefits and challenges / Panel on the Functionality and Usability of Data from the American Community Survey, Constance F. Citro and Graham Kalton, editors ; Committee on National Statistics, Division of Behavioral and Social Sciences and Education.

p. cm.

Includes bibliographical references and index.

ISBN-13: 978-0-309-10672-6 (pbk. perfect bound : alk. paper)

ISBN-10: 0-309-10672-9 (pbk. perfect bound : alk. paper) 1. American community survey. 2. Household surveys—United States. 3. United States—Census. I. Citro, Constance F. (Constance Forbes), 1942- II. Kalton, Graham. III. National Research Council (U.S.). Panel on the Functionality and Usability of Data from the American Community Survey. IV. National Research Council (U.S.). Committee on National Statistics.

HA37.U55U85 2007

317.3—dc22

2007024090

Additional copies of this report are available from the National Academies Press, 500 Fifth Street, NW, Washington, DC 20001; (800) 624-6242 or (202) 334-3313 (in the Washington metropolitan area); Internet, http://www.nap.edu

Printed in the United States of America

Suggested citation: National Research Council. (2007). *Using the American Community Survey: Benefits and Challenges.* Panel on the Functionality and Usability of Data from the American Community Survey, Constance F. Citro and Graham Kalton, Editors. Committee on National Statistics, Division of Behavioral and Social Sciences and Education. Washington, DC: The National Academies Press.

THE NATIONAL ACADEMIES

Advisers to the Nation on Science, Engineering, and Medicine

The **National Academy of Sciences** is a private, nonprofit, self-perpetuating society of distinguished scholars engaged in scientific and engineering research, dedicated to the furtherance of science and technology and to their use for the general welfare. Upon the authority of the charter granted to it by the Congress in 1863, the Academy has a mandate that requires it to advise the federal government on scientific and technical matters. Dr. Ralph J. Cicerone is president of the National Academy of Sciences.

The **National Academy of Engineering** was established in 1964, under the charter of the National Academy of Sciences, as a parallel organization of outstanding engineers. It is autonomous in its administration and in the selection of its members, sharing with the National Academy of Sciences the responsibility for advising the federal government. The National Academy of Engineering also sponsors engineering programs aimed at meeting national needs, encourages education and research, and recognizes the superior achievements of engineers. Dr. Charles M. Vest is president of the National Academy of Engineering.

The **Institute of Medicine** was established in 1970 by the National Academy of Sciences to secure the services of eminent members of appropriate professions in the examination of policy matters pertaining to the health of the public. The Institute acts under the responsibility given to the National Academy of Sciences by its congressional charter to be an adviser to the federal government and, upon its own initiative, to identify issues of medical care, research, and education. Dr. Harvey V. Fineberg is president of the Institute of Medicine.

The **National Research Council** was organized by the National Academy of Sciences in 1916 to associate the broad community of science and technology with the Academy's purposes of furthering knowledge and advising the federal government. Functioning in accordance with general policies determined by the Academy, the Council has become the principal operating agency of both the National Academy of Sciences and the National Academy of Engineering in providing services to the government, the public, and the scientific and engineering communities. The Council is administered jointly by both Academies and the Institute of Medicine. Dr. Ralph J. Cicerone and Dr. Charles M. Vest are chair and vice chair, respectively, of the National Research Council.

www.national-academies.org

PANEL ON THE FUNCTIONALITY AND USABILITY OF DATA FROM THE AMERICAN COMMUNITY SURVEY

GRAHAM KALTON (*Chair*), Westat, Rockville, MD
PAUL P. BIEMER, RTI International, Research Triangle Park, NC
NANCY DUNTON, School of Nursing, University of Kansas, Kansas City
MARTIN R. FRANKEL, Zicklin School of Business, Baruch College, New York
D. TIM HOLT, University of Southampton, United Kingdom (emeritus)
SHARON LOHR, Department of Mathematics and Statistics, Arizona State University, Tempe
CHARLES L. PURVIS, Metropolitan Transportation Commission, Oakland, CA
JOSEPH J. SALVO, New York City Department of City Planning
HAL S. STERN, Department of Statistics, University of California, Irvine

CONSTANCE F. CITRO, *Co-Study Director*
MICHAEL L. COHEN, *Co-Study Director*
DANIEL L. CORK, *Senior Program Officer*
BARBARA A. BAILAR, *Consultant*
F. JAY BREIDT, *Consultant*
MEYER ZITTER, *Consultant*
AGNES E. GASKIN, *Senior Program Assistant*

COMMITTEE ON NATIONAL STATISTICS
2006-2007

Acknowledgments

The Panel on the Functionality and Usability of Data from the American Community Survey (ACS) wishes to thank the many people who contributed to the panel's work.

As the sponsor of the project, the U.S. Census Bureau—under the leadership of director Louis Kincannon and then-deputy director Hermann Habermann—provided consistent and strong support as we reviewed the utility of ACS estimates and data products and related issues. As associate director for decennial census, and in his new role as deputy director, Preston J. Waite set the basic direction for the 2010 census and the replacement of the traditional census long-form sample with the new ACS; he provided considerable advice during the panel's meetings and also served as a discussant at a session at the 2006 Joint Statistical Meetings in Seattle describing the panel's work. The communication between the panel and Census Bureau throughout the study was greatly facilitated by the efforts of Philip Gbur as contracting officer and David Hubble (now of Westat) as lead technical liaison. Both were always readily accessible and extremely helpful in providing answers to questions. Before his retirement from the Census Bureau, Rajendra Singh ably assisted in interactions with the panel, for which we are grateful.

Further, a number of Census Bureau staff made very informative presentations to the panel or provided useful materials, including Teresa Angueira, Lisa Blumerman, Robert Fay, Deborah Griffin, Douglas Hillmer, David Hubble, Lawrence McGinn, J. Gregory Robinson, and Signe Wetrogan. We are also greatly indebted to Mark Asiala, Alfredo Navarro, Michael

Starsinic, and Lynn Weidman for answering detailed methodological questions and for carrying out computations that were requested by the panel. Finally, Census Bureau staff made available an early draft of a technical report on ACS design and methodology (U.S. Census Bureau, 2006, see http://www.census.gov/acs/www), which the panel found very useful in its work.

Unfortunately, our panel did not have the ability to directly interact with the late Charles (Chip) Alexander, the key architect of the data collection program that would become the ACS. His death in 2003 left a void that the Census Bureau—indeed, the entire federal statistical system—still struggles to fill. Throughout our work, though, we have benefited from the ideas embodied in his writings and greatly appreciate them.

As consultant to the panel, F. Jay Breidt (Department of Statistics, Colorado State University) prepared two extremely useful papers and associated presentations on alternative estimands from the ACS multiyear data and the use of population controls for ACS estimates at various levels of aggregation. These papers helped the panel develop its ideas on these important topics, and we are greatly pleased to include them as appendixes to this report.

In April 2005, the panel convened a special meeting on user perspectives, emphasizing the current uses of census long-form-sample data by state and local organizations, as well as the media, and the prospects for use of ACS data by these constituencies. Panel members Nancy Dunton, Chuck Purvis, and Joe Salvo were particularly instrumental in assembling this group. We thank the participants in that meeting for their time and insightful comments: Sarah Breshears (State Data Center, University of Arkansas at Little Rock), Warren Brown (Cornell Institute for Social and Economic Research), Nathan Erlbaum (New York State Department of Transportation), Linda Gage (California Department of Finance, Demographic Research Unit), Jeff Hardcastle (University of Nevada, Reno), John McHenry (Demographic Data for Decision-Making, Inc.), Paul Overberg (*USA TODAY*), Richard Rathge (Department of Agribusiness and Applied Economics, North Dakota State University), Ed Schafer (San Diego Association of Governments), and David Swanson (Department of Sociology and Anthropology, University of Mississippi). Subsequent discussions with Gage, Hardcastle, and Overberg were also very helpful to the panel.

Over the course of our regular meetings, we benefited from presentations and discussions from a wide variety of data users from federal and state agencies. We thank Chris Chapman (National Center for Education Statistics), George Hough (Oregon State Data Center and Portland State University), Sandra Mason (Bureau of Labor Statistics), Elaine Murakami (Federal Highway Administration), Thomas Nardone (Bureau of Labor Statistics), Donald Oellerich (Office of the Assistant Secretary for Planning and Evaluation, U.S. Department of Health and Human Services), William O'Hare (Annie E. Casey Foundation), Alan Pisarski (consultant), Ann Poliska (Bureau of Labor Statistics), Susan Schechter Bortner (then of the U.S. Office of Management and Budget), Marilyn Seastrom (National Center for Education Statistics), and Ronald Sepanik (U.S. Department of Housing and Urban Development). Another of our panel meetings included presentations by Jay Breidt on his commissioned papers and very helpful comments and suggestions from four discussants: David Binder (Statistics Canada, retired), Wayne Fuller (Iowa State University), Eric Slud (University of Maryland), and Alan Zaslavsky (Harvard Medical School Department of Health Care Policy). We also appreciate the comments offered by Allen Schirm (Mathematica Policy Research, Inc.) as a discussant on the panel's work at a session of the 2006 Joint Statistical Meetings.

Our panel was one of three concurrent panel studies conducted by the Committee on National Statistics (CNSTAT) on aspects of the decennial census. Though the panels covered substantially different subject areas, we benefited from interactions with members of our sister panels, the Panel on Residence Rules in the Decennial Census and the Panel on Correlation Bias and Coverage Measurement in the 2010 Census. Paul Voss (Department of Rural Sociology, University of Wisconsin-Madison, emeritus) merits particular credit; during his period of service as chair of the residence rules panel, he also participated in several of our panel's meetings (including the April 2005 special meeting on user perspectives) and provided very important comments throughout the process.

The panel is especially indebted to Constance Citro, director of CNSTAT, who drafted most of the chapters of the report and provided much of the insight on the use of census long-form-sample data products for which the ACS is serving as a replacement. Her wide experience on census issues, her extremely clear writing style, and the clarity of her reasoning were of essential importance to the panel's success. Michael Cohen, assisted by Daniel Cork and Meyer Zitter, organized the work of the panel's meetings and interactions with Census Bureau staff, data users, and others. Barbara Bailar, serving as consultant to the panel, took the lead in the initial drafting of two chapters of the report. Christine McShane provided expert technical editing of the draft report. Finally, Agnes Gaskin provided all of the administrative support for the panel, smoothly arranging travel and meetings, including two off-site meetings, and Bridget Edmonds assisted in preparation of the manuscript.

In addition to a session at the 2006 Joint Statistical Meetings, the panel made use of other opportunities to discuss the general nature of its work and to solicit ideas. In particular, we benefited from interaction with the Association of Public Data Users; at their 2004 annual meeting, members of the panel discussed general themes and issues for its work and received extremely helpful feedback in response. An update was provided at the association's 2006 annual meeting. Similarly, we gained insight from comments by Andrew Reamer (The Brookings Institution) and others on presentations at a November 2006 Washington Statistical Society seminar and at meetings of the Federal State Cooperative Program for Population Estimates (FSCPE). We also appreciated the opportunity to mention the panel's work at a meeting of the Census Information Center/FSCPE/State Data Center steering committee in early 2007.

Most importantly, I am indebted to the members of the Panel on the Functionality and Usability of Data from the American Community Survey. They were extremely hard working, providing draft text and comments on several rounds of drafts on a difficult subject. Special kudos go to Tim Holt, who happily traveled across the Atlantic for the work of the panel, and Joe Salvo, who took the lead on sections of the report dealing with user education and applications of the data. He was assisted in the preparation of a case study (in Chapter 3 of the report) by Jennifer Jensen of the New York City Department of City Planning.

This report and appended papers have been reviewed in draft form by individuals chosen for their diverse perspectives and technical expertise, in accordance with procedures approved by the Report Review Committee of the National Research Council (NRC). The purpose of this independent review is to provide candid and critical comments that will assist the institution in making the published report as sound as possible and to ensure that the report meets institutional standards for objectives, evidence, and responsiveness to the study charge. The review comments and draft manuscript remain confidential to protect the integrity of the deliberative process.

We thank the following individuals for their participation in the review of the report or the papers: Katharine G. Abraham, Joint Program in Survey Methodology, University of Maryland; Patricia C. Becker, President, APB Associates, Southfield, MI; David Binder, Methodology Branch, Statistics Canada (retired); Manning Feinlieb, Bloomberg School of Public Health, Johns Hopkins University; Ken Hodges, Data Research and Development, Claritas Inc., Ithaca, NY; James Lepkowski, Institute for Survey Research, University of Michigan; Elaine Murakami, Office of Planning, Federal Highway Administration, U.S. Department of Transportation; Jean Opsomer, Department of Statistics and Center for Survey Statistics and Methodology, Iowa State University; Eric V. Slud, Department of Mathematics, University of Maryland; David A. Swanson, Department of Sociology and Anthropology and Center for Population Research, University of Mississippi; John H. Thompson, Executive Vice President, The National Opinion Research Center, Chicago, IL; and Alan M. Zaslavsky, Department of Health Care Policy, Harvard Medical School.

Although the reviewers listed above provided many constructive comments and suggestions, they were not asked to endorse the conclusions or recommendations nor did they see the final draft of the report or the papers before their release. The review of the report was overseen by Douglas Massey, Department of Sociology, Princeton University. Appointed by the NRC, he was responsible for making certain that an independent examination of the report was carried out in accordance with institutional procedures and that all review comments were carefully considered. Responsibility for the final content of this report rests entirely with the authoring panel and the institution.

Graham Kalton, *Chair*
Panel on the Functionality and
Usability of Data from the
American Community Survey

Contents

APPENDIXES

Tables

Boxes

Executive Summary

The American Community Survey (ACS), after a decade of testing, is a reality: the first set of ACS data products, released in August–November 2006, reports on the social, demographic, economic, and housing characteristics in 2005 of cities, counties, and other areas with 65,000 or more people. With the advent of the ACS, there will no longer be a long-form sample as part of the decennial census.

The Census Bureau asked a panel of the Committee on National Statistics to assess the usability of ACS data. The report advises users on making the transition from the long-form sample to the ACS. It identifies areas for research and development by the Census Bureau so that the ACS can realize its full potential to improve the nation's information on people and communities.

THE ACS IN BRIEF

The ACS has similar content on people and housing as the 2000 long-form questionnaire, but its design is different. It is a continuing monthly survey, in which sampled housing unit addresses receive a questionnaire each month, cumulating to about 2 million responding households each year and 10–11 million over 5 years. ACS data products are *period* estimates that average 12, 36, and 60 months of data, respectively, for 1-year, 3-year, and 5-year periods. In contrast, the 2000 long-form sample of over 16 million responding households pertained to a fixed time—Census Day, April 1.

The ACS has three major benefits compared with the long-form sample:

- The first benefit is timeliness: ACS data products are released 8–10 months, instead of 2 years, after data collection.
- The second, and an even more important, benefit is frequency: ACS data products are updated every year instead of every 10 years, which will make it possible in many areas to track trends in such important population characteristics as educational attainment, employment, poverty, diversity, and others.
- A third benefit is higher quality of the data in terms of completeness of response to the survey items: the much more complete response to the ACS compared with the 2000 long-form sample is achieved by the use of computer-assisted telephone and personal interviewing of households that do not respond by mail. The ACS interviewers are experienced and highly trained in contrast to the lightly trained temporary enumerators that were used for nonresponse follow-up in the 2000 census. In addition, ACS telephone interviewers contact mail respondents to obtain answers to missing items, a step not done in 2000.

A weakness of the ACS compared with the long-form sample is the significantly larger margins of error in ACS estimates, even when cumulated over 5 years. The primary reason is the much smaller sample size of the ACS. Another important reason is the greater variation in the ACS sample weights resulting from the subsampling for field interviewing of households not responding by mail or telephone. Also, the postcensal population and housing estimates used as survey controls are less effective than the full census controls used with the long-form sample: they are subject to unmeasured estimation error, they are applied at a less detailed level than the census controls, and they are not directly related to the ACS in the way that the census controls are related to the long-form sample.

The larger ACS sampling errors are a particular problem for small cities, counties, and other governmental jurisdictions; they also apply to small neighborhoods in large cities, but neighborhoods can often be combined satisfactorily into larger user-defined areas for analysis. For small areas for which 1-year period estimates are not available or sufficiently precise, users must learn to work with 3-year and 5-year period estimates, which are very different from point-in-time estimates.

The census long-form sample was heavily used by federal, state, and local government agencies, researchers, the private sector, the media, and the public. The ACS continuous design will initially challenge many such users in adapting their applications based on the long-form sample to the new data. Yet this design also provides the platform for important new applications that the long-form sample could not support.

This summary provides the panel's general guidelines for using the ACS for such applications as fund allocation, program planning by federal, state, and local governments, transportation modeling, private-sector decision making, research on population and housing trends, and general public understanding. It then presents the panel's recommendations to the Census Bureau for investment in the ACS and in user education and outreach that will be necessary to make the most effective use of the new data.

GENERAL GUIDELINES FOR ACS USE

The panel encourages users to follow the general guidelines below in working with the ACS period estimates.

a. Always examine margins of error before drawing conclusions from a set of estimates.
b. Review the available information about nonsampling errors for estimates of interest and use this information in interpreting findings from the ACS.
c. Carefully consider the pros and cons of alternative strategies for extracting value from ACS 5-year period estimates for very small areas, such as aggregating small-area estimates into estimates for larger, user-defined areas.
d. When using ACS data to estimate shares of some total, compare estimates among areas or population groups, or assess trends over time, use ACS estimates that pertain to the same time period (1-year, 3-year, or 5-year) for all geographic areas or population groups that are being compared. Do not use a mixture of different period estimates.
e. When analyzing trends over time for an area or population group, use ACS 1-year period estimates whenever they are available and sufficiently precise for the purpose of interest and be cognizant of changes in geographic area boundaries that may affect comparability. Keep in mind that the sampling error for the estimate of the difference between pairs of 1-year period estimates will be larger than the sampling error of either estimate.
f. If only 3-year or 5-year period estimates are available or sufficiently precise, use them with care for analyzing trends over time for an area or population group. In general, avoid analyses of changes over time that are based on overlapping period estimates (for example, 5-year period estimates for 2010–2014 and 2011–2015).

g. Take advantage of the availability of 1-year and 3-year period estimates for public use microdata areas, which include about 100,000 people, to assist with analyses for smaller areas.
h. Take care to label ACS estimates, including those for 1 year, 3 years, and 5 years, as *period* estimates.
i. Use ACS 3- and 5-year period estimates for income, housing value, and housing costs with care. To compensate for the differing time periods for which dollar amounts are collected, those amounts are adjusted to a common calendar year by the change in the national consumer price index. This inflation adjustment expresses all of the reported dollar amounts in a comparable manner with regard to purchasing power as of the most recent calendar year in the period. However, the resulting estimates should not be interpreted as current-year estimates.
j. Use care in comparing ACS estimates with estimates from other data sources, including the 2000 long-form sample and other surveys, and be cognizant of the differences that could affect the comparisons. Such differences may include population coverage, sample size and design, reference periods, residence rules, and interview modes.

OVERARCHING RECOMMENDATIONS

The panel strongly supports the ACS, but it does not underestimate the challenges facing the Census Bureau, which must produce a flood of data products every year, or the challenges facing the user community. The continuous ACS design will ultimately support not only current applications, but also new applications requiring innovative data products. However, there will be a learning curve. For a successful transition that leads to the full use of the ACS, the panel makes five overarching recommendations (identified by chapter number) to the Census Bureau on investment in the ACS, increasing the precision of ACS estimates, a user education and outreach program, priorities for research and development, and looking to the future.

Recommendation 7-1: The Census Bureau should continue to make sufficient funding of the ACS one of its top priorities. It should seek adequate funding on a continuing basis, not only for data collection and production, but also for ongoing programs of methodological research and evaluation and user outreach and education.

Recommendation 4-4: The Census Bureau should identify potential ways to increase the precision of ACS estimates for small geographic areas, particularly small governmental jurisdictions, through reallocation of the sample and through increases in the overall sample size. Cost savings should be sought to support such increases, although increases that could significantly improve the precision of estimates will require additional funding from Congress. Sample reallocation should also be considered to minimize anomalies across areas (for example, jurisdictions with very similar populations that fall into different sampling rate categories).

Recommendation 7-2: The Census Bureau should develop a comprehensive program of user education, outreach, and feedback for the ACS. Two goals of the program should be (1) to educate users in the basics of the ACS, how it differs from the census long-form sample and other data sources, and appropriate methods to use the data; and (2) to develop paths for systematic feedback from users to improve the training materials, identify potential problems with the data, and suggest ways to improve data products and documentation to maximize the utility of the data and facilitate data use.

Recommendation 7-9: The Census Bureau should assign priority to the following topics for research and development: sample size and allocation; the Master Address File (MAF); population controls; residence rules; estimates of change with multiyear averages; comparisons with other surveys and administrative records; and the development of automated tools for data quality review of ACS products.

Recommendation 7-10: As part of its research and development program for the ACS, the Census Bureau should dedicate a portion of resources to pursue innovative, longer term projects. While short-term research and development must focus on the ACS as a replacement for the census long-form sample, research must also address how the ACS can improve the nation's information on population and housing in ways that were not possible with the long-form sample and may not even be envisioned today.

ADDITIONAL RECOMMENDATIONS

The panel's additional recommendations to the Census Bureau address areas for research and development for the ACS, including the sample frame, data collection for housing units, sampling and data collection for group quarters, data products, data quality review, period estimation, and survey operations. These are followed by recommendations that address user education and outreach and data quality monitoring and improvement.

Sample Frame (Master Address File)

Recommendation 4-1: Given the centrality of the MAF to the ACS, the Census Bureau should ensure that adequate resources are provided to maintain the highest possible completeness and accuracy of MAF address information on a continuous basis.

Recommendation 4-2: The Census Bureau should plan now for programs to follow the 2010 census to ensure that the MAF is updated on a continuous basis more completely than is being done prior to 2010. These programs should include not only the current updates from the Delivery Sequence File and the Community Address Updating System, but also such initiatives as continuing local review, the use of ACS field interviewers to investigate address problems, and the use of address information from the Census Bureau's e-StARS database of linked administrative records.

Recommendation 4-3: The Census Bureau should support a continuing research program on the quality of the MAF and the cost-effectiveness of the various operations that are designed to update the MAF. This program should include periodic field checks on MAF addresses, comparisons with housing unit estimates for specific areas, comparisons with the e-StARS database, and comparisons with the results of the 2009 complete block canvass that will be used to prepare the 2010 census MAF. The program should also include studies of methods to improve the listing of small multiunit addresses in urban areas, characteristics of duplicate housing units, and characteristics of undeliverable mail addresses. In addition, the program should examine the effectiveness of the Community Address Updating System and explore ways to improve its performance.

Data Collection for Housing Units

Recommendation 4-5: The Census Bureau should conduct experimental research on the effects of the different data collection modes used in the ACS—mailout-mailback, computer-assisted telephone interviewing (CATI), and computer-assisted personal interviewing (CAPI)—on ACS estimates and, when possible, on response errors for questionnaire items. In addition, the Census Bureau should assess how different patterns of responding by mail, CATI, and CAPI among population groups and geographic areas affect comparisons of ACS estimates and inform data users of consequential differences.

Recommendation 4-6: The Census Bureau should conduct experiments to determine the extent to which ACS respondents give different answers to the decennial census usual residence rule and the ACS 2-month residence rule and the extent to which they apply the specific ACS residence rules (for example, reporting commuter workers at the family residence, applying the 2-month rule prospectively). To help clarify residence according to the census and ACS concepts, the experimental questionnaire should ask about other residences at which respondents spend time. The Census Bureau should assess the implications of the experimental results for ACS population estimates for different geographic areas and population groups. Depending on the results, the Census Bureau should consider appropriate changes in the ACS questionnaire instructions on residence or in the residence rules themselves.

Sampling and Data Collection for Group Quarters

Recommendation 4-7: The Census Bureau should discuss with data users their requirements for detailed information from the ACS for residents of institutions and other types of group quarters, particularly at the local level. The discussions should assess benefits against costs, and the results should be used to determine any changes to the group quarters component of the ACS—for example, the possible deletion of institutions from the ACS universe—that would be cost-beneficial for users and stakeholders.

Data Products—Confidentiality Protection

Recommendation 4-8: Because of the potential value of month of data collection for analysis of the ACS public use microdata samples, the Census Bureau should revisit its decision to omit this variable as a confidentiality protection measure. If further research determines that including exact month of data collection would significantly increase disclosure risk, the Census Bureau might consider perturbing the month of data collection or taking other steps to protect confidentiality. Similarly, the Census Bureau should consider developing selected summary tables that identify the season of collection (such as summer or winter) for geographic areas for which such information would be useful.

Recommendation 4-9: The Census Bureau should undertake research to develop confidentiality protection rules and procedures for tabulations from the ACS that recognize the protection afforded to respondents by pooling the data over many months. Whenever possible, the

Census Bureau should prefer confidentiality protection procedures that preserve the ability to aggregate smaller geographic areas into larger, user-defined areas.

Data Products—Collapsing Cells for Large Sampling Errors

Recommendation 4-10: The Census Bureau should monitor the extent of collapsing of cells that is performed in different tables to meet minimum precision standards of 1-year and 3-year period tabulations from the ACS and assess the implications for comparisons among geographic areas and over time. After sufficient information has been gleaned about the extent of data collapsing, and its impact on users, the Census Bureau, in consultation with data users, should assess whether its collapsing rules are sound or should be modified for one or more subject areas.

Data Products—Inflation Adjustments

Recommendation 4-11: The Census Bureau should provide users with a full explanation of its inflation adjustment procedures and their effects on multiyear ACS estimates of income, housing costs, and housing value. It should consult with users about other kinds of income and housing amount adjustments they may need and conduct research on appropriate estimation methods (for example, methods to produce latest-year amounts from multiyear averages). It should consider publishing selected multiyear averages in nominal dollars as well as inflation-adjusted dollars.

Data Products—Tabulation Specifications

Recommendation 4-12: If some or all group quarters residents continue to be included in the ACS, the Census Bureau should consult with users regarding the most useful population universe for tabulations, which, depending on the table, could be the entire population, the household and group quarters populations separately, or the noninstitutional and institutional populations separately.

Recommendation 4-13: The Census Bureau should consider expanding the geographic areas for ACS tabulations in order to afford users greater flexibility for aggregating small areas into larger user-defined areas. Two possibilities to investigate are to lower the population threshold for 1-year period estimates to, say, 50,000, and to produce 3-year (and possibly 1-year) period estimates for user-defined statistical subareas of large cities (aggregations of census tracts or block groups) and counties (aggregations of places and towns).

Data Quality Review

Recommendation 4-14: The Census Bureau should increase its research and development on automated tools and standardized procedures to facilitate timely review and quality control of the large volume of ACS data products.

Period Estimation

Recommendation 5-1: The Census Bureau should conduct an in-depth review of the weighting scheme used for producing ACS 1-year period estimates and assess a range of alternative schemes that might improve the quality of the estimates.

Recommendation 5-2: The Census Bureau should evaluate the quality of the postcensal housing unit estimates and the MAF sampling frame in relation to one another. In the light of this evaluation, the Census Bureau should assess the suitability of the current housing unit control factor adjustment and modify it as necessary.

The Census Bureau should attempt to identify areas in which improvements can be made to the postcensal housing unit estimates and to the MAF sampling frame. In particular, it should investigate an integrated approach for developing the postcensal housing unit estimates and for continuously updating the MAF that would benefit both and reduce the variability in the housing unit control factor.

Recommendation 5-3: As a high priority, the Census Bureau should undertake research to evaluate the effect of the postcensal population controls on ACS estimates and to examine alternative methods of making the adjustment that may be superior to the one currently used (including dispensing with the population controls entirely). The Census Bureau should make users aware in ACS documentation that biases in the ACS estimates caused by errors in the population controls are not reflected in the margins of error reported with the estimates and should conduct research to examine the effects of these errors on ACS estimates.

The Census Bureau should also give priority to research on ways to improve the postcensal population estimates at the county level, including estimates of internal migration and international immigration and the classification of race and ethnicity.

Recommendation 6-1: The Census Bureau should conduct research to examine the bias and variance properties of the planned multiyear weighting scheme and compare these properties with those of some alternative schemes.

Recommendation 6-2: The Census Bureau should consult users about the utility of the currently proposed multiyear period estimates—particularly for estimates of totals—for areas that change markedly in population size. It should investigate whether there are other forms of estimates that could be produced and would better serve user needs.

User Education and Outreach

Recommendation 7-3: As an integral part of its education, outreach, and feedback program for the ACS, the Census Bureau should establish a dedicated ACS user staff. That staff should partner with organizations that will assist end users, including the State Data Center network as a key partner and many other organizations and groups. The staff should work with the media to help them understand ACS data so that they can explain and showcase the value of the data to communities in an effective and accurate way.

Recommendation 7-4: The Census Bureau should establish an ongoing advisory group of experienced data users with whom to interact about user education materials, web site design, table content, and other aspects of the data products and education and outreach program for the ACS.

Data Quality Monitoring and Improvement

Recommendation 7-5: The Census Bureau, in collaboration with user education partners, should carry out research on ways to facilitate understanding of the quality measures provided on the ACS web site. The Census Bureau and its partners should also consider what additional quality indicators—for example, some of the indicators presented at a finer level of geographic detail—would be useful to provide for the 2005 ACS and subsequent 1-year period estimates and what indicators to provide for the 3-year and 5-year period ACS estimates when those become available.

Recommendation 7-6: The Census Bureau, in consultation with data users and statistical methodologists, should evaluate its presentation of sampling errors of estimates that are published on the ACS web site and its descriptions of methods for computing approximate estimates of sampling errors for estimates for which sampling errors are not published. Steps should be identified to improve the usability and ease of comprehension of information on sampling errors.

Recommendation 7-7: The Census Bureau should develop and publish an ongoing quality profile for the ACS to inform users of the survey's data quality, to guide the development of a continuing program of data quality assessments, and to identify areas for survey improvement. The Census Bureau should seek input from users on priority topics for assessment and design reports that they would find to be useful additions to the technical reports.

Recommendation 7-8: The Census Bureau should continue to seek funding with which to implement methods panels (large samples of households) for experimentation with questionnaire design, question wording, residence rules, data collection mode, and other features of the ACS. The methods panels should be conducted annually so that the survey can be kept current in meeting data needs and collecting responses in the most efficient and effective ways.

1

Introduction

For decades, policy makers, planners, researchers, the media, and the public have looked to the decennial census as the source of detailed information on the numbers and characteristics of the U.S. population for the nation, states, metropolitan areas, counties, cities, towns, school districts, and neighborhoods. The census provides complete counts of people by such basic characteristics as age, sex, race, and ethnicity for areas as small as city blocks. It has also provided estimates from a very large sample—called the long-form sample—for areas as small as groups of blocks on people's education, employment, income, disability, commuting, and other characteristics and about the housing in which they live. Other household surveys, such as the Current Population Survey, the Survey of Income and Program Participation, the American Housing Survey, and the National Health Interview Survey, provide more frequent, detailed information on a variety of topics, but estimates from these surveys are generally available only for the nation as a whole or, sometimes, for states or large metropolitan areas.

In late summer 2006, the U.S. Census Bureau released 2005 data from a major new continuous survey designed to provide small-area data, the American Community Survey (ACS). Each month, the ACS questionnaire—similar in content to the census long form—is mailed to 250,000 housing units across the nation; as with the long-form sample, response to the ACS is required by law. Two big differences from the long-form sample are that the ACS is conducted on a continuous basis instead of once every 10 years and the data are released every year. Over the course of time, the ACS will provide detailed data for all of the small areas covered by the long-form sample. With the advent of the ACS, the long-form sample will be dropped from the 2010 and successive censuses; consequently, the decennial census will now include only a short form with basic questions on age, race, sex, ethnicity, household relationship, and housing tenure (owner or renter).

Each summer and fall from 2006 forward, the Census Bureau will release ACS statistics for the previous calendar year for areas with 65,000 or more people. In addition, by 2008, enough responses will have accumulated over the 3-year period 2005–2007 for the Census Bureau to release statistics in the fall for areas with at least 20,000 people. Finally, by 2010, enough responses will have accumulated over the 5-year period 2005–2009 for the Census Bureau to release statistics in the fall for all areas, including very small places and neighborhoods. Each year, the 1-year, 3-year, and 5-year estimates will be updated to reflect newer data.[1]

The implementation of the ACS represents a seismic shift in the landscape of small-area data on the U.S. population. This shift promises important benefits to users in terms of much more timely and up-to-date information than the long-form sample could provide. However, there will inevitably be a learning curve and costs in time and other resources of users to make the transition from the once-a-decade long-form sample to the continuous ACS.

1-A PANEL CHARGE

Recognizing the need to assist users in the transition from the long-form sample to the ACS, in 2004 the Census Bureau asked the Committee on National Statistics of the National Academies to convene a Panel on the Functionality and Usability of Data from the American Community Survey. The charge to the panel was to study the effects of using small-area ACS estimates that are based on multiyear measurements released every year for applications that previously used static, one-time estimates from the long-form sample. The major goals of the panel's work are to provide a base of information to ease the transition from the long-form sample to the ACS for a wide variety of data uses and to explore methodological issues raised by the use of this survey.

The panel undertook a range of activities to respond to this charge. We listened to groups of small-area data users on several occasions, including at meetings we organized with major federal agency users and experienced state and local government users and at a special session of the October 2004 conference of the Association of Public Data Users. The panel also commissioned papers on the properties of different types of multiyear

[1]Similar data products will be available for areas in Puerto Rico from the Puerto Rico Community Survey (see Box 2-1).

estimates and the effects of using census-based population estimates as controls to smooth the ACS estimates. Subgroups of panel members met with Census Bureau staff to learn about as many aspects as possible of the ACS data collection and estimation process, including the construction of the Master Address File (MAF), from which ACS sample addresses are drawn; the mailing and nonresponse follow-up procedures; and the weighting, estimation, and data release procedures. From these activities and our deliberations, the panel developed the findings and recommendations presented in this report.

1-B HISTORICAL BACKGROUND

1-B.1 Evolution of the Long-Form Sample

The U.S. decennial census, conducted every 10 years beginning in 1790, serves a constitutionally mandated purpose to provide the number of people for each state for reapportionment of the U.S. House of Representatives. A closely linked purpose, to redraw the boundaries of congressional, state, and local legislative districts after each reapportionment, requires that the census provide data at the block level for demographic groups.

Beginning as early as 1820, the census obtained additional information beyond the basic head count on the characteristics of the population to respond to the needs of policy makers and the public for a better understanding of the growing new nation. Censuses from 1820 to 1860 included questions on such topics as school attendance, literacy, industry and occupation of employment, and citizenship. Censuses from 1870 through 1930 included a large number of questions asked of everyone.

The 1940 census saw the first application of newly developed methods for population sampling to reduce census costs and the burden on the public (Citro, 2000c). In this census, about one-sixth of the total of about 50 questions were asked of only a 5 percent sample. In 1950 about two-fifths of the total of about 50 questions were asked on a sample basis. This approach to obtaining a wide range of information about the population can be termed a "paired strategy," in which a sample survey that asks a large number of questions is embedded in an enumeration of the entire population on basic characteristics.

Prior to 1960, all census data were collected by census enumerators. The 1960 census saw the introduction of the mails to assist the enumeration, with separate short-form and long-form questionnaires. In this and subsequent censuses, the long forms contained the small number of questions asked of everyone—as on the short form—as well as additional questions asked of only a sample (see Citro, 2000b). Long-form-sample sizes varied across censuses. In the 2000 census, the long form was sent to about one-sixth of housing units overall (17 percent, or about 18 million housing units), although sampling rates varied from 13 to 50 percent depending on the population size of the area. Estimates from the long-form sample were released for areas as small as census tracts and block groups. Estimates for the entire population from the data collected for everyone were released at the individual block level, but this was not the case for long-form-sample estimates, both because the estimates were not sufficiently precise at that level and out of concern to protect individual confidentiality.

The long-form questions have changed over time to reflect changing needs for small-area data to implement federal legislation and administer federal programs. In addition to the basic items asked on the short form and depending on how one counts items with multiple parts, the 2000 census long form included 54 sample items about people, covering such topics as marital status, educational attainment, place of birth, citizenship, language spoken at home, English proficiency, ancestry, military service, year moved into residence, various types of disability, responsibility for grandchildren in the home, current and prior year employment status, occupation and industry, transportation to work, and income by type. The 2000 census long form also included 30 sample items about housing, covering such topics as market value of owned home, rent, cost of utilities, characteristics of house or apartment, year structure built, ownership finances, and number of vehicles.

These data have been used by the federal government for such purposes as implementation of sections of the Voting Rights Act, allocation of billions of dollars of federal funds to states and localities, assessment of charges of employment discrimination, and planning, monitoring, and evaluating federal programs. They have also been used by state and local governments for fund allocation, program planning and evaluation, facility planning, and economic development and marketing. Private-sector organizations (retail establishments, restaurants, banks, advertising firms, utility companies, health care providers, etc.) have used long-form-sample data for site location, the targeting of advertising and services, workforce development, and the assessment of compliance with government requirements. Finally, researchers have used long-form-sample data to help understand key social processes, such as internal migration and the correlates of poverty (see National Research Council, 2004b:Ch. 2; National Research Council, 1995:Apps. C, D, E, F, G, H, M, for details).

1-B.2 Why Seek an Alternative to the Long-Form Sample?

On one hand, the paired strategy used in modern censuses through 2000 of embedding a long-form sample in the basic decennial census had at least three advantages compared with using a separate survey to collect long-form-type data on a continuous basis: point-in-time reference periods; close agreement for small areas between the sample estimates and the complete counts by age, sex, race, ethnicity, and housing type; and—under some assumptions—cost savings from taking advantage of the census infrastructure. On the other hand, the paired strategy had at least three disadvantages: decreasing relevance of the data the longer the period between Census Day and the time for which estimates are desired; impaired data quality because of the priority given to completing the head count; and infrequent opportunities to revise the questionnaire. In addition, the paired strategy introduced inefficiencies into the census operations and—perhaps—impaired the completeness of the census head count.

1-B.2.a Advantages of the Paired Strategy

The long-form-sample data, like all census data, referred to a single point in time, which in 2000 was Census Day, April 1 (even though questionnaires were mailed in mid-March and follow-up operations to complete the enumeration spread out over several months). For some economic characteristics, such as income, the data referred to a single reference period, which was the preceding calendar year. Accompanying this point-in-time reference period was a concept for assigning people to a specific "usual residence," which was defined as the place where the person lived or stayed most of the time. These concepts were easy for users to understand and work with. In contrast, a continuous survey requires users to become accustomed to somewhat different and more complex concepts of reference periods and residence. In the case of the ACS, the estimates for a calendar year are based on aggregating data over the 12 months of data collection; the reference period is either the time when a household fills out the questionnaire or the preceding 12 months (for income, weeks and hours worked, and some housing costs); and residence is defined using a 2-month residence concept.

The paired strategy had the advantage that estimates for head counts and basic demographic characteristics from the long-form sample could be controlled to conform to the full census figures for small areas, using a statistical procedure called a raking ratio adjustment. This procedure reduced sampling and nonsampling error in the long-form-sample data products and produced a high level of consistency between estimates for demographic groups and small areas from the complete enumeration and the sample. Postcensal estimates are used to control the ACS, but they contain more error than the census counts and are not available for very small areas.

The "piggy-backing" of the long-form sample on the existing infrastructure for the short-form census may have had the advantage for the Census Bureau of reducing the costs of administering the long form. "Infrastructure" includes the MAF or comprehensive list of housing units in the United States; the mailout of forms to most households; the follow-up of nonrespondents and the associated local enumeration office structure; and data capture, editing, imputation, and preparation of data products. Under some assumptions, the marginal cost of the long-form sample was primarily that of the additional data capture and processing; hence, the cost savings through use of the paired strategy. Another source of cost savings is that, with the paired strategy, the MAF needed to be updated only once a decade, whereas the ACS requires the MAF to be constantly updated. An ongoing survey such as the ACS requires separate continuous data collection and processing, as well as continuous estimation, tabulation, and publication operations. Moreover, a continuous survey is never likely to obtain as high a mail response rate as the long form (which benefited from the massive publicity surrounding the decennial census) and hence requires more costly field follow-up.

1-B.2.b Disadvantages of the Paired Strategy

From the user perspective, a major disadvantage of the paired strategy was that the data collected became less relevant and more out of date the farther one was from Census Day. Typically, because of the need to give priority to the head count processing, the long-form-sample data were not released until 2 years or more after Census Day, and they were not updated for another 10 years. How much error resulted from the use of out-of-date information depended on how quickly the population of an area changed over time.

For example, consider estimates of the immigrant (foreign-born) population for the cities of Los Angeles, San Diego, and San Francisco, California, provided by the ACS test surveys conducted in 2000–2004. The data show complex patterns of stability and change within and among the three cities across this short time span—patterns that would be difficult to predict if the only data available were from a long-form sample every 10 years. The percentages of immigrants in the population of all three cities did not change between 2000 and 2004, but the percentages of the immigrant population who arrived after 2000 and who became citizens grew significantly. In Los Angeles, the percentage of people aged 5 and older who spoke a language other than English increased significantly between 2000 and 2004, but in San Diego and San Francisco, this percentage did not change.[2]

The effects of Hurricanes Katrina and Rita on the populations of New

[2]At http://factfinder.census.gov, specify "get data" for the ACS and go to "Data Profiles, Selected Social Characteristics" for a specific year. All web addresses are current as of March 2007.

Orleans and other affected Gulf Coast areas offer a striking example of the way in which long-form-sample data from a census could become irrelevant. Even though increasingly out of date, the 2000 long-form-sample data may have approximated the characteristics of the population in these areas reasonably well through the summer of 2005. However, after the hurricanes hit and so many people fled the area, moved into temporary housing, lost their jobs, or experienced other major changes in their living situations, the 2000 long-form-sample data no longer came close to approximating the numbers and characteristics of the remaining residents. In contrast, in June 2006 the Census Bureau was able to issue a special product from the 2005 ACS for these areas, providing separate estimates for the period January-August 2005 and the period September-December 2005.[3]

Not only were the long-form-sample products delayed because of the priority given to completing the head count, but also in 2000, even more than in prior censuses, there was not a dedicated effort to collecting long-form information during nonresponse follow-up. Furthermore, in 2000 there was no operation to follow up households that mailed back incomplete long-form questionnaires. Consequently, nonresponse rates were very high in 2000 for many long-form items, particularly those obtained in follow-up operations by temporary, minimally trained field staff. They were high absolutely and in comparison with the 2000 ACS test survey (known as the Census 2000 Supplementary Survey or C2SS), for which the Census Bureau used permanent, highly trained interviewers in nonresponse follow-up (Schneider, 2004:Appendix Tables 1, 2). Measurement error may also have been greater for long-form-type information collected as part of the census than is the case for the ACS.

The paired strategy limited the opportunities to revise the questionnaire to respond to emerging data needs or to improve response quality. Although it is unclear how often ACS questions can be revised, the strategy of obtaining long-form-type information in the continuous measurement design of the ACS should allow for additions to the questionnaire to address current issues of interest more frequently than once every 10 years.

An important consideration for the Census Bureau is that the infrastructure needed to administer a short-form census can be much more efficient than the infrastructure needed to administer both the short form and the long form. Indeed, the current design for carrying out the short-form-only census in 2010 envisions the use of handheld computing devices

[3]Available at http://www.census.gov/acs/www/Products/Profiles/gulf_coast/index.htm. The special ACS estimates were not without problems—the hurricanes disrupted postal service delivery, dislocated the interviewer workforce, and made it difficult to complete interviews from sample households. Moreover, post-hurricane population and housing unit estimates were not available to use as controls for the ACS post-hurricane estimates, so no controls were used.

to collect and transmit data during nonresponse follow-up operations and to assist in managing the large, temporary enumerator workforce. These handheld devices are considered crucial by the Census Bureau to lowering the costs of field data collection by reducing the time needed to find the next address in an enumerator's daily assignments, by automatically capturing and transmitting data, by reducing the amount of paper to be managed, and by helping to better manage enumerators' assignments and compensation.

Not having to collect long-form data facilitates the use of handheld devices by the census enumerators because many fewer data items must be asked and recorded. Also, not having the additional long-form questions should reduce the amount of paper that the centers responsible for processing the mail returns will need to handle by about 20 percent. These considerations certainly reduce the cost-efficiency argument of having the long-form sample use the short-form infrastructure, although whether they would completely overcome the cost advantages of the paired strategy compared with a separate continuous survey is not clear.

Finally, it is possible that dropping the long form from the census will improve the completeness and accuracy of the basic head count. In 2000, adverse publicity about the perceived intrusiveness of questions on the long form helped reduce the mail return rate for long forms, which was 9 percentage points below that for short forms (National Research Council, 2004b:Box 4.1). A short-form-only census, other things being equal, will probably have a higher mail return rate than in 2000. The higher the return rate, the less is the likelihood that people will be missed or doublecounted because they moved between Census Day and the time (several weeks or months later) that census enumerators visit nonresponding households or because the enumerators will find no one at home and obtain information from a neighbor or landlord.

1-B.3 Evolution of the ACS

Considerations of data needs and census efficiencies drove the development of the ACS as a replacement for the long-form sample. Several European countries have moved to continuous measurement as well—see Box 1-1.

The need for more frequent small-area data on the social and economic conditions of the population was discussed at least as far back as a 1941 proposal by Philip Hauser, then deputy director of the Census Bureau, for an "annual sample census" (see Alexander, 2001). In 1981, Leslie Kish of the Institute for Social Research at the University of Michigan proposed a "continuous measurement" or "rolling sample" design in place of the census, in which one-tenth of the population would be surveyed each year, cumulating the estimates over 1, 2, or more years to increase their precision

BOX 1-1
Continuous Measurement in Three Countries

The United States is not alone among developed countries in exploring the advantages of a rolling sample design for some or all of the content of a traditional census. Such investigation in other countries has usually assumed that a rolling design would replace a traditional census and not just long-form-type characteristics and that extensive use would be made of administrative records.

France

In 2004 France replaced its census with a form of rolling sample design (Desplanques, 2003). Every year a sample is drawn that comprises all households in one-fifth of the smallest communes (local administrative units) together with one-fifth of households in larger-sized communes. The data are cumulated over 5 years; in addition, data from administrative records are used to supplement the yearly samples to produce more reliable 1-year estimates.

Germany

Germany has not conducted a census since 1987 due to privacy concerns. In its place, Germany conducts a microcensus, involving every year 1 percent of all households in Germany, each of which stays in the sample for 4 years (see http://www.destatis.de/micro/e/micro_c1.htm). The questionnaire includes mandatory and supplementary items. Data from the microcensus and administrative records are used to develop population estimates.

Great Britain

The national statistical agency of Great Britain has explored the pros and cons of a rolling census. The Office for National Statistics (2003) reached a conclusion that a rolling design could be feasible to implement following the next full census (scheduled for 2011) if population and address registers are sufficiently developed to augment the sample-based data.

for small geographic areas (National Research Council, 1995:71).[4] Daniel Horvitz (1986) proposed a design that rolled by geographic area: a full census, including short-form and long-form content, would be conducted every year of one-tenth of the nation's counties. In 1988, Roger Herriot, then chief of the Population Division of the Census Bureau, proposed an ongoing "Decade Census Program" that mixed aspects of the Kish and Horvitz designs.

[4]Kish wrote a series of papers on continuous measurement, advocating its use for a variety of purposes, particularly in developing countries (see Alexander, 2001).

After the 1990 census, members of Congress expressed concern about the perceived adverse effect of the long-form questionnaire on the completeness of census response. In 1990, 29 percent of households that received the long form failed to mail back their form, compared with 24 percent of households that received the short form (National Research Council, 2004b:100), and some thought that this differential contributed to the poorer coverage of the population in 1990 in comparison with 1980. Other members of Congress were interested in more frequent estimates for small areas of such long-form-sample statistics as the percentage of school-age children in poverty for use in allocating federal funds to states and localities. Consequently, in 1994, the Census Bureau formed a staff to implement a continuous measurement design similar to that proposed by Kish in a few test sites, so that it could be evaluated as a replacement for the long-form sample in 2000. Renamed the American Community Survey, a questionnaire similar to the long form was fielded in four counties in 1996.[5]

The decision was made in the mid-1990s to retain the long-form sample in 2000 and to implement the ACS on a gradual basis, so that it could be further tested and compared with the 2000 census results. The goal was to conduct a short-form-only census in 2010 and to fully implement the ACS so that it could provide estimates for small areas that were about as precise as long-form-sample estimates for small areas by accumulating samples over 5 years. However, very early in the development process, rising costs led to a decision to scale back the originally planned sample size of 500,000 housing units per month to a sample of 250,000 housing units per month (National Research Council, 1995:127).

The original 4 ACS test sites were gradually increased to include 36 counties in 31 sites for the years 1999–2004 (see U.S. Census Bureau, 2003). In addition, the C2SS was fielded as an experiment in more than one-third of U.S. counties by using the monthly ACS data collection design and a questionnaire very similar to the long form. Cumulated over the 12 months of 2000, the C2SS, including the test sites, obtained responses from about 587,000 housing units; it demonstrated the feasibility of conducting an ACS nationwide and at the same time as the decennial census (U.S. Census Bureau, 2001). Similar nationwide test surveys were fielded in 2001–2004. In January 2005, full implementation of the ACS was inaugurated for the first month's sample of 250,000 housing units; and in January 2006, group quarters residents were added to the ACS. Plans are to proceed with a short-form-only census in 2010.

[5]The late Charles ("Chip") Alexander played a pivotal role in designing the ACS and moving it toward full implementation (see, for example, Alexander, 1993, 1997, 1998).

1-C ISSUES FOR THE PANEL

Two basic features of the ACS data products raised key questions for the panel's consideration: (1) the plan to use multiyear estimates from aggregating monthly samples from the ACS to replace point-in-time estimates from the decennial census long-form sample and (2) the plan to use census-based population estimates at the level of counties (or groups of small counties) for July 1 of each year to calibrate the sample-based ACS estimates. The purpose of using the population estimates, which are developed from the previous census updated with births, deaths, and migrant flows from administrative records, as controls is to reduce the effects of sampling error and compensate for any incompleteness of coverage in the population surveyed in the ACS. There are also separate controls for housing units.

The use of multiyear averages is central to the ACS design. The question it raises is the extent to which users can easily apply the ACS data products to the important and varied uses that, until now, were met through the long-form-sample data products for the census year. What does it mean to have a 5-year estimate for an area of, say, the poverty level or the average length of time to commute to work cumulated from 60 months' worth of data? How should users interpret differences in estimates for the same geographic area, such as a large county, that will be available from data cumulated over 1, 3, and 5 years? Turning to the detailed procedures for producing multiyear estimates, does the Census Bureau's plan to weight the data to the average population over the period represent the optimal approach, or are there other approaches that might on balance have better properties?

The decision to use controls for counties (or groups of small counties) for specific population groups (defined by age, sex, race, and ethnicity) for July 1 each year raises several questions. One question concerns the effect of controlling the ACS monthly samples spread over a year to point-in-time population estimates that are updated from April 1 population counts from the last census. Another question concerns whether the county population estimates by age, sex, race, and ethnicity are of adequate quality to be used for this purpose. Might, for instance, controlling to population estimates at higher levels of geographic aggregation or using fewer population groups as controls offer advantages over the current plan?

In addition to these two primary areas of investigation, the panel needed to examine the ACS as a data collection and production system to answer questions concerning the functionality and usability of ACS data products. This examination led the panel to address four areas of ACS operations: (1) sampling for housing units, including initial sampling from the MAF and subsampling of nonrespondents for follow-up by computer-assisted personal interviewing; (2) data collection for housing units, including the mode of collection (mail, telephone, personal visit) and residence rules; (3) sampling and data collection for group quarters; and (4) treatment of the data, including disclosure avoidance, collapsing of tables to improve precision, inflation adjustments for income and housing costs, the population universe and geographic areas for tabulations, and data quality review.

A last broad issue concerns the ultimate role of the ACS in the federal statistical system. Since the ACS has been developed to replace—and improve on—the decennial long-form sample, it is first necessary to assess the quality and usability of the ACS data products in comparison with the long-form-sample data products. The panel thinks, however, that focusing on these comparisons limits one's perspective concerning the ultimate utility of the ACS. Instead of viewing the ACS data products as simply analogous to or as a substitute for decennial census long-form-sample data products, the panel thinks that the ACS should be viewed as an entirely new and unique source of information to support public and private decision making. ACS data can be used as input for analyses that would not be feasible with long-form-sample data products. In that context, the panel evaluated the various characteristics of the ACS and its data products with a longer term view of what the potential for the ACS will be over time, and how it can be helped toward achieving its full promise as a key component of the federal collection of individual and household demographic, social, and economic data.

1-D OVERVIEW OF THE REPORT

Following this introduction, the panel's report has three parts: using the ACS (Part I), technical issues (Part II), and priorities for user outreach and continued development of the ACS (Part III).

Chapters 2 and 3 in Part I are addressed to users of long-form-sample data who want to use the ACS for their applications. Chapter 2 reviews key features of the ACS design; the major advantages of the survey in terms of timeliness, frequency, and quality of data; and the major challenges of using the ACS data in terms of period estimates replacing point-in-time estimates and high levels of sampling error for small areas. This chapter is essential reading for users who will work with the ACS data. It ends with an assessment of the usefulness of the ACS.

Chapter 3 provides guidance for applying the ACS data. Through examples of key applications that currently use long-form-sample data, the chapter considers how users can work with the various ACS products for those applications and the considerations they need to take into account in deciding which products to use for their applications. Both Chapters 2 and 3 take as given the Census Bureau's announced plans for ACS data products. Both chapters are written to be as user-friendly as possible, although, given the complexity of the ACS, they will be most helpful to two groups: people who expect to use the ACS for repeated, in-depth applications and people who expect to serve as intermediaries helping other users with their applications.

The next three chapters in Part II are addressed to technical users and the Census Bureau. They critically review key features of the ACS design, operations, and data products and offer recommendations for areas of further research and possible modification in the future. Chapter 4 addresses features of the ACS sample design and operations that are particularly relevant to the quality of the data and hence their usability for various applications. Chapter 5 reviews the procedures to weight the 1-year data so that totals agree with the Census Bureau's population estimates for major demographic groups, as well the Bureau's housing estimates, in large counties and groups of smaller counties. The chapter also discusses adjustments that are made to the data to account for nonresponse. Chapter 6 reviews the construction, interpretation, and possible alternative methods for producing multiyear period estimates.

Chapter 7 in Part III brings the user and technical strands together: it reviews and makes recommendations for three key areas of continuous research and development for the ACS. The three are (1) education and outreach activities to various user communities to help ease the transition from the census long-form-sample data products to those from the ACS and to ensure a continuous feedback loop between the Census Bureau and data users; (2) priorities for continuous methodological and operational improvement of the ACS; and (3) a vision of the future in which the ACS contributes in new and innovative ways to expanding information on the nation's people and communities.

The appendixes include a list of abbreviations and acronyms used in the report (Appendix A) and two papers written for the panel by F. Jay Breidt: Appendix B is on population controls and Appendix C is on multiyear period estimates. The report concludes with biographical sketches of panel members and staff in Appendix D.

PART I

Using the American Community Survey

2

Essentials for Users

This chapter addresses users of the decennial census long-form sample who want to know, in general terms, what benefits—and challenges—the new American Community Survey (ACS) presents to them. The chapter first summarizes the basics that every user should know about the ACS and key ways in which it is similar to and differs from the decennial census long-form sample. It then addresses two central issues: (1) why users should care about the ACS in terms of the benefits it offers and (2) some of the challenges those benefits present for users. Finally, it offers the panel's assessment of the value of the ACS to users based on the available knowledge about its properties.

2-A ACS DESIGN BASICS

To work with data from the ACS, users should be acquainted with the following features of its design and operations: the population or universe covered, rules for assigning people to a place of residence, questionnaire content and reference periods, sample size and design, data collection procedures, data products, and data-processing procedures to generate the products. The key factor to keep in mind is that, unlike the census long-form sample, the ACS is *continuous*: a fresh sample of addresses is surveyed every month, and data products represent *cumulations* of monthly data for 1-year, 3-year, and 5-year periods. The discussion below pertains to the ACS

BOX 2-1
The Puerto Rico Community Survey (PRCS)

The Puerto Rico Community Survey (PRCS) is identical in most respects to the American Community Survey in the 50 states and the District of Columbia. There were no PRCS test surveys or test sites for Puerto Rico in 2000–2004, so that 2005 PRCS data are the first post-2000 long-form-type data available for Puerto Rico.

Following the same basic design as the ACS, the initial sample size of the PRCS is 3,000 housing units each month or 36,000 housing units each year—about 2.4 percent of the total of about 1.5 million residential addresses in Puerto Rico. Initial sampling rates for blocks vary by the estimate of occupied housing units in the governmental jurisdiction or census tract (see Table 2-3, Part A), although the PRCS rates are slightly higher than the ACS rates (see U.S. Census Bureau, 2006:Table 4.1).

Data collection in the PRCS uses mailout, CATI, and CAPI, like the ACS. However, because of low mail response, all mailout/CATI nonrespondents are sampled at a 50 percent rate for the CAPI follow-up as of June 2005.

Areas for which PRCS products are published include:

- 78 municipios (county equivalents): 12 will have 1-year estimates; 65 will have 3-year estimates
- 455 barrios (subdivisions of municipios, similar to minor civil divisions): 5 will have 1-year estimates; 34 will have 3-year estimates (based on 2000 census counts)
- 225 zonas urbanas (census designated places that are governmental centers of municipios) and communidads (other census designated places): 9 will have 1-year estimates; 20 will have 3-year estimates (based on 2000 census counts)
- 871 census tracts and 2,477 block groups (5-year estimates only)

The PRCS is explained in greater detail by U.S. Census Bureau (2006).

in the United States; see Box 2-1 for a brief overview of the Puerto Rico Community Survey (PRCS).[1]

2-A.1 Population Coverage (Universe)

The ACS for 2005 covered the household population. The 2006 ACS covered not only the household population, but also people who live in college dormitories, armed forces barracks, prisons, nursing homes, and other group quarters. The 2006 ACS population coverage was the same as the census long-form-sample coverage, except that the ACS did not in-

[1]All of Section 2-A draws heavily on U.S. Census Bureau (2006).

clude people found at soup kitchens or street locations frequented by the homeless and a few other transient situations. Table 2-1 lists the types of residences included in the 2006 ACS.

2-A.2 Residence Rules

The ACS instructs the respondent for a household to provide data on all people who, at the time of filling out the questionnaire, are living or staying at the household address for more than 2 months (including usual residents who are away for less than 2 months). In contrast, the long-form sample asked household respondents to report all people who *usually* lived at the address as of Census Day, April 1, meaning they lived or stayed there most of the time. People whom the ACS samples in group quarters beginning in 2006 are counted at the group quarters location, in effect applying a *de facto* residence rule regardless of how long an individual has lived or expects to live in the group quarters. The long-form sample also in effect generally applied a *de facto* residence rule for group quarters residents, although residents of some types of group quarters were allowed the option of indicating another usual place of residence. (An unduplication process was used to determine the correct enumeration for people listed at the group quarters and the other residence; such a process would not be possible for the ACS because it is not embedded in a census.)

For many people, their ACS residence will be the same as their long-form-sample residence. However, some people may report a different residence: for example, people who live in a house or apartment in New York most of the year but reside in Florida in December through March should report Florida as their address if sampled for the ACS in Florida in the winter, whereas their Census Day address is in New York.

2-A.3 Content and Reference Periods

The 2005 ACS includes about 55 questions for every person and 30 questions for every housing unit in the sample—approximately the same content as in the 2000 census long-form sample. There are some differences:

- The ACS mail questionnaire uses a matrix layout for questions on sex, age, race, ethnicity, and household relationship, compared with a person-by-person format in the long-form questionnaire.
- The ACS mail questionnaire provides room to respond for 5 household members compared with 6 on the long-form questionnaire (telephone follow-up is used to obtain information on additional household members).

TABLE 2-1 Types of Residences in the American Community Survey (ACS)

Residence Type	2000 Population (Percentage)	
Housing Unit Residence[a]	**97.2**	
Single-family, detached		64.6
Single-family, attached		5.4
2-or-more-unit structure		20.4
In an apartment building (including condominium or co-op)		
In an assisted living facility with separate apartments		
In a group quarters (e.g., house master's residence)		
In a home (e.g., basement apartment, upstairs apartment)		
In multi-unit military family housing on or off base		
Mobile home that is occupied or, if vacant, that is permanently sited		6.7
Boat at a mooring, RV, or occupied van		0.1
Institutional Group Quarters Residence (beginning in 2006 ACS)	**1.4**	
Nursing home or other long-term care facility		0.6
Correctional institution (for example, prison or jail)		0.7
Other institutions (for example, hospital or residential school for people with disabilities, long-term care home for juveniles)		0.1
Noninstitutional Group Quarters Residence (beginning in 2006 ACS)	**1.3**	
College dormitory		0.7
Military quarters (in barracks on a base; on a ship assigned to home port)		0.1
Other noninstitutional group quarters		0.5[b]
Residence in the ACS and the 2000 Long-Form Sample		
Convent, monastery		
Group home		
Halfway home		
Hospice		
Job Corps center		
Migrant worker quarters		
Shelter, emergency shelter		
YMCA, YWCA, hostel		
Residence NOT in the ACS but in the 2000 Long-Form Sample		
Circus quarters		
Crews on merchant ships		
Domestic violence shelter		
Recreational vehicle in a campground		
Soup kitchen or mobile food van site		
Street location for the homeless		

[a]Housing units are separate living quarters with direct access from the outside or through a common hall (U.S. Census Bureau, 2006:D-17).

[b]Includes 170,706 people (0.06 percent of the population of 281.4 million in 2000) living in emergency shelters for the homeless, shelters for runaway children, transitional shelters, and hotels and motels used to provide shelter for people without conventional housing (U.S. Census Bureau, 2001).

SOURCES: Types of residences adapted from U.S. Census Bureau (2006:Ch. 8, Attachment A); population percentages from http://factfinder.census.gov, Summary File 1, Table P37; Summary File 3, Table H33.

- Many ACS items refer to a time period different from that of the corresponding items on the long-form questionnaire: for example, usual hours worked per week, weeks worked per year, and income items on the ACS refer to the 12 months prior to the day when the household filled out the questionnaire, whereas these items on the long form always referred to the previous calendar year (1999 for the 2000 census).
- The ACS currently includes three items not on the 2000 long form: (1) whether the household received food stamps in the previous 12 months and their value; (2) the length of time and main reason for staying at the address (for example, permanent home, vacation home, to attend school or college); and, for women ages 15–50, whether they gave birth to any children in the past 12 months.

Table 2-2 compares the items on the 2005 ACS questionnaire with the items on the 2000 census long form. The Census Bureau is proposing several changes to the ACS questionnaire beginning in 2008. These changes, if approved, will include the addition of three new questions on marital history, health insurance coverage, and veterans' service-related disability, the deletion of the question on length of time and main reason for staying at the address, changes to the basic demographic items for consistency with the 2010 census questionnaire, and changes in wording and format to improve reporting of several other questions as determined by a 2006 test. A question on field of bachelor's degree will be tested in 2007 and may be added to the ACS beginning in 2009.

2-A.4 Sample Design and Size

The ACS sends out questionnaires to about 250,000 housing unit addresses every month that have been sampled from the Census Bureau's Master Address File (MAF; see Chapter 4 for details of the sampling operation). Each month's sample includes addresses in every one of the nation's 3,141 counties. The monthly samples cumulate to about 3 million addresses over a year, or about 2.3 percent of the total number of about 129.5 million housing unit addresses in the United States in 2005. The sample is constructed so that no housing unit address will be included more than once every 5 years.

The ACS sample is very large compared with the samples for major national household surveys. However, the long-form sample was even larger: in 2000, the long form was sent to about 18 million addresses, or one-sixth of total housing unit addresses in the United States at the time, and 16.4 million usable long-form questionnaires were included in the final edited data file. The ACS monthly and even yearly samples cannot be as large as

TABLE 2-2 Items on the 2005 ACS Questionnaire and the 2000 Census Long Form

2005 ACS Item (in question order)	Asked in 2000 Census?
Person Items	
Sex	Yes (short-form item)
Age (at interview)	Yes (as of April 1, short form)
Date of birth (month, day, year)	Yes (short form)
Relationship to household reference person (person 1)	Yes (more detail, short form)
Marital status	Yes
Hispanic origin	Yes (short form)
Race (option for multiple races)	Yes (short form)
Place of birth	Yes
Citizenship	Yes
Year of immigration	Yes
Attended school in last 3 months	Yes (since February 1, 2000)
Grade attending	Yes
Highest degree completed	Yes
Ancestry or ethnic origin	Yes
Language spoken at home	Yes
How well speaks English	Yes
Place of residence 1 year ago (city or town, county, state)	Yes (5 years ago)
Disability involving sight or hearing	Yes
Disability limiting physical activity	Yes
Difficulty learning, remembering due to disability of 6 or more months	Yes
Difficulty dressing, bathing, or getting around the home	Yes
For people ages 15 and older	
Difficulty going outside the home alone to shop, etc.	Yes (ages 16 and older)
Difficulty working at a job or business	Yes (ages 16 and older)
Given birth in past 12 months (women ages 15–50)	No
Responsible for own grandchildren in the home	Yes
How long responsible for grandchildren	Yes
Veteran status (active duty)	Yes
Period of active military service	Yes
Number of automobiles, vans, trucks for use by household members	Yes
Years of active military service (less than 2 years, 2 years or more)	Yes
Working last week for pay or profit	Yes
Place of work (address)	Yes
Usual means of transportation to work last week	Yes
If by car, truck, or van, how many people used it	Yes
Time left home for work	Yes
Minutes to work	Yes
On layoff last week	Yes
Temporarily absent from work last week	Yes
Whether will be recalled to work	Yes
Looking for work last 4 weeks	Yes
Could have worked last week	Yes

TABLE 2-2 Continued

2005 ACS Item (in question order)	Asked in 2000 Census?
When last worked	Yes
Weeks worked, last 12 months	Yes (1999)
Hours usually worked per week, last 12 months	Yes (1999)
Class of worker of current or most recent employment	Yes
Industry of current or most recent employment	Yes
Occupation of current or most recent employment	Yes
Wage and salary income, last 12 months	Yes (1999)
Self-employment income (farm and nonfarm), last 12 months	Yes (1999)
Interest, dividend, net rent, royalty, and trust income, last 12 months	Yes (1999)
Social Security income, last 12 months	Yes (1999)
Supplemental Security Income, last 12 months	Yes (1999)
State or local public assistance income, last 12 months	Yes (1999)
Retirement, survivor, or disability pension income, last 12 months	Yes (1999)
Any other regular income, last 12 months	Yes (1999)
Total income, last 12 months	Yes (1999)
Housing Items	
Type of building/number of units in structure	Yes
Year building built	Yes
When household reference person (person 1) moved in	Yes
Number of acres on property (single-family or mobile home)	Yes
Agricultural sales, last 12 months (single-family or mobile home on 1 or more acres)	Yes (1999)
Whether business on property (single-family or mobile home)	Yes
Rooms in unit	Yes
Bedrooms in unit	Yes
Complete plumbing facilities	Yes
Complete kitchen facilities	Yes
Telephone service available	Yes
Number of automobiles, vans, trucks for use by household members	Yes
Heating fuel most used	Yes
Electricity cost, last month	Yes (annual cost)
Gas cost, last month	Yes (annual cost)
Water and sewer cost, last 12 months	Yes (annual cost)
Oil, coal, kerosene cost, last 12 months	Yes (annual cost)
Receive food stamps, value last 12 months	No
Monthly condominium fee	Yes
Owner or renter (tenure)	Yes (short-form item)
Monthly rent (and whether includes various utilities)	Yes
Whether rent includes meals	Yes
Value of property if were for sale	Yes

continued

TABLE 2-2 Continued

2005 ACS Item (in question order)	Asked in 2000 Census?
Annual real estate taxes	Yes (last year)
Annual hazard insurance	Yes
Monthly mortgage payment	Yes
Whether mortgage payment includes taxes	Yes
Whether mortgage payment includes insurance	Yes
Whether a second mortgage and/or home equity loan	Yes
Second mortgage/home equity loan monthly payment	Yes
Annual costs for mobile home and site (personal property taxes, site rent, fees and licenses)	Yes (last year, also includes installment loans)
Whether any household members live here year round	No
Number of months members live here	No
Main reason members stay at this address	No

NOTES: The 2005 ACS and 2000 census long-form sample provided room on the mailback questionnaire for characteristics of up to 5 and 6 household members, respectively. Questionnaires should be consulted for precise question wording.

SOURCES: See http://www.census.gov/acs/www/SBasics/SQuest/SQuest1.htm for the 2005 ACS; Anderson (2000:388-399) for the 2000 long form.

the long-form sample because the costs would be too great. Accumulated over 5 years, the ACS sample will total about 15 million housing unit addresses, but the ACS sample is then reduced by the subsampling for in-person follow-up of households not responding to the mail and telephone data collection procedures (see below). This subsampling may reduce the ACS 5-year sample to 10–11 million housing unit addresses.

Because data on governmental jurisdictions will be an important output of the ACS and because many governmental units are very small in population size, the ACS oversamples housing unit addresses in small governmental units relative to other areas similar to the 2000 long-form-sample design. Oversampling provides more precise estimates for small counties, places, townships, school districts, and American Indian and Alaska Native areas than would otherwise be possible. In a similar manner, for the personal visit follow-up operation, the ACS oversamples mail and telephone nonrespondents in census tracts that are expected to have low mail and telephone response rates relative to other census tracts. In order to afford the costs for the additional follow-up, not only are smaller subsamples followed up in person in census tracts that are expected to have high mail and telephone response, but also the initial sample for these tracts is reduced by 8 percent. Tables 2-3a, 2-3b, and 2-3c provide initial annual and 5-year sampling rates for governmental units and census tracts of different popu-

TABLE 2-3a Housing Unit Addresses, 2005 ACS and 2000 Census Long-Form Sample: Approximate Initial Block-Level Sampling Rates

	2005 American Community Survey		
Type and Size of Smallest Area Containing a Block	Annual Initial Sampling Rate	Cumulative 5-Year Initial Sampling Rate	2000 Long-Form-Sample Census Day Sampling Rate
Governmental unit (county, place, township in 12 states, school district, American Indian or Alaska Native area)			
With < 200 occupied housing units (fewer than about 500 people)	10.0% (1 in 10)	50.0% (1 in 2)	50.0% (1 in 2)
With 200–800 occupied housing units (about 500-2,000 people)	6.9% (1 in 14)	34.5% (1 in 3)	50.0% (1 in 2)
With 800–1,200 occupied housing units (about 2,000-3,000 people)	3.5% (1 in 28)	17.5% (1 in 6)	25.0% (1 in 4)
Census tract with > 2,000 occupied housing units (more than about 5,000 people)[a]	1.7% (1 in 59) or 1.6% (1 in 63)	8.5% (1 in 12) or 8.0% (1 in 13)	12.5% (1 in 8)
Other area[a]	2.3% (1 in 44) or 2.1% (1 in 48)	11.5% (1 in 9) or 10.5% (1 in 10)	16.7% (1 in 6)
Overall	2.3% (1 in 44)	11.5% (1 in 9)	16.7% (1 in 6)

NOTES: Number of occupied housing units is estimated from the MAF. Because the initial ACS sample size will be kept at approximately 3 million residential addresses per year, the initial sampling rates shown will be slightly reduced as the number of occupied housing units grows. Townships and other minor civil divisions are recognized for sampling purposes in 12 states where they are functioning governments: Connecticut, Maine, Massachusetts, Michigan, Minnesota, New Hampshire, New Jersey, New York, Pennsylvania, Rhode Island, Vermont, and Wisconsin.

[a]The smaller of the two ACS sampling rates shown applies for blocks in census tracts with predicted mail/CATI response rates greater than 60% (see Table 2-3b).

SOURCE: Adapted from U.S. Census Bureau (2006:Tables 4.1, 4.2) for the ACS.

TABLE 2-3b Housing Unit Addresses, 2005 ACS and 2000 Census Long-Form Sample: Census Tract-Level CAPI Subsampling Rates in the 2005 ACS for Mail/CATI Nonrespondents[a]

Subsampling Rate Category	CAPI Subsampling Rate	Illustrative Completed Sample Cases as Percentage of Initial Sample
Census tracts with predicted mail/CATI response rate less than or equal to 35%	50% (1 in 2)	If, say, 20% mail/CATI response, then completed sample will be 60% of initial sample—20% mail/CATI plus 40% CAPI (1/2 of 80%)
Census tracts with predicted mail/CATI response rate between 36 and 50%	40% (2 in 5)	If, say, 40% mail/CATI response, then completed sample will be 64% of initial sample—40% mail/CATI plus 24% CAPI (2/5 of 60%)
Census tracts with predicted mail/CATI response rate between 51 and 60%	33% (1 in 3)	If, say, 55% mail/CATI response, then completed sample will be 70% of initial sample—55% mail/CATI plus 15% CAPI (1/3 of 45%)
Census tracts with predicted mail/CATI response rate greater than 60%[b]	33% (1 in 3)	If, say, 80% mail/CATI response, then completed sample will be 87% of initial sample—80% mail/CATI plus 7% CAPI (1/3 of 20%)

[a]Nonmailable addresses are followed up in the CAPI data collection phase at a rate of 67% (2 in 3).

[b]In census tracts outside oversampled governmental units with high predicted mail/CATI response rates, the initial sample is reduced by a factor of 0.92 (see Table 2-3a). This reduction is implemented to satisfy a budget constraint for personal interviewing.

SOURCE: Adapted from U.S. Census Bureau (2006:Tables 4.1, 4.2) for the ACS.

TABLE 2-3c Housing Unit Addresses, 2005 ACS and 2000 Census Long-Form Sample: Illustrative Rates of Completed Sample Cases[a]

Type and Size of Area	2005 ACS Cumulative 5-Year Rate of Completed Sample Cases (Percent)				2000 Long-Form-Sample Census Day Rate of Completed Sample Cases
Mail/CATI Response Rate for CAPI Subsampling	20	40	55	80	
Governmental unit (county, place, township in 12 states, school district, American Indian or Alaska Native area)					
With < 200 occupied housing units (fewer than about 500 people)	30.0	32.0	35.0	43.5	50.0
With 200–800 occupied housing units (about 500–2,000 people)	20.7	22.1	24.2	30.0	50.0
With 800–1,200 occupied housing units (about 2,000–3,000 people)	10.5	11.2	12.3	15.2	25.0
Census tract with > 2,000 occupied housing units (more than about 5,000 people)	5.1	5.4	6.0	7.0[b]	12.5
Other area	6.9	7.4	8.1	9.1[b]	16.7

NOTES: The ACS rate of completed sample cases for an area is the initial sampling rate (see Table 2-3a, second column under ACS) times the percentage calculated in the last column of Table 2-3b to illustrate the effect of CAPI subsampling. The illustrative rates of completed sample cases shown will be reduced by unit nonresponse, which was 3% in the 2005 ACS and 7% in the 2000 long-form sample.
CAPI: computer-assisted personal interviewing; CATI: computer-assisted telephone interviewing.

[a]For nonmailable addresses, the illustrative rate of completed sample cases (before nonresponse) will be 67% of the initial sampling rates in Table 2-3a, or, reading from top to bottom, 33.5%, 23.0%, 11.7%, 5.7% (using the higher of the two initial sampling rates in Table 2-3a), and 7.7% (using the higher of the two initial sampling rates in Table 2-3a).

[b]Uses the lower of the two initial sampling rates in Table 2-3a.

SOURCE: Adapted from U.S. Census Bureau (2006:Tables 4.1, 4.2) for the ACS.

lation sizes, subsampling rates for addresses that must be followed up in person, and estimated 5-year rates of sample cases after subsampling; Table 2-4 provides counts of governmental units by type and population size.

2-A.5 Data Collection

Each month the residential housing unit addresses in the ACS sample with mailable addresses—about 95 percent of each month's sample of 250,000 addresses—are sent a notification letter followed 4 days later by a questionnaire booklet. A reminder postcard is sent out 3 days after the questionnaire mailing. Whenever a questionnaire is not returned by mail within the following 3 weeks, a second questionnaire is mailed to the address. If there is still no response and if the Census Bureau is able to obtain a telephone number for the address, then trained interviewers conduct telephone follow-up using computer-assisted telephone interviewing (CATI) equipment. Telephone follow-up is also used to obtain missing information from households that mailed back incomplete questionnaires. About 33 percent of mail questionnaires in 2005 required telephone follow-up because key items were missing or because the household reported more members than there was room to provide information (U.S. Census Bureau, 2006:7-9).

For samples of addresses for which no mail or CATI responses are received after 2 months, or the postal service returned the questionnaire because it could not be delivered as addressed, or the address is not in street name and number format and so was not mailed out in the first place (for example, post office box or rural route addresses), interviewers are sent into the field with laptop computers. They visit housing units in person (or, in about 20 percent of cases, make contact by telephone) and collect the ACS data through computer-assisted personal interviewing (CAPI). The personal interview follow-up is conducted on a sample basis in order to save costs: about two-thirds of nonmailable addresses and between one-third and one-half of mailable addresses in each census tract—depending on the expected mailback and CATI response rate for the census tract—are followed up in person. Interviewers also visit group quarters in person to collect data for residents, using paper-and-pencil questionnaires.

An important difference between the ACS and 2000 census long-form-sample data collection procedures is that all nonresponding housing units were included in the long-form follow-up operations. Long-form-sample questionnaires were sent out in mid-March 2000, preceded by a notification letter and followed by a reminder postcard. For every address for which a questionnaire was not returned by mail, temporary interviewers (enumerators) went into the field to try to obtain responses in the period of late April-June. The enumerators were often not successful in obtaining

TABLE 2-4 Types of Governmental Units by Population Size in 2000

	Counties[a]		Places[b]		Minor Civil Divisions[c]		School Districts[d]	
Population Size Category	No.	Percentage Total Pop.	No.	Percentage Total Pop.	No.	Percentage Total Pop.	No.	Percentage Total Pop.
Total	3,141	100.00%	19,849	76.57%	8,414	12.46%	14,125	100.00%
Under 500 people	6	<0.01	6,230	0.54	1,966	0.18	854	0.07
500–2,000	71	0.03	6,194	2.33	3,226	1.27	2,380	0.93
2,000–3,000	72	0.06	1,498	1.30	847	0.74	1,151	0.91
Areas containing blocks that are oversampled, subtotal[e]	149	0.09	13,922	4.17	6,039	2.19	4,385	1.91
3,000–5,000	143	0.20	1,498	2.07	868	1.19	1,638	2.05
5,000–10,000	404	1.08	1,677	4.23	719	1.78	2,505	5.76
10,000–20,000	652	3.40	1,147	5.70	448	2.25	2,370	10.75
20,000–50,000	879	10.09	971	10.69	280	2.90	2,009	19.61
50,000–65,000	167	3.41	189	3.82	31	0.62	326	5.89
65,000–100,000	223	6.41	189	5.35	19	0.55	379	9.58
100,000–250,000	293	15.88	175	9.17	6	0.34	380	18.32
250,000–500,000	119	14.80	45	5.85	3	0.38	87	9.53
500,000–1,000,000	78	19.67	19	4.29	1	0.26	32	6.79
1,000,000–2,500,000	28	15.36	10	4.66	0	0.00	11	4.85
2,500,000 or more	6	9.60	7	16.57	0	0.00	3	4.96
Other areas, subtotal	2,992	99.91	5,927	72.40	2,375	10.27	9,740	98.09

[a]Source for counties: 2000 Census Gazetteer File (extract of Summary File 1) (http://www.census.gov/geo/www/gazetteer/); total 2000 population of 281.4 million.

[b]Places include all incorporated places plus census-designated places in Hawaii; census-designated places in the other states will have estimates published for them (see Table 2-5) but are not recognized for purposes of oversampling; see note *a* above for source of estimates.

[c]Minor civil divisions (not coterminous with places) are recognized in 12 states for purposes of oversampling (Connecticut, Maine, Massachusetts, Michigan, Minnesota, New Hampshire, New Jersey, New York, Pennsylvania, Rhode Island, Vermont, and Wisconsin); they are recognized in additional states for publication but do not benefit from oversampling in these states; see note *a* above for source of estimates.

[d]Population size for school districts is as of 2003; some school districts (for example, elementary and secondary districts) overlap in population, so the population base for percentages is 314.6 million and not the 2003 population of 290.8 million. Source for school districts: 2003 population estimates (http://www.census.gov/hhes/www/saipe/school/sd03layout.html).

[e]See Table 2-3a for oversampling rates for blocks in small jurisdictions of fewer than 500 people, 500-2,000 people, and 2,000-3,000 people.

complete information. Moreover, there was no telephone follow-up to fill in missing items on mailed-back questionnaires. Consequently, missing data rates were very high for many long-form questionnaire items in 2000—considerably higher than in the ACS to date (see Section 2-B.2 below).

2-A.6 Data Products

The ACS will produce data products that resemble those from the 2000 long-form sample. The products will be available primarily through the American FactFinder on the Census Bureau's web site: http://factfinder.census.gov (see Box 2-2). Because of the continuous design of the ACS, its data products will differ in two important respects from the long-form-sample products. First, the ACS data will be available every year instead of once a decade. As was just done for the ACS 2005 data, each year's data products will be released in waves from August through November.[2] Second, the ACS data that are released each year will not pertain to a point in time, like the long-form-sample data for Census Day; instead, they will be cumulated over a 12-, 36-, or 60-month period for governmental and statistical units depending on population size in order to provide sufficiently precise estimates for publication (see Table 2-5).

ACS products will include tables and profiles of characteristics for governmental and statistical areas; see Box 2-2. The confidentiality of these products will be protected by various means. One method is to combine detailed categories into broader categories when the individual categories contain too few sample cases. Another method is termed "data swapping," in which computer programs may swap the data for an entire household that is at risk of being identified (for example, the only minority household in a block group) with the data for another similar household in a different area. Only a small percentage of records, which are never identified, are swapped. In addition to the various procedures that are implemented to protect confidentiality for the 1-year, 3-year, and 5-year period estimates, the Census Bureau will combine individual categories and even delete entire tabulations from 1-year and 3-year period products when the sampling errors are very large (see Chapter 4 for details).

ACS products will also include public use microdata samples (PUMS). PUMS files comprise samples of individual records, suitably processed to protect confidentiality by such means as:

- deleting names and addresses from the records;
- limiting geographic identification to large areas known as public

[2]See http://www.census.gov/acs/www/Special/Alerts/Latest.htm for the release schedule for data products from the 2005 ACS.

use microdata areas (PUMAs), which are defined to include about 100,000 people; and

- limiting the detail that is provided for sensitive variables—for example, assigning a catchall code to income amounts over a certain threshold, such as $100,000 or more, and not identifying the specific amount.

2-A.7 Data Processing—1-Year Period Estimates

The computer programs to generate the ACS 1-year period data products each summer operate on 12 months of data collected in the preceding calendar year. These data include all of the mailed-back, CATI, and CAPI responses that were received in January through December of that year (including additional information obtained by telephone for incompletely filled out mail questionnaires). The major data processing steps are described briefly below.

2-A.7.a Coding, Editing, and Imputation

As in the long-form sample, the first data-processing step for the ACS is to assign codes for write-in responses for such items as ancestry, industry, and occupation by using automated and clerical coding procedures. Coding is performed on a monthly basis. Then once a year, the raw data, the codes assigned to write-in items, and various operational data for the responses for the preceding January–December are assembled into an "edit-input file." Computer programs review the records on this file for each household to determine if the data are sufficiently complete to be accepted for further processing and to determine the best set of records to use in instances when more than one questionnaire was obtained for a household.

Computer programs then edit the data on the accepted, unduplicated records in various ways. Computer programs also supply values for any missing information that remains after editing, using data from neighboring households with similar characteristics in a process called "hot-deck imputation." The goal of editing and imputation is to make the ACS housing and person records complete for all persons and households

Because of the varying reference periods in the ACS, dollar amounts of income are adjusted for inflation using the national consumer price index for all urban consumers research series (CPI-U-RS) so that every amount reflects the average value of the dollar for the calendar year. For example, a person who reported an income of $20,000 in February 2005 for the period February 2004–January 2005 would have this amount inflated by a figure of 1.031 to give an amount of $20,620. The figure of 1.031 comes from dividing the average annual CPI-U-RS for 2005 (which has the value

BOX 2-2
Data Products from the American Community Survey

Tabulations

- *Base or detailed tables, for 1-year, 3-year, and 5-year periods, all geographies that meet the relevant population size cutoff*—Hundreds of tables that cross-classify two or more characteristics for a wide variety of subjects (for example, employment by sex and age); race and Hispanic-origin iterations for key characteristics; tables providing item imputation rates. For 1-year and 3-year periods, collapsed tables may be provided when categories in a detailed table are suppressed because the estimates do not meet minimum precision criteria.

  Similar to the tabulations in Summary File 3 from the 2000 long-form sample; 5-year period ACS estimates will also include tabulations of journey-to-work items for traffic analysis zones (one or more blocks, block groups, or census tracts) that, in 2000, were produced on a special tabulation basis (known as the Census Transportation Planning Package).

- *Subject tables, for 1-year, 3-year, and 5-year periods, all geographies as above*—Over 60 single-topic tables of frequently requested information, with distributions and medians.

  Comparable to the Quick Tables from the 2000 long-form sample but with more detail.

- *Population profiles for selected race, ethnicity, and ancestry groups, for 1-year, 3-year, and 5-year periods, areas with 1 million or more people*—Key distributions (availability of 1-year and 3-year profiles depends on the size of the group).

  New data product for the ACS. Profiles will be produced for most of the groups tabulated in Summary File 4 from the 2000 long-form sample.

- *Data profiles (single year), all geographies with 65,000 or more people*—Four profiles of demographic, social, economic, and housing characteristics and one narrative.

  Comparable to profiles from the 2000 long-form sample with more geographic detail; narrative profile, which covers all four subject areas, is new.

- *Multiyear profiles, all geographies with 65,000 or more people*—Four profiles for the current year and four prior years, indicating differences for a specified year from the current year that are statistically significant with 90 percent margin of error.

  New data product for the ACS; multiyear profiles are available for the 2000–2004 ACS test surveys; the first release of multiyear profiles from the full ACS will be in 2008.

- *Ranking tables and charts, for 1-year periods*—86 ranking tables for states that enable the user to determine which differences among states are statistically significant with 90 percent confidence.

Expanded from 19 subjects from the 2000 long-form sample.

Downloadable Tables

These tables are accessible from the ACS File Transfer Protocol (FTP) site (http://www.census.gov/acs/www/Special/acsftp.html).

- *Base tables, single-year profiles, multiyear profiles, and ranking tables*—Permits user analysis, such as summing categories, computing percentages, etc.
- *ACS summary files for 1-year, 3-year, and 5-year period estimates (under development)*—In response to users, the Census Bureau is developing an ACS product similar to Summary File 3 from the 2000 long-form sample. ACS summary files will contain all of the base tables for 1-year, 3-year, and 5-year period estimates and will be readily analyzable for such uses as comparing areas and population groups.

Public Use Microdata Sample (PUMS) Files

- *1 percent sample file, available for each calendar year of ACS data*—Each year's file will contain about 1.25 million household and 3 million person records selected from the final realized sample of about 2 million housing units, or 40 percent of the final sample.

  Content: All housing and person items, together with imputation flags.

  Geographic identification: states, within-state areas of 100,000 or more population (public use microdata areas or PUMAs).

  Comparable, when cumulated over 5 years, to the 5 percent sample file from the 2000 long-form sample; the 1 percent ACS sample file provides finer geographic identification than the 1 percent long-form-sample file (PUMAs of 100,000 or more population compared with super-PUMAs of 400,000 or more population). Permits multivariate, microlevel analysis.

Geographic Products

- *Geographic comparison tables*—86 tables (same topics as ranking tables, but not in rank order) for geographic components of the nation and states (for example, a table of median age for counties in Alabama).
- *Thematic maps*—86 maps (same topics as ranking tables) for geographic components of the nation and states.

NOTE: See Table 2-5 for types of geographic areas and population sizes for which 1-year, 3-year, and 5-year period estimates are available. Unless otherwise noted, products are available from http://www.census.gov/acs/www and provide 90 percent margins of error for estimates.

TABLE 2-5 Major Types of Geographic Areas for Which 1-Year, 3-Year, and 5-Year Period Estimates Are Available from the American Community Survey

	Estimate Type		
Area Type	1-Year Period	3-Year Period	5-Year Period
States and District of Columbia	51	51	51
Congressional districts	436	436	436
Public use microdata areas (PUMAs) (these areas have at least 100,000 people)	2,071	2,071	2,071
Metropolitan and micropolitan statistical areas	492	905	936
Urban areas	363	809	3,607
Counties and county equivalents	775	1,812	3,141
Cities, towns, and census-designated places	492	2,062	25,112
Townships and villages (minor civil divisions) (recognized for publication in 28 states)	186	984	21,200
School districts (elementary, secondary, and unified)	878	3,257	14,394
American Indian and Alaska Native areas	14	36	603
Census tracts	0	0	65,433
Block groups	0	0	208,790

NOTES: 1-year period estimates are available for governmental and statistical areas with at least 65,000 people; 3-year period estimates are available for governmental and statistical areas with at least 20,000 people; 5-year period estimates are available for all governmental and statistical areas, including census tracts (statistical areas of about 4,000 people) and block groups (statistical areas of about 1,500 people). Other areas for which estimates are provided (not shown) include combined statistical areas, Hawaiian Home Lands, urban and rural territory, areas inside and outside the principal city of a metropolitan or micropolitan statistical area, areas outside metropolitan and micropolitan areas.

SOURCE: Tabulation provided by the U.S. Census Bureau, February 21, 2007. Because of changes in population and geographic boundaries, the actual numbers of areas with estimates published may differ from the numbers shown.

of 284.3 relative to the base index of 100 in December 1977) by the average of the monthly CPIs for the 12 months from February 2004 to January 2005 (which is 275.8).[3]

2-A.7.b Weighting

The edited, filled-in data records for the 12 months in a calendar year are weighted in a series of steps to produce 1-year period estimates that represent the entire population. Chapter 5 provides a complete description of the nine steps in the weighting process for housing units and their members; four key steps (1, 3, 5, and 6) are briefly described here. (Similar steps will be used to weight the sample records for residents of group quarters beginning in 2006.)

Step 1, Base Weights Initially, the ACS housing unit and person records are assigned "base" weights that reflect the rate at which the unit was originally sampled from the MAF and, for CAPI responses, the rate at which it was subsampled for follow-up. Housing unit base weights can vary from as low as 10 (for housing units selected at a rate of 1 in 10 that mailed back their questionnaire or responded by telephone) to as high as 180 (for housing units selected at a rate of 1 in 60 that were followed up by CAPI at a rate of 1 in 3).

Step 3, Nonresponse Adjustment An important adjustment is made to the base weights for occupied housing units to account for unit nonresponse; in this adjustment, the weights are inflated to account for the failure to interview all housing units in the sample.

Steps 5 and 6, Housing and Population Controls Two other key adjustments are made to the weights to improve the precision of the survey estimates and to compensate for the fact that some people are overlooked in sample households and some addresses are left off the MAF. Each of these adjustments is performed for estimation areas, which are large counties or groups of small counties. First (step 5), the weights for housing units are adjusted (controlled) to agree with independently derived estimates of total housing units in the applicable estimation area as of July 1 of the year being processed. The housing controls are derived by updating the previous census counts with information on new construction building permits, shipments of mobile homes, and estimates of housing loss (see Chapter 5). Second (step 6), the weights for persons are adjusted to agree with independently derived estimates of people in age, sex, race, and ethnic groups in the

[3]See http://www.bls.gov/cpi/cpiurstx.htm.

applicable estimation area as of July 1. The population controls are derived by updating the previous census counts with information on births, deaths, and net migration (see Chapter 5). In a similar, but much more detailed, procedure for 2000, long-form-sample responses were weighted up to agree with the census complete counts for age, sex, race, and ethnic groups and type and size of household in subcounty areas as of April 1, 2000 (see U.S. Census Bureau, 2003:Ch. 8).

The multistep weighting process is designed to produce estimates of people and housing units that are as complete as possible and that take into account the various aspects of the complex ACS design. A key point for users to keep in mind is that the weights will vary—sometimes a great deal—both within and across many governmental units. On one hand, this variation in weights will make the estimates less reliable than they would be with an equal probability sample of the same size. On the other hand, the variations in initial sampling and CAPI subsampling rates are intended to serve specific purposes, such as improving the precision of estimates for small governmental units within a budget constraint that limits the total sample size.

2-A.7.c Tabulation

The final data-processing steps for the 1-year period estimates are to generate tabulations, profiles, and other products, such as PUMS. At this stage, procedures are implemented to protect data confidentiality and to combine categories (or delete entire tables) to meet precision standards.

Throughout, the ACS data for a calendar year are processed as a whole and not month by month. The only exception to date is that for areas in states affected by Hurricane Katrina and Hurricane Rita, the Census Bureau issued two sets of 2005 data products in early June 2006—one set for January–August 2005 and the other set for September–December 2005, reflecting conditions before and after the hurricanes.[4]

2-A.8 Data Processing—3-Year and 5-Year Period Estimates

The computer programs to produce 3-year period products use the fully processed records for the 3 preceding calendar years, containing 36 months' worth of responses; the programs to produce 5-year period products use the fully processed records for the 5 preceding calendar years, containing 60 months' worth of responses. The only new steps are to modify the inflation adjustments and the weights.

The income inflation adjustments are modified so that every amount

[4]Available at http://www.census.gov/acs/www/Products/Profiles/gulf_coast/index.htm.

is expressed in terms of the average value of the dollar for the most recent calendar year. For example, for 3-year or 5-year period estimates released in 2010, covering 2007–2009 and 2005–2009, respectively, all income amounts would be adjusted to reflect the average value of the dollar for 2009. Amounts for housing value and costs are also inflated to reflect the average value of the dollar for the most recent calendar year.

While the Census Bureau has not worked out all of the details of the weighting for 3-year and 5-year period data products, the general procedure will be to remove the adjustments to the 1-year period weights for housing unit nonresponse and agreement with housing unit and population controls and to make new adjustments. Unit nonresponse adjustments will be implemented for all occupied housing units for which data were obtained in the relevant 36 months or 60 months. Averages of the independent housing unit and population estimates for 3 years or 5 years, as applicable, will be used to adjust the weights of each housing unit and person for whom data were obtained during the relevant 36 months or 60 months.

The final data-processing steps for the 3-year and 5-year period estimates are to generate the various data products. At this stage, procedures are implemented to protect data confidentiality; also, for the 3-year period estimates, procedures are implemented (as for the 1-year period estimates) to combine categories (or delete entire tables) to meet precision standards. No screening for precision is applied to 5-year period estimates, as they are considered to be the building blocks for user-defined areas, such as groups of census tracts or block groups in a city.

2-B ACS BENEFITS

Two paramount benefits that users will gain from the ACS in comparison with the census long-form sample are the more timely issuance of the data and the greater frequency with which the data will be released. Timeliness refers to the speed with which estimates are produced after the data are collected; frequency refers to how often the estimates are produced. A third important benefit will very likely be improved data quality in that the ACS data will likely be more complete and accurate than the long-form-sample data.

2-B.1 Timeliness and Frequency

Instead of producing point-in-time estimates once a decade for governmental and statistical areas, every year the ACS will produce period estimates—5-year period estimates for all areas, including small neighborhoods (census tracts and block groups) and small governmental units; 3-year period estimates for all areas with at least 20,000 people; and 1-year period estimates for areas with at least 65,000 people. Table 2-6 shows how the various period estimates will become available over the period 2005–2009, as sufficient months of data are accumulated, and how the ACS estimates will continue to be produced from that point onward.

The Census Bureau's data release schedule calls for each set of estimates to become available 8–10 months after all the data needed to produce the estimates are collected. Long-form-sample tabulations typically required 2 years or more after Census Day to become available. Even more important than the faster schedule for data processing is that the ACS estimates released each summer and fall will provide a continual flow of updated

TABLE 2-6 Release Year and Calendar Year of Period Estimates from the ACS

Type of Period Estimate	Release Year (Late Summer-Fall)									
	2006	2007	2008	2009	2010	2011	2012	2013	2014	2015
	Calendar Year(s) of Data									
1-year period estimates for areas with 65,000 or more people	2005	2006	2007	2008	2009	2010	2011	2012	2013	2014
3-year period estimates for areas with 20,000 or more people			2005–2007	2006–2008	2007–2009	2008–2010	2009–2011	2010–2012	2011–2013	2012–2014
5-year period estimates for block groups, census tracts, and all other areas					2005–2009	2006–2010	2007–2011	2008–2012	2009–2013	2010–2014

NOTE: See Table 2-5 for major types of governmental and statistical areas for which 1-year, 3-year, and 5-year period estimates are available.

information, enabling users to analyze recent data for an area or population group and compare it with other areas and groups.

Moreover, the annual ACS products will enable users to construct time series for analyzing change. To the extent permitted by sampling error, these data will make it possible to detect trends in the percentage of people in an area who are employed, live in poverty, or have attained a college degree, and whether the trends for an area mirror or deviate from national trends. Similarly, the data will show changes in the ethnic makeup of an area, housing costs for homeowners and renters, and many other characteristics of interest to data users. Of course, several factors can interrupt a time series for an area of interest, such as a change in geographic boundaries, a change in the wording of the question used to measure a characteristic of interest, or, occasionally, a revision to the county population estimates that are used to control the ACS estimates.

2-B.2 Data Quality

Another major benefit of the ACS over the census long-form sample should be higher quality of the data in terms of the completeness and accuracy of response. Missing and inaccurate responses are components of nonsampling error that can result in bias in survey estimates, as distinct from the variable error due to the use of a sample (discussed in Section 2-C below). Both kinds of error are important: sampling variability can be so large as to render an unbiased estimate of little use for decision making, while even a very precise estimate in terms of sampling error can be misleading if the bias in the estimate is large (see Box 2-3 for brief descriptions of elements of bias and variability).

The assessment of the likely higher quality of the ACS rests primarily on comparisons of estimates from the Census 2000 Supplementary Survey (C2SS) and the 2000 long-form sample. The C2SS was a full-scale test of the ACS questionnaire and data collection procedures. It included about 587,000 responses from a nationwide sample in 1,203 counties plus samples in 36 counties that were ACS test sites.

By comparing the C2SS and the 2000 long-form sample, the Census Bureau was able to evaluate the relative quality to be expected from the ACS in terms of unit (household) weighted response rates, population coverage, item response rates, and quality control processes. See Box 2-4 for indicators of sample size, household response, population coverage, and item response for each year of the ACS.[5] The C2SS equaled or outperformed

[5]The indicators can be accessed from the main ACS web site (http://www.census.gov/acs/www) under "Using the Data" or by selecting a subject area for which ACS data are desired, clicking on "survey methodology" and then on "quality measures": http://www.census.gov/acs/www/UseData/sse/index.htm.

BOX 2-3
Sources of Sampling and Nonsampling Error in Survey Estimates

Multiple sources of error can affect the estimates from a survey such as the ACS. In this context, *error* has a statistical meaning; namely, it refers to the difference between an estimate and the (unknown) true population parameter (for example, the true percent poor school-age children). Despite the colloquial meaning of the word, such an error is not necessarily an indication that a mistake has been made.

Survey methodologists generally classify statistical errors into two major categories—*variability*, or errors that lead to variation in the survey estimates across hypothetical repetitions of the survey process under identical survey conditions; and *bias*, or the systematic component of errors that results in a difference between the average of the survey estimates across these hypothetical repetitions and the true value of the parameter being estimated. Some sources of error (for example, differences among interviewers) can substantially affect both variability and bias. Much survey research is devoted not only to the measurement of variability and bias, but also to the development of procedures to reduce their effects on survey estimates. (For categorizations of sources of error, see Groves et al., 2004, and Biemer and Lyberg, 2003.)

Variability

- The estimates from a survey are never precise but vary to a greater or lesser extent from their average over hypothetical repetitions of the survey under identical survey conditions.
- For most surveys, the major source of imprecision is sampling variance that arises when estimates are based on a sample and not a complete census of the universe. Other things equal, sampling error decreases as the size of the sample

the 2000 long-form sample on all of these attributes, as did the C2SS-like ACS test surveys conducted in 2001–2004 and the full 2005 ACS. Population coverage and unit and item response rates have also been higher in the ACS than in the Current Population Survey Annual Social and Economic Supplement (CPS ASEC), which is the nation's premier household survey of income, participation in government programs, employment, and family relationships.[6]

In addition to examining basic quality measures, the Census Bureau also compared the distributions of responses to individual items for the C2SS and the long-form sample, which mainly identified consistencies be-

[6]In comparing the ACS and the CPS ASEC, users should bear in mind the many differences between the two surveys (see Nelson, 2006, and Section 3-F.2).

increases, and vice versa. (See Box 2-5 for explanations of statistical terms associated with the measurement of sampling error.)

- Other elements of a survey can increase the imprecision of estimates, including variability introduced by respondents, interviewers, and coders, and by procedures to impute values for missing responses. Errors arising from such other sources are referred to as nonsampling errors. For example, the same respondent may give different answers to a question about income or race when interviewed more than once due to random factors, such as how the respondent interprets the question. For large-scale surveys for estimates of totals, these other sources of variance may contribute much more to imprecision than sampling variance.

Bias

- The estimates from a survey may differ systematically from the true value for any number of reasons. Nonsampling error sources often give rise to bias.
- Sources of bias include that the question wording elicits responses that differ from the construct intended by the survey designer; that respondents consistently overestimate or underestimate the true value (for example, the amount of their income last year); that imputation and weighting adjustment procedures may not compensate adequately for nonresponse and noncoverage; and that the weighting adjustment controls used to correct for coverage errors are inaccurate for certain areas and population groups.

Some variability and bias in survey estimates is inevitable (bias and some sources of variability also affect censuses). The challenge for users is, with the help of methodologists, to understand enough of the extent and nature of sampling and nonsampling errors in survey estimates to assess the utility of the estimates for the user's purpose and identify possible strategies for ameliorating the effects of these errors on survey inferences.

tween the two surveys, as well as a few differences. These comparisons were performed for the nation as a whole and for individual counties in the ACS test sites, which were oversampled relative to the other C2SS counties. The finding of consistency between estimates from the C2SS and the long-form sample cannot prove that the C2SS estimates are unbiased. Consistency, however, does offer reassurance that the C2SS—and, by extension, the ACS—are measuring items in the same way.

The highlights of the overall and individual item evaluations of the C2SS compared with the 2000 long-form sample are summarized below; the complete findings are available in seven reports issued by the Census Bureau (U.S. Census Bureau, 2002b, 2004a-f; see also National Research Council, 2004b:Ch. 7; Schneider, 2004).

BOX 2-4
Four Quality Measures Available for the American Community Survey

The Census Bureau currently provides four indicators of nonsampling errors for the nation and states. They can be accessed from the main ACS web site under "Using the Data" (http://www.census.gov/acs/www, middle of the page) or by selecting a subject area for which ACS data are desired, clicking on "survey methodology" and then on "quality measures" (http://www.census.gov/acs/www/UseData/sse/index.htm).

Sample Size

Sample size is critical for estimating the level of sampling error in survey estimates. The ACS web site provides two sample sizes for the year in question: (1) initial sample addresses, or the total number of addresses selected from the MAF to receive a questionnaire and (2) final interviews, or the total number of questionnaires received by mail, CATI, or CAPI. The second measure is smaller than the first—for example, 2.9 million addresses were initially selected for the 2005 ACS, but the number of final interviews was only 1.9 million. The principal reason is because CAPI is used to follow up only a subsample of addresses that lack a mail or CATI response; in addition, some sampled addresses turn out to be nonexistent or nonresidential, and some households do not respond even after follow-up.

Coverage Error

Coverage error occurs in the ACS as in other household surveys. Undercoverage occurs when the sampling frame does not include all addresses and when not all people in sampled addresses are included on the questionnaire; overcoverage occurs when households or individuals are duplicated or otherwise erroneously included. Net coverage is defined relative to decennial census–based population estimates.

The ACS web site provides net coverage rates for men and women for states and the United States and for six race/ethnicity categories for the United States. These rates are the weighted ACS estimate for the year in question for the relevant demo-

2-B.2.a High Unit Response Rate in the C2SS

The Census Bureau never expected that the mail response would be as high in the ACS as in the census, nor has it been: it was 56 percent in the C2SS compared with 71 percent in the 2000 long-form sample. To save on costs, the Census Bureau specified that only about one-third of ACS mail and telephone nonrespondents would be interviewed in person. Hence, to obtain a final household response rate that can be compared to the long-form-sample rate as a measure of public cooperation, the subsample of households sent for CAPI (in-person) interviewing in the ACS must be weighted to account for the subsampling before they are added to the mail and CATI (telephone) respondents.

graphic group and geographic area before being controlled to population estimates, divided by the corresponding census-based population estimate.

The population and housing unit estimates used to adjust the ACS estimates for coverage errors (see Sections 5-C and 5-D) pertain to only a few characteristics (age, sex, race, ethnicity, and total housing), are only applied for large counties and groups of small counties (estimation areas), and themselves contain errors. To the extent that the controls are flawed and that noncovered or duplicated people differ from correctly covered people, then estimates from the ACS may be biased.

Unit Nonresponse

Unit nonresponse relates to the number of final interviews from a survey. To the extent that responding and nonresponding units differ, estimates from a survey may be biased. The ACS web site provides unit response rates and unit nonresponse rates by type (refusal, unable to locate, no one home, temporarily absent, language problem, other, and insufficient data from an interview to be included in the data set). The numerator for unit response rates is the number of mail, CATI, and CAPI responses for the year in question, weighted to account for different initial sampling and CAPI subsampling rates. The denominator is a similarly weighted estimate of the number of cases eligible to be interviewed. The intent in estimating the denominator is to exclude that fraction of the sample of addresses that turn out to be nonexistent, nonresidential, or otherwise ineligible for inclusion in the ACS.

Item Nonresponse

Item nonresponse occurs when interviews are complete enough to include in the estimation but answers are missing (or invalid) for one or more questions. Item nonresponse rates indicate the potential for measurement error due to differences between the values imputed for missing responses and the actual values. The ACS web site provides nonresponse rates for individual tabulated items. The numerator for each rate is the number of allocated responses (imputations that use reported information from other persons or housing units); the denominator is the total number of responses, including valid responses, allocations, and assignments (assignments use other information for the same person to fill in or correct a response).

Using a weighted response rate, the C2SS compares favorably with the 2000 long-form-sample rate. The 2000 final edited long-form-sample data file included 93 percent of sampled households; the remaining 7 percent were dropped because the information collected for them was too scant (National Research Council, 2004b:Table 7.7). By comparison, the C2SS weighted household response rate was 95 percent. The weighted household response rates in the 2001–2003 ACS test surveys were higher than the C2SS rate, averaging 97 percent. The 2004 rate was lower (93 percent) because of a funding reduction that necessitated dropping telephone and personal follow-up operations for January 2004. The 2005 weighted household response rate (the first year under full implementation) was 97 percent.[7] The 2005 ACS mail response rate declined to 51 percent from 56 percent in the C2SS, but the CATI and CAPI operations more than made up the difference.

2-B.2.b Good Population Coverage

When weighted to account for sampling and unit nonresponse, the ACS estimates of the population, like those from other household surveys, typically fall short of census counts (or census-based population estimates). More people are missed in surveys than in the census, either because the sampling frame of addresses is less complete or because larger numbers of people are not reported by sampled households. People may also be duplicated or included erroneously in surveys as in the census, but surveys more often miss people, resulting in greater net undercoverage of the population.

The process of controlling the survey weights to the population estimates attempts to compensate for coverage errors, but the controls are available for only a few characteristics (see Section A.7.b above), so that achieving high coverage rates to begin with is important. Before applying controls, the C2SS covered 97 percent of the household population; subsequent supplementary surveys covered 94 percent of the population, and the 2005 ACS covered 95 percent of the household population.[8] By comparison, the 2004 CPS ASEC covered only 88 percent of the population age 16 and over (Nelson, 2006:Table B).

2-B.2.c More Complete Item Response in the C2SS

Imputation rates for questionnaire items—that is, the percentage of item responses for households (for housing questions) or household residents (for person questions) for which an answer had to be imputed from another household's responses because the item was missing—are a commonly used measure of missing data. By this measure, the C2SS significantly outperformed the 2000 long-form sample. The C2SS had lower imputation rates for household members for 26 of 27 housing items and 48 of 54 population items that were included on both questionnaires (Schneider, 2004: Appendix). For example, 19.3 percent of household residents in the 2000 long-form sample were imputed a response for the question on number of weeks they worked last year, compared with only 9.6 percent of household residents in the C2SS who were imputed a response for the question on number of weeks they worked in the past 12 months.

---

[7]See Quality Measure 3 at http://www.census.gov/acs/www/UseData/sse/index.htm.

[8]See Quality Measure 2 at http://www.census.gov/acs/www/UseData/sse/index.htm.

For most items, the 2000 census imputation rates for enumerator returns exceeded those for mailed-back returns, sometimes by large margins. In contrast, for many items, the C2SS imputation rates were lower for interviewer returns than for mailed-back returns, indicating the higher quality of the follow-up effort in the C2SS (National Research Council, 2004b:Table 7.5).

Moreover, there is evidence suggesting that the quality of the ACS data collection improved after 2000. The imputation rates for 20 of 36 housing items and 51 of 57 population items were lower in the 2001–2004 ACS test surveys compared with the C2SS—sometimes substantially lower—and no rate was higher.[9] For example, only 18.8 percent of household members age 15 and over had some or all of their income imputed in the 2004 supplementary survey, compared with 23.9 percent in the C2SS and 29.7 percent in the 2000 long-form sample. Item imputation rates remained at low levels in the 2005 ACS.

2-B.2.d *Greater Quality Control in the C2SS and the ACS*

For the C2SS the Census Bureau implemented quality assurance procedures that were not included in the 2000 long-form-sample procedures because of cost and timing constraints. These same procedures are being used in the ACS.

An important operation that contributes to quality for the ACS is the telephone follow-up of mailed-back questionnaires that do not meet standards for completeness of coverage of household members or content. Moreover, telephone and in-person follow-up of nonrespondents is conducted by experienced, highly trained interviewers who are assisted by computerized questionnaires with built-in edit checks and skip patterns. The 2000 census lacked telephone follow-up for mailed-back questionnaires that were missing several items, and the enumerators who conducted the in-person follow-up of households that did not mail back their forms used paper-and-pencil questionnaires. Moreover, the temporary, lightly trained enumerators focused on obtaining the answers to the basic questions and not on the additional long-form-sample questions.

Another quality assurance procedure in the ACS is not to allow proxies for household respondents, such as neighbors or landlords. In 2000, 6.2 percent of long forms were obtained by proxy, and three-fifths of these had to be dropped from the final tabulation file because the data were so incomplete (National Research Council, 2004b:291).

---

[9]See Quality Measure 4 at http://www.census.gov/acs/www/UseData/sse/index.htm.

2-B.2.e *Consistency of Responses for Many Items*

Comparisons of the C2SS and the 2000 long-form-sample estimates for individual items cannot establish which is closer to a measure of truth—additional research would be required to examine this issue, using such techniques as matches with consistently defined administrative records and reinterviews of households. Yet such comparisons can identify the extent to which items are broadly consistent between the two surveys, thereby giving users confidence in the ACS as a replacement for the long-form sample. Complete consistency should not be expected even with the same questions because of differences in reference periods and residence rules, question formatting, editing, interview mode, and other survey procedures, yet the finding of major differences would be cause for disquiet and suggest further needed research.

The comparisons of the C2SS and the 2000 long-form sample for household residents found that most items were broadly consistent between the two sources at the national level. Individual comparisons performed for 18 of the 36 counties in the ACS test sites also found a high degree of consistency for most items. (It was not possible to perform comparisons for group quarters residents because they were not included in the C2SS.) "Driving to work alone" was an instance of a category in which national estimates were similar between the C2SS and the long-form sample, but estimates differed substantially for some of the counties that were examined individually. For this category, the C2SS estimates were appreciably higher in three counties and appreciably lower in three counties than the corresponding long-form-sample estimates (U.S. Census Bureau, 2004b:23).

National-level differences between the two surveys that were statistically and substantively significant occurred for race, ancestry, vacancy status, tenure (owner/renter), number of rooms in the housing unit, disability status for people ages 5–64, employment status, and median income. Users should always exercise care when comparing estimates from any of the ACS data sets (2005, 2001–2004 test surveys, C2SS) with the 2000 long-form sample because of differences in the ACS and long-form-sample design and operations. They should be particularly careful when making comparisons for the items discussed below.

Race and Ethnicity The C2SS estimated a higher percentage of "white alone" and a lower percentage of "some other race alone" compared with the 2000 long-form sample (77.5 versus 75.3 percent for "white alone" and 3.9 versus 5.5 percent for "some other race alone"). These results appear due in part to differences in wording and format of the race and ethnicity questions, which the Census Bureau is investigating (U.S. Census Bureau, 2004a:32). Another contributing factor is that census enumerators, who were not as well trained as C2SS interviewers, often failed to ask the race question appropriately (Martin and Gerber, 2003:42-43).

Ancestry The C2SS estimated higher percentages in many ancestry groups compared with the 2000 long-form sample, particularly for Germans (17.0 versus 15.4 percent), English (10.3 versus 8.8 percent), and Irish (12.1 versus 11.0 percent). The reason is that the Census Bureau does not impute an ancestry when none is reported, and the C2SS had more complete reporting of this item than the long-form sample: 88 percent of people reported at least one ancestry in the C2SS compared with only 81 percent in the 2000 long-form sample (U.S. Census Bureau, 2004e:45-47).

Vacancy Status The C2SS estimated a higher percentage of vacant housing units (9.7 percent) than the 2000 long-form sample (9.0 percent). The higher estimated vacancy rate in the C2SS applied not only to the nation as a whole, but also to most of the counties examined individually in spite of wide variations among them in vacancy rates. This result is contrary to the expectation that the ACS 2-month residence rule and 3-month data collection period implemented in the C2SS would lead to a lower estimated percentage of vacant housing. (For example, a vacant unit to which a questionnaire is mailed in the first month of collection could well be occupied by the time it is visited by an interviewer in month 3.) Why the C2SS estimated a lower vacancy rate than the 2000 long-form sample is not clear; it may be that census enumerators were more apt to classify a vacant unit as an occupied unit than the C2SS interviewers (U.S. Census Bureau, 2004a:vi-vii, 4-41). Continued assessment of vacancy status estimates in comparison with the American Housing Survey and other data sources will be required to evaluate the accuracy of this item in the ACS.

Tenure (Owner/Renter) Status The C2SS estimated a lower percentage of owner-occupied housing units (65.4 percent) than the 2000 long-form sample (66.2 percent), which may be due to a high rate of imputation for this item in the census, so that the C2SS estimates may be more accurate (U.S. Census Bureau, 2004a:43, 45).

Rooms in Housing Unit The C2SS produced different estimates of numbers of rooms in the housing unit compared with the 2000 long-form sample—specifically, the C2SS estimated fewer small units of 1–2 rooms (5.5 versus 7.0 percent), more mid-sized units of 4–5 rooms (39.7 versus 36.9 percent), and fewer large units of 7 or more rooms (26.4 versus 27.9 percent). The extent to which these differences may be due to differences in question wording, format, and sequencing, differences in data capture and editing, inconsistent definitions of what constitutes a separate room by interviewers and respondents, and other factors is not known (U.S. Census Bureau, 2004f:36-41).

Disability Status The C2SS estimated a substantially lower percentage of disabled people ages 21–64 than the 2000 long-form sample (13.8 versus 19.1 percent), as well as a lower percentage of disabled people ages 5–20 (6.8 versus 8.0 percent). The problem seems to involve the nonresponse follow-up phase of the 2000 census. It appears that people who were visited by census enumerators misunderstood the questions about employment disability and difficulty going outside the home alone. An indicator that supports this hypothesis is that 75 percent of people in the census who reported an employment difficulty to an interviewer were actually employed, compared with only 21 percent in the C2SS (U.S. Census Bureau, 4004e:33-36; see also Israel, 2006).

Employment Status The C2SS estimated a higher percentage of employed people than the 2000 long-form sample (62.3 versus 61.4 percent). This difference may be due to several factors, including different reference periods. The C2SS and the 2000 long-form-sample estimates of unemployment were comparable (3.5 versus 3.4 percent) but significantly lower than the Current Population Survey estimate of 4.0 percent civilian unemployment for 2000 (U.S. Census Bureau, 2004b:18-20). Over the period 2000–2003, the CPS persistently estimated not only substantially higher annual unemployment rates than the ACS test surveys, but also somewhat higher employment rates (Palumbo, 2005). The CPS is the official source of unemployment figures, and its estimates are presumably more accurate because they are based on responses to a more detailed set of questions obtained by trained interviewers using CATI and CAPI interviewing. In addition, the CPS uses a fixed reference period for reporting employment status (see Bureau of Labor Statistics and U.S. Census Bureau, 2002:Ch. 16).

Median Income The C2SS estimated lower median household and family income than the 2000 long-form sample—$40,137 versus $41,994 for household median income and $48,014 versus $50,046 for family median income. Estimates of families and people in poverty were virtually the same, although the C2SS estimated higher percentages of poor children and poor single-woman families with children than the 2000 long-form sample (16.8 versus 16.1 percent poor children, 35.4 versus 34.3 percent poor single-woman families with children) (U.S. Census Bureau, 2004b:33-38).

The reasons for these differences are not known, although they may be due to differences in completeness of reporting and reference periods. The income reference periods for the C2SS spanned January–December 1999 (for people interviewed in January 2000) to December 1999–November 2000 (for people interviewed in December 2000), while the income reference period for the 2000 long-form sample was a uniform January–December 1999. It is possible that the C2SS captured the onset of recession in a way that the 2000 long-form sample could not. The inflation adjustment procedure for the C2SS could also be a factor. For the Census Bureau's comparative analysis, the C2SS data were backward adjusted to reflect the average inflation experience for 1999, not 2000. Analysis by Turek, Denmead, and James (2005) suggests that inflation adjustments to the ACS may not accurately reflect economic growth (or decline) over a year.

2-C ACS CHALLENGES

The ACS will benefit users by providing more timely and frequent data for small areas that are likely to be based on responses of higher quality than the census long-form sample. However, there are challenges to using the data that stem principally from the continuous design of the ACS. Two major challenges are (1) the period nature of the ACS 1-year, 3-year, and 5-year estimates in contrast to the point-in-time nature of the long-form-sample estimates (Section C.1) and (2) the greater sampling error of the ACS estimates compared with the long-form-sample estimates (Section C.2).

2-C.1 Period Estimates

All estimates from the ACS will be period estimates—that is, the estimates will represent averages of months of data—12 months for 1-year estimates, 36 months for 3-year estimates, and 60 months for 5-year estimates. The issue is how to interpret the period estimates from the ACS, which are not the same as the (approximately) point-in-time estimates for Census Day (April 1) from the census long-form sample.

Consider first the 1-year period estimates, which include responses in all 12 months of the calendar year for sampled housing units that existed on the MAF as of January of the year (see Section 4-A). An independent estimate of total housing for July 1 is used to control the estimated number of housing units in an estimation area, but the reported characteristics of the units may vary throughout the year. Consequently, for housing characteristics (utility costs, value, rent, number of rooms, and others), the 1-year period estimates are 12-month averages, which may often differ from long-form-type point-in-time estimates.

Similarly, even though independently derived census-based population estimates for July 1 for major age, sex, race, and ethnicity groups are used to control the 1-year period estimates of people, such characteristics as education, income, and others may vary during the year. The 1-year period estimates are consequently 12-month averages of such population characteristics as education, income, veterans status, and others. A 1-year period estimate for an area will correspond to a point-in-time estimate for July

1 only if the population and its characteristics are stable during the year, which will not be true of areas with distinct seasonal populations, such as summer and winter residents who differ appreciably not only in numbers, but also in socioeconomic characteristics (see Section 3-C.3).

The 3-year and 5-year period estimates have similar attributes to the 1-year estimates. They do not represent the characteristics of the population for either the end year or the middle year—interpretations that may appeal to users but are misleading. Rather, they are period estimates, or averages, over 36 or 60 months. Such period estimates have lower sampling error than other types of estimates that could be constructed from the data, such as a middle-year estimate (see Chapter 6). They place more of a burden on the user, however, in interpreting them individually and in interpreting trends in them over time (see Section 3-C.1.b).

To become comfortable with the 3-year and 5-year period estimates, users need to think of them as pertaining to a *period of time*, not to a specific year or date. Thus, the poverty estimate for a small town from averaging the ACS data for the 5-year period from 2010 to 2014 could be termed "the average poverty estimate for our town for the first half of the decade," while the estimate from averaging the ACS data for the 5-year period from 2015 to 2019 could be termed "the average poverty estimate for our town for the second half of the decade." Similarly, poverty rates based on 3 years of ACS data could be assigned such terms as "the average poverty estimate for our city for the early [or middle, or later] part of the decade." This kind of description will not work when a 3-year or a 5-year period estimate does not neatly correspond to a readily identifiable portion of a decade (for example, an estimate for 2012–2015). Yet the general point remains, which is the need for users to develop descriptive phrases and other ways to reinforce the idea that all ACS estimates pertain to a period of time.

Once the ACS has been fully operational for a sufficient number of years, many large areas will have estimates available each summer and fall for more than one period, such as 1-year, 3-year, and 5-year estimates for areas with 65,000 or more people (refer back to Table 2-6). Unless very little or no change has occurred in the area's population or characteristics, these 1-year, 3-year, and 5-year estimates are likely to differ. Users who are only interested in the estimates for an area as a whole, and not for any smaller components, can decide which set of period estimates best suits the goals of their analysis, considering such factors as the likely variability in the characteristic over time and the level of sampling error that is tolerable for their application (see Section 2-C.2 below). Users who want to look at larger areas and also their components—for example, a city and its planning areas made up of groups of census tracts—will need to use the same period estimates throughout to ensure comparability. Most likely, users will have to work with the 5-year estimates, which will be the only estimates available for the smallest areas. Alternatively, they will need to develop methods to relate estimates for different time periods, such as using 1-year period estimates for a state, large county, or PUMA to update 5-year period estimates for small governmental jurisdictions (see Section 3-B.3).

In using the ACS data to study trends and changes over time, users will need to keep in mind the implications of changes in an area's geographic boundaries and population size for their analysis. With regard to population size, a governmental unit may gain or lose population so that it crosses a population size threshold for the publication of estimates. For example, a small city may grow from 60,000 to 70,000 over a 5-year period. Beginning in the year when the city achieves 65,000, it will have 1-year as well as 3-year and 5-year period estimates produced. Significant population decline, however, if sustained, could cause an area to be dropped from the 1-year or even the 3-year period estimates series. Population changes may also increase or decrease the initial sampling rate for an area.

With regard to boundaries, the Census Bureau will continue to update regularly the geographic boundaries of most types of governmental units every year—for example, to reflect an annexation or a combination or splitting up of units. It will update school district boundaries every 2 years and update the boundaries of statistical areas, including metropolitan areas, urbanized areas, PUMAs, census tracts, and block groups every decade in conjunction with the census. For ACS estimates for such governmental units as counties and cities, the Census Bureau will use the geographic boundaries as of January 1 of the most recent year of data that figure into the particular set of estimates. Consider a large city that annexed territory in late 2008 for which the user is working with 1-year, 3-year, and 5-year period estimates pertaining to 2010, 2008–2010, and 2006–2010, respectively. All these estimates will include data for the current enlarged city boundaries. The Census Bureau will not, however, revise estimates that precede the most recent 5 years to reflect boundary changes.

2-C.2 Sampling Error

The use of a sample rather than a complete enumeration introduces sampling error that affects the precision of the estimates from a survey. Such error is related to the variability of the characteristic in the population, the size of the sample, and the sample design. For a given estimate and sample design, the larger the sample size, the smaller is the sampling error.

2-C.2.a Design Factors

Overall, the ACS 5-year period estimates for an area will exhibit greater sampling error than the 2000 census long-form-sample estimates for the same area. (The sampling errors for 3-year and 1-year period estimates will be greater yet.) Two reasons are that the ACS cumulative 5-year initial sample size is only about three-fourths that of the long-form sample and that the ACS then uses subsampling for the CAPI interviews. For planning purposes, the Census Bureau estimated that the sampling errors (known as the standard errors) of the ACS 5-year period estimates would be about 20 percent greater than the errors of the long-form-sample estimates, but recent work (Starsinic, 2005) suggests that the ACS sampling errors will exceed the long-form-sample errors by about 50 percent.

The ACS design, like the long-form-sample design, oversamples very small governmental jurisdictions (refer back to Table 2-3a). This oversampling reduces the sampling errors of estimates for those units, but it increases the errors for larger areas that are undersampled, as well as somewhat increasing the errors for larger units that include some oversampled and some undersampled areas relative to a design with the same sampling rate for all areas.

The subsampling used for CAPI interviews in the ACS increases sampling error for two reasons: first, the subsampling reduces the final sample

BOX 2-5
Brief Descriptions of Statistical Terms Used in This Report

- *Standard error of an estimate:* A commonly used statistic that expresses the imprecision in an estimate that is due to sampling. This imprecision is known as sampling error. It is to be distinguished from nonsampling errors from such sources as misreporting and nonresponse, which are often systematic in nature and result in biased survey estimates (see Box 2-3).
- *Coefficient of variation (CV) or relative standard error:* The standard error expressed as a percentage of the estimate. CVs of 10–12 percent or less are often accepted as a reasonable standard of precision for an estimate.
- *90 percent margin of error (MOE):* Plus or minus 1.65 times the standard error of an estimate.
- *90 percent confidence interval (CI):* The 90 percent MOE expressed as a range around the estimate.

Example Calculations

Consider the example of MEDIUM CITY, 5-Year Period ACS Estimate (see Tables 2-7a, 2-7b, and 2-7c). Assume that MEDIUM CITY has a population of 100,000 with an estimated 20,000 school-age children, of whom 3,000 (15 percent) are estimated to be poor. For a 15.0 percent poverty rate for school-age children with a 1.28 percentage point standard error:

- CV = 8.5 percent (1.28/15.0)

size; second, the additional weighting that is needed to compensate for the subsampling increases the sampling error relative to a design without subsampling. The reduction in sample size would be particularly severe for areas in which households are less likely to mail back their questionnaires, but, to ameliorate this effect, the Census Bureau follows up somewhat higher proportions of nonresponding households in census tracts with lower expected mail and telephone nonresponse rates than other areas (refer back to Table 2-3b).

2-C.2.b Illustrative, Approximate Sampling Error Estimates for the ACS

The sampling error in an estimate may be measured by its standard error (see Box 2-5 for definitions of statistical terms). In the case of a percentage estimate, the estimated standard error depends on the size of the percentage and on the sample size for the relevant population that is used as the base for estimating the percentage. Sample size is affected not only

- 90 percent MOE = ± 2.1 (1.28 × 1.65)
- 90 percent CI = 12.9–17.1 percent poor children

Interpretation of Example

What does it mean to say that the 90 percent MOE for this estimate is plus or minus 2 percent, which translates into a CI of 13 to 17 percent poor children? This interval provides a measure of the uncertainty in the estimate due to taking a sample rather than measuring the city's entire population. A different sample would give a slightly different estimate—perhaps 14 percent or 16 percent poor school-age children. If one could look at all the possible samples that could be selected for the city using the ACS sample selection method and construct a 90 percent CI from each sample, one would expect 90 percent of these intervals to include the true percentage of poor school-age children in the city.

Another way to look at this is to consider the 90 percent CI for the percent poor school-age children for all U.S. cities. One would expect 90 percent of the city CIs to include the true percent for their respective cities. However, if the city samples are selected independently, one would expect 10 percent of the cities to have samples for which the percentage of poor school-age children is far enough away from the true value that their 90 percent CIs do not include the true value.

NOTE: The ACS data products show 90 percent MOEs. This practice is not standard in survey research. The standard 95 percent MOE (1.96 times the standard error of an estimate) results in wider CIs, which are more likely to cover the true value.

by the original design, but also by nonresponse and, in the case of the ACS, by the extent of CAPI subsampling that is done for personal visit follow-up to contain costs.

Tables 2-7a, 2-7b, and 2-7c provide rough, approximate estimates of sampling error for an estimated 15 percent poor school-age children from the ACS (1-year, 3-year, and 5-year period estimates) and the 2000 long-form sample for areas ranging in population from 500 to 2.5 million people. The calculations assume that school-age children are 20 percent of the total population and that areas with 3,000 or fewer people are oversampled. The calculations take account—for both the ACS and the long-form sample—of the added sampling error from household nonresponse but not the added error from item nonresponse.

Specifically, Table 2-7a shows relative standard errors—that is, the standard error as a percentage of the estimate, also called the coefficient of variation (see Box 2-5). Table 2-7b shows approximate 90 percent margins of error (MOEs) plus or minus the estimate of 15 percent poor school-age children for each size area (90 percent MOEs are 1.65 times the corresponding standard error). Finally, Table 2-7c translates the MOEs into 90 percent confidence intervals surrounding the 15 percent school-age poverty estimates.

The tables and text use 90 percent MOEs and confidence levels to follow the long-standing practice of Census Bureau publications; however, this practice is not standard in statistical work. It gives smaller MOEs and confidence intervals than is the case when the 95 percent standard is used: with the 95 percent standard, the MOEs and confidence intervals would be about 20 percent larger.

The panel developed the sampling error estimates in the tables by starting with a generalized variance estimation function provided by the Census Bureau for the 2000 long-form sample; we then computed the sampling error estimates for the ACS as multiples of the long-form-sample estimates (see notes at the end of Table 2-7c). The multiplication factors are derived from Census Bureau research with the ACS test sites, the C2SS, and the 2001–2004 ACS test surveys.

For the 2005 ACS, the Census Bureau directly estimated the sampling errors for specific estimates, including not only school-age poverty, but also other characteristics, using a repeated replication method (U.S. Census Bureau, 2006:Ch.12). The 2005 ACS data were only recently released, however, and the panel was not able to analyze their sampling errors; moreover, these estimates pertain only to areas with 65,000 or more people. Nevertheless, an unsystematic examination of the sampling errors for selected 2005 ACS poverty estimates suggests that they are similar to those shown in Tables 2-7a, 2-7b, and 2-7c.

2-C.2.c Assessment of Sampling Error from Illustrative Estimates

Looking at Table 2-7a for the ACS 5-year period estimates, the relative standard errors, or coefficients of variation, for all but the smallest governmental units are half again as large (51 percent) as the corresponding relative standard errors for the 2000 long-form sample. For the ACS 3-year period estimates, the relative standard errors are, in turn, almost 30 percent larger than those for the ACS 5-year estimates and 95 percent larger than those for the long-form-sample estimates. For the ACS 1-year period estimates, the relative standard errors are more than 2 times larger than those for the ACS 5-year period estimates and more than 3 times larger than those for the long-form-sample estimates.

To illustrate, consider first the best case shown in Tables 2-7a, 2-7b, and 2-7c, which is the long-form-sample estimate of 15 percent poor school-age children for an area with 2.5 million people. For this estimate, the relative standard error is only 1.1 percentage points, the 90 percent MOE is only ±0.3 percentage points, and the 90 percent confidence interval is quite narrow—14.7 to 15.3 percent. In other words, the estimate is very precise and provides useful information for a variety of applications, such as fund allocation and program planning. For the same estimate for the same size area from ACS data accumulated over 5 years, the relative standard error is only somewhat larger at 1.7 percentage points, and the ACS data have the advantage of being more up to date.

At the other extreme, the worst case is for estimates of 15 percent poor school-age children for areas with 500 people. These areas are oversampled in both the ACS and the long-form sample, but the sample sizes are so small that the estimates are very imprecise. The 90 percent confidence interval for the estimate of 15 percent poor school-age children from the long-form sample ranges from 5.8 to 24.2 percent poor (90 percent MOE of ±9.2 percentage points), while that from the ACS 5-year period estimates ranges from 3.9 to 26.1 percent poor (90 percent MOE of ±11.1 percentage points). Intervals this wide are not helpful to users, and the range would be wider yet for areas with 500 people that are not oversampled—for example, a township in one of the 38 states for which the Census Bureau does not recognize townships as functioning governments for purposes of oversampling (refer back to Table 2-3), or a block group in a large area.

What constitutes an acceptable level of precision for a survey estimate depends on the uses to be made of the estimate. A commonly used standard for many uses is that a sample estimate should have a relative standard error, or coefficient of variation, of 10 percent or less—sometimes increased to 12 percent or less for a characteristic like poverty, which is clustered within a household or family. This standard does not apply in some instances: specifically, for estimates of proportions that are less than 5 percent

TABLE 2-7a Illustrative, Approximate Relative Standard Errors (Coefficients of Variation, or CVs) for an Estimate of 15 Percent Poor School-Age Children from the ACS and the 2000 Census Long-Form Sample, by Population Size of Area

	Children Ages 5–17		ACS 1-Year Period Estimate		ACS 3-Year Period Estimate		ACS 5-Year Period Estimate		2000 Long-Form Sample Estimate	
Population Size of Area (1)	Total (20% of total pop.) (2)	Poor (15% of ages 5–17) (3)	CV(%) (4a)	(Sample Cases) (4b)	CV(%) (5a)	(Sample Cases) (5b)	CV(%) (6a)	(Sample Cases) (6b)	CV(%) (7a)	(Sample Cases) (7b)
2,500,000	500,000	75,000	***3.8%***	(7,900)	***2.2%***	(23,700)	***1.7%***	(39,550)	***1.1%***	(77,500)
1,000,000	200,000	30,000	***6.0***	(3,150)	***3.5***	(9,500)	***2.7***	(15,800)	***1.8***	(31,000)
500,000	100,000	15,000	***8.5***	(1,600)	***4.9***	(4,750)	***3.8***	(7,900)	***2.5***	(15,550)
250,000	50,000	7,500	***12.1***	(800)	***7.0***	(2,350)	***5.4***	(3,950)	***3.6***	(7,750)
100,000	20,000	3,000	19.1	(300)	***11.0***	(950)	***8.5***	(1,600)	***5.6***	(3,100)
65,000	13,000	1,950	23.7	(200)	13.6	(600)	***10.6***	(1,050)	***7.0***	(2,000)
50,000	10,000	1,500	27.0	(150)	15.6	(450)	***12.1***	(800)	***8.0***	(1,550)
25,000	5,000	750	38.2	(80)	22.0	(250)	17.1	(400)	***11.3***	(800)
20,000	4,000	600	42.7	(60)	24.6	(200)	19.1	(300)	12.6	(600)
10,000	2,000	300	60.4	(30)	34.8	(90)	27.0	(150)	17.9	(300)
5,000	1,000	150	85.4	(20)	49.2	(50)	38.1	(100)	25.2	(150)
3,000	600	90	95.6	(10)	55.0	(40)	42.7	(80)	28.3	(150)
1,500	300	45	72.8	(10)	41.9	(40)	32.5	(70)	21.5	(150)
500	100	15	100.2	(10)	57.7	(20)	44.7	(30)	37.3	(50)

NOTES: The coefficient of variation (CV) is the standard error (SE) of an estimate expressed as a percentage of the estimate (see Box 2-5). CVs that indicate often acceptable levels of precision are in ***bold italics***. ACS estimates are not published below the solid line in column 4 (below 65,000 people) and column 6 (below 20,000 people).

Column 1: Assumed population size of an area.

Column 2: Assumed number of school-age children (ages 5–17); assumed to be 20 percent of column 1.

Column 3: Assumed number of poor school-age children; assumed to be 15 percent of column 2.

Column 4a: ACS 1-year period estimate CV; based on SE estimated as 2.24 times ACS 5-year period estimate SE (see column 6a).

Column 4b: Approximate number of completed sample cases of school-age children (column 2) for 1 year of ACS; estimated as 20 percent of the ACS 5-year number of completed sample cases in column 6b; rounded to nearest 50 (nearest 10 when fewer than 100 cases).

Column 5a: ACS 3-year period estimate CV; based on SE estimated as 1.29 times ACS 5-year period estimate SE (see column 6a).

Column 5b: Approximate number of completed sample cases of school-age children (column 2) for 3 years of ACS; estimated as 60 percent of the ACS 5-year number of completed sample cases in column 6b; rounded to nearest 50 (nearest 10 when fewer than 100 cases).

Column 6a: ACS 5-year period estimate CV; for areas with 1,500 or more people (column 1), based on SE estimated as 1.51 times 2000 long-form-sample SE (from Starsinic, 2005); for areas with 500 people (column 1), based on SE estimated as 1.2 times 2000 long-form-sample SE (see column 7a). The factor of 1.51 accounts for the smaller initial 5-year ACS sampling rate compared with the 2000 long-form-sample rate, as well as CAPI subsampling and nonresponse in the ACS. The factor of 1.2 takes into account that the ACS initial sampling rate is the same as the long-form sampling rate for areas of this small size (see Table 2-3, Part A).

Column 6b: Approximate number of completed sample cases of school-age children (column 2) for 5 years of ACS; estimated as 0.51 and 0.73 times the long-form number of completed sample cases (column 7b) for areas with 1,500 or more people and 500 people, respectively, times 0.97 to allow for nonresponse in the ACS; rounded to nearest 50 (nearest 10 when fewer than 100 cases). The 0.51 and 0.73 factors are based on the ratio of ACS 5-year period cumulative rates of completed sample cases to the 2000 long-form-sample rate (see Table 2-3, Part C). These factors assume a 60 percent mail and CATI response rate from the initial sample as in the 2005 ACS.

Column 7a: 2000 long-form sample CV; based on SE estimated according to the formula in U.S. Census Bureau (2005:8-23), which is: $SE(p) = F(\sqrt{(5/b)\, p\, (100-p)})$, where b is the population base of the estimated percentage, p, and F is a design factor. The base, b, is the number of school-age children in column 2; p is 15 percent poor school-age children; and F varies by the characteristic estimated (poverty) and the assumed long-form-sample sizes for different size areas (from U.S. Census Bureau, 2005:Table C):

- 1.5 design factor for areas of 5,000 or more people, with assumed sample sizes of about 15 percent, instead of 16.7 percent, of school-age children (allowing for unit nonresponse);
- 1.3 design factor for oversampled areas of 3,000 people, with assumed sample sizes of about 20 percent instead of 25 percent of school-age children; and
- 0.7 design factor for oversampled areas of 1,500 people or fewer, with assumed sample sizes of about 45 percent instead of 50 percent of school-age children.

SEs for areas of 3,000 or fewer people that are ***not*** oversampled (including census tracts in larger governmental units and townships not in one of the 12 states in which they are recognized as functioning governments for purposes of oversampling—see Table 2-3, Part A) will be larger than those calculated.

Column 7b: Approximate number of completed sample cases of school-age children (column 2) for 2000 long-form sample; estimated using 2000 long-form sampling rates from Table 2-3 times 0.93, which is the percentage of usable cases of the total sample in 2000. These estimates do not enter into the SE and CV calculations, which are based on design factors estimated for the actual 2000 long-form sample.

TABLE 2-7b Illustrative, Approximate 90 Percent Margins of Error (MOEs), Plus or Minus an Estimate of 15 Percent Poor School-Age Children from the ACS and the 2000 Census Long-Form Sample, by Population Size of Area

Population Size of Area	Children Ages 5–17: Total (20% of total pop.)	Children Ages 5–17: Poor (15% of ages 5–17)	ACS 1-Year Period Estimate 90% MOE	ACS 3-Year Period Estimate 90% MOE	ACS 5-Year Period Estimate 90% MOE	2000 Long-Form Sample Estimate 90% MOE
2,500,000	500,000	75,000	±0.9%	±0.5%	±0.4%	±0.3%
1,000,000	200,000	30,000	±1.5	±0.9	±0.7	±0.4
500,000	100,000	15,000	±2.1	±1.2	±0.9	±0.6
250,000	50,000	7,500	±3.0	±1.7	±1.3	±0.9
100,000	20,000	3,000	±4.7	±2.7	±2.1	±1.4
65,000	13,000	1,950	±5.9	±3.4	±2.6	±1.7
50,000	10,000	1,500	±6.7	±3.8	±3.0	±2.0
25,000	5,000	750	±9.5	±5.4	±4.2	±2.8
20,000	4,000	600	±10.6	±6.1	±4.7	±3.1
10,000	2,000	300	±14.9	±8.6	±6.7	±4.4
5,000	1,000	150	(±21.1)	±12.2	±9.4	±6.2
3,000	600	90	(±23.6)	±13.6	±10.6	±7.0
1,500	300	45	(±18.0)	±10.4	±8.0	±5.3
500	100	15	(±24.8)	±14.3	±11.1	±9.2

NOTES: The 90 percent margin of error (MOE) is plus or minus (±) the standard error of an estimate times 1.65 (see Table 2-7a notes). The MOEs in parentheses are inexact. They are intended simply to indicate the substantial level of imprecision when the area has a very small population. In these cases, the subtraction of the MOE from the 15 percent estimate yields a negative value, which is an impossible result. Although the standard procedure for deriving the MOE is applied throughout the table, the underlying assumption of that procedure—that the sampling distribution of the estimate is approximately the normal distribution—is not applicable in these cases.

TABLE 2-7c Illustrative, Approximate 90 Percent Confidence Intervals (CIs) Around an Estimate of 15 Percent Poor School-Age Children from the ACS and the 2000 Census Long-Form Sample, by Population Size of Area

	Children Ages 5–17					
Population Size of Area	Total (20% of pop. total)	Poor (15% of ages 5–17)	ACS 1-Year Period Estimate 90% CI	ACS 3-Year Period Estimate 90% CI	ACS 5-Year Period Estimate 90% CI	2000 Long-Form Sample Estimate 90% CI
2,500,000	500,000	75,000	14.1–15.9%	14.5–15.5%	14.6–15.4%	14.7–15.3%
1,000,000	200,000	30,000	13.5–16.5	14.1–15.9	14.3–15.7	14.6–15.4
500,000	100,000	15,000	12.9–17.1	13.8–16.2	14.1–15.9	14.4–15.6
250,000	50,000	7,500	12.0–18.0	13.3–16.7	13.7–16.3	14.1–15.9
100,000	20,000	3,000	10.3–19.7	12.3–17.7	12.9–17.1	13.6–16.4
65,000	13,000	1,950	9.1–20.9	11.6–18.4	12.4–17.6	13.3–16.7
50,000	10,000	1,500	8.3–21.7	11.2–18.8	12.0–18.0	13.0–17.0
25,000	5,000	750	5.5–24.5	9.6–20.4	10.8–19.2	12.2–17.8
20,000	4,000	600	4.4–25.6	8.9–21.1	10.3–19.7	11.9–18.1
10,000	2,000	300	0.1–29.9	6.4–23.6	8.3–21.7	10.6–19.4
5,000	1,000	150	(0.0–36.1)	2.8–27.2	5.6–24.4	8.8–21.2
3,000	600	90	(0.0–38.6)	1.4–28.6	4.4–25.6	8.0–22.0
1,500	300	45	(0.0–33.0)	4.6–25.4	7.0–23.0	9.7–20.3
500	100	15	(0.0–39.8)	0.7–29.3	3.9–26.1	5.8–24.2

NOTES: The 90 percent confidence interval (CI) ranges from an estimate minus the 90 percent margin of error to the estimate plus the 90 percent margin of error (see Table 2-7b). The 90 percent confidence intervals in parentheses are inexact. The lower limit of the confidence interval calculated in the standard way is a negative number, which is not possible. For simplicity, the lower limit has been set to 0 in these cases. See also the notes for Table 2-7b.

of a population group in an area. The formula for estimating the coefficient of variation is very unstable for estimates of small proportions, and the estimated coefficients can be misleadingly large.

Table 2-7a shows that estimates from the 2000 long-form sample of 15 percent poor school-age children meet the 12 percent standard of precision for areas with a minimum population between 20,000 and 25,000 people (4,000–5,000 school-age children), but estimates from accumulated ACS 5-year data meet this standard only for areas with at least 50,000 people (10,000 school-age children). Estimates from the ACS 3-year and 1-year data meet this standard only for areas with at least 80,000 people (16,000 school-age children) and 250,000 people (50,000 school-age children), respectively.

The relative standard errors in Table 2-7a are calculated for estimates of 15 percent poor children among all school-age children. The latter group, in turn, is assumed to be 20 percent of the total population, so that poor school-age children are only 3 percent of the total population. If, instead, the table were to provide relative standard errors for estimates of 15 percent poor people—including all children and adults—among the total population, then the levels of precision shown would be considerably improved (see Table 2-8). Thus, the long-form sample would provide estimates that meet the 12 percent or less precision standard for areas as small as 1,500 people, while estimates from accumulated ACS 5-year data would meet this standard for areas as small as 10,000 people. Estimates from accumulated ACS 3-year and 1-year data would meet this standard for areas as small as about 15,000 and 50,000 people, respectively (see Table 2-8). In other words, simple one-way tabulations from the ACS may meet common standards for precision for relatively small areas, although that is not likely to be the case once another variable is introduced, such as age or race.

Users should not simply rely on commonly cited precision standards in deciding whether to use particular estimates. They also need to take into account the specific requirements of their application. For example, deciding which subset of school districts should receive additional funding directed to low-income students may require a narrower confidence interval than the standard. Thus, a 90 percent confidence interval of 12 to 18 percent poor school-age children, which corresponds to a 12 percent relative standard error for an estimate of 15 percent poor school-age children, may be too wide an interval for purposes of fund allocation. Still, for some applications, a ballpark estimate with an even wider confidence interval may suffice.

In deciding which set of ACS estimates is best suited for a particular application, users will need to make trade-offs between timeliness and sampling error. For example, a user could decide that a 3-year period estimate is preferable to a 1-year period estimate for a large city or county in order to achieve a greater level of precision. Alternatively, a user could decide that

TABLE 2-8 Illustrative, Approximate Relative Standard Errors (Coefficients of Variation, or CVs) for an Estimate of 15 Percent Poor People from the ACS and the 2000 Census Long-Form Sample, by Population Size of Area

Population Size of Area (1)	Poor People (15% of total pop.) (2)	ACS 1-Year Period Estimate CV(%) (3a)	ACS 1-Year Period Estimate (Sample Cases) (3b)	ACS 3-Year Period Estimate CV(%) (4a)	ACS 3-Year Period Estimate (Sample Cases) (4b)	ACS 5-Year Period Estimate CV(%) (5a)	ACS 5-Year Period Estimate (Sample Cases) (5b)	2000 Long-Form-Sample Estimate CV(%) (6a)	2000 Long-Form-Sample Estimate (Sample Cases) (6b)
2,500,000	375,000	*1.7%*	(39,550)	*1.0%*	(118,600)	*0.8%*	(197,650)	*0.5%*	(387,500)
1,000,000	150,000	*2.7*	(15,800)	*1.6*	(47,450)	*1.2*	(79,050)	*0.8*	(155,000)
500,000	75,000	*3.8*	(7,900)	*2.2*	(23,700)	*1.7*	(39,550)	*1.1*	(77,500)
250,000	37,500	*5.4*	(3,950)	*3.1*	(11,850)	*2.4*	(19,750)	*1.6*	(38,750)
100,000	15,000	*8.5*	(1,600)	*4.9*	(4,750)	*3.8*	(7,900)	*2.5*	(15,500)
65,000	13,000	*10.6*	(1,050)	*6.1*	(3,100)	*4.7*	(5,150)	*3.1*	(10,100)
50,000	7,500	*12.1*	(800)	*7.0*	(2,350)	*5.4*	(4,050)	*3.6*	(7,750)
25,000	3,750	17.1	(400)	*9.8*	(1,200)	*7.6*	(2,000)	*5.1*	(3,900)
20,000	3,000	19.1	(300)	*11.0*	(950)	*8.5*	(1,600)	*5.6*	(3,100)
10,000	1,500	27.0	(150)	15.6	(450)	*12.1*	(800)	*8.0*	(1,550)
5,000	750	38.2	(80)	22.0	(250)	17.1	(400)	*11.3*	(800)
3,000	450	42.7	(70)	24.6	(200)	19.1	(350)	*12.6*	(700)
1,500	225	32.5	(70)	18.7	(200)	14.5	(350)	*9.6*	(700)
500	75	44.8	(30)	25.8	(100)	20.0	(150)	16.7	(250)

NOTES: See Notes for Table 2-7a—columns 3a–6b in Table 2-8 correspond to columns 4a–7b, respectively, in Table 2-7a. Population sizes for calculating standard errors are in column 1. To obtain an approximate 90 percent margin of error, multiply 15 percent by the estimated coefficient of variation (CV) above to obtain the estimated standard error and multiply the result by 1.65. For example, the 90 percent margin of error for an ACS 1-year period estimate of 15 percent poor people in an area of 65,000 total population is 15 times 0.106 equals 1.6, times 1.65 equals ±2.6, which, in turn, gives a 90 percent confidence interval of 12.4–17.6 percent poor people.

several years of 1-year period ACS estimates will be informative regarding trends and the current situation for the city, even though the estimates are less precise (see discussion in Chapter 3).

2-C.2.d Documentation of Sampling Error

The Census Bureau commendably is trying to impress upon users the extent of sampling error in the ACS estimates. Originally, for data products issued through mid-2005 from the C2SS and the ACS test surveys for 2001–2004, the Census Bureau published upper and lower 90 percent confidence interval bounds (for example, 13–17 percent for a 15 percent estimate of poor school-age children). In response to users, who are more accustomed to the MOE concept (as reported in the media for public opinion polls), the Census Bureau decided to replace the upper and lower bounds in tables with the 90 percent MOEs for specific estimates (such as ±0.2 percentage points). In addition, the Census Bureau will not publish 1-year or 3-year estimates when their imprecision is deemed to be too great. In these instances, the standard tabulation categories will be combined to the point at which the tabulations meet the Census Bureau's threshold for a minimally acceptable level of precision. The 5-year period estimates will not be treated in this manner, even for very small areas for which they are highly imprecise, because the 5-year small-area estimates are the building blocks for a wide range of user applications similar to how the long-form-sample data were used (see Section 4-D.2).

In contrast, the sampling error of the long-form-sample estimates was not highlighted, but instead was contained in footnotes and auxiliary documentation. Moreover, margins of error were not provided for specific estimates; instead, users were provided with general formulas for making their own computations of sampling error. As a result, many users have been unaware of the sampling error in the long form-sample estimates they have been using.

2-D SUMMARY ASSESSMENT

The ACS promises to be of great benefit to many users for a wide range of applications for which they previously relied on information from the decennial census long-form sample. The three major benefits of the ACS are its timeliness, frequency, and the improved quality of the responses when compared with the long-form sample. Not only will the ACS information be released within 8–10 months of completion of data collection, compared with 2 years or more for the long-form sample, but it will also be updated every year instead of once a decade. Moreover, there is strong evidence that the ACS will provide data with reduced nonsampling error because of such factors as the use of trained interviewers to collect the information from nonrespondents. In tests of the ACS, improvements in quality are evident in more complete response to almost every item compared with the long-form sample. Furthermore, in personal interviews, some items will be more accurately reported because the computer-assisted interviewing can more readily correct respondent misperceptions about what is being asked and resolve inconsistent responses.

A complication for users of switching from the census long-form sample to the ACS is the continuous fielding and processing of the ACS. This design produces estimates that pertain to periods of time—averages over 12, 36, or 60 months—instead of the traditional point-in-time estimates with which users are familiar from the long-form sample and other household surveys. Users will need to work together and with the Census Bureau to develop strategies for application of the ACS information that take account of the survey's continuous design. In Chapter 3 we outline some of these strategies.

Sampling error or imprecision of the estimates is a problematic aspect of the ACS, although users should remember that many long-form-sample estimates did not meet common standards of precision for small areas, either (see Tables 2-7a, 2-7b, 2-7c, and 2-8). When the data are averaged over 5 years, it appears that the ACS will provide reasonably precise estimates for small population groups, such as poor school-age children, for areas with 50,000 or more people but not for smaller areas. The ACS 1-year estimates for such a small population group will have low precision unless the area has at least 250,000 people. For larger population groups, such as total poor, the ACS 5-year estimates will likely provide reasonably precise estimates for areas of at least 10,000 people, while the ACS 1-year estimates will meet that standard for areas of at least 50,000 people.

ACS estimates for census tracts, which average 4,000 people, and block groups, which average 1,500 people, will be very imprecise. Indeed, they were not precise from the long-form sample for other than large population groups. However, these areas can be combined in various ways by users who want to compare planning districts, wards, or other components of large cities, counties, and other areas.

The bottom line for large geographic areas—such as states, congressional districts, and large metropolitan areas, cities, and counties—is that the ACS estimates will be a great asset to data users. The data will be timely, up to date, of good quality, and reasonably precise. The 5-year data for census tracts and block groups, while not precise in and of themselves, will provide building blocks that should enable detailed analyses of the populations of large geographic areas.

Estimates from the ACS for small governmental units, even with oversampling, are the most problematic from the perspective of sampling error. Consider a place of 1,500 people and 300 school-age children, of whom 45 children or 15 percent are estimated to be poor. Table 2-7c shows a 90 percent confidence interval of 7 to 23 percent poor school-age children from 5 years of ACS data. Based on the calculations used to derive Table 2-7c, the margin of error of the ACS estimate is 51 percent greater than that from the 2000 long-form sample, which already has a high margin of error, and this increase may be somewhat underestimated. Moreover, the option of combining small governmental units into larger analytical units in order to improve the precision of estimates is less applicable than in the case of combining census tracts or block groups within a larger jurisdiction.

Chapter 3 discusses possible strategies for data users who are interested in very small governmental units to make effective use of the ACS estimates. It will also be imperative to maintain the planned sample sizes for the ACS over time and, furthermore, for the Census Bureau, in cooperation with users, to seek ways to improve the precision of the estimates for small areas (see Section 4-A.5).

3

Working with the ACS: Guidance for Users

The American Community Survey (ACS) can benefit decision makers, planners, and analysts in virtually every type of setting, including federal executive and legislative agencies, state and local government agencies, metropolitan planning organizations, nonprofit organizations, professional associations, universities, think tanks, and private businesses in many sectors. The ACS will also be invaluable to educators, students, the media, and the public.

This chapter addresses how users can work with the various ACS products that are planned to become available and the factors to consider when deciding which products to use for particular purposes. Because not every potential application can be included (or indeed foreseen), the chapter highlights key applications for federal, state, and local government agencies, transportation planners, researchers, the media, and the public who currently use long-form-sample data. The specific users and applications that are discussed include:

- Federal agency users (Section 3-A). Highlighted applications include the use of ACS 1-year, 3-year, and 5-year period estimates for fund allocation to states and localities (3-A.1) and to update the U.S. Department of Housing and Urban Development's income limits for housing assistance programs (3-A.2).
- State agency users (Section 3-B). A highlighted application is the use of ACS 5-year period estimates for fund allocation and grants to localities.

- Local government users (Section 3-C). Uses of different ACS estimates are discussed separately for big cities (3-C.1) and small, oversampled jurisdictions (3-C.2). Also discussed is the special case of jurisdictions with large seasonal populations (3-C.3).
- Transportation planners (Section 3-D). Their applications will rely heavily on the ACS 5-year period estimates and also the public use microdata sample (PUMS) files.
- Academic and other researchers (Section 3-E). Researchers will make heavy use of the PUMS files and of an ACS summary file, similar to Summary File 3 from the 2000 long-form sample that is currently under development.
- Media outlets and the public (Section 3-F). These groups will likely make the most use of the ACS 1-year period summary estimates provided in profiles, ranking tables, and change tables.

Whatever their category (federal agency, local government, other), users should review other sections in addition to the one addressed to them. Many of the specific applications discussed—each of which illustrates some but not all issues regarding use of ACS data products to replace long-form-sample data products—pertain to more than one category of user.

Section 3-G discusses an issue that affects all users—namely, the fact that new population and housing numbers from the decennial census every 10 years will likely interrupt the time series of ACS estimates. The reason is that the ACS estimates for calendar years ending in 0 through 9 each decade will be calibrated at the level of an individual county (or a group of small counties) to annual population estimates updated from the previous census by records of births and deaths and estimates of net migration. A similar calibration will be made to housing unit totals updated from the previous census. When a new census is taken, the census counts will not necessarily coincide with the updated estimates, thereby producing discontinuities in the ACS time series.

The chapter concludes (Section 3-H) by summarizing the panel's general guidelines for effective use of the ACS and suggesting ways in which users who expect to work extensively with the ACS small-area data can prepare during the ramp-up period from 2006 to 2010, as the first sets of 1-year, 3-year, and 5-year period estimates become available. Many users are rightly concerned, first and foremost, with how well the ACS can serve as a replacement source of useful and usable estimates for planning, research, public education, and a host of other applications that currently rely on the long-form sample. The examples in this chapter serve principally to address this underlying concern about the functionality of ACS data to meet current needs.

The decoupling of long-form-type data from the once-a-decade census, however, promises to allow the ACS to develop in ways that, while not clear today, will allow this new survey to become much more powerful than the long-form sample could ever be. We urge users to take a long view of the ACS and be open to new uses that were not possible with the long-form sample but that the continuously updated ACS data can support.

The Census Bureau, for its part, needs to provide as much guidance and training as possible to users to help them maximize the upside and minimize the downside of working with this complex data set. As discussed in Chapter 7, the Census Bureau should proactively identify ways to assist the occasional user who will not be in a position to master the ins and outs of the ACS data—for example, by highlighting estimates that meet reasonable standards for precision. The Census Bureau should also support an ongoing education and outreach program for users who plan to work extensively with ACS data, including the staffs of state data centers and other groups whose mission is to assist the broad user community. As discussed in Chapter 4, the Census Bureau should consider the development of new data products that would help many users, such as 3-year period estimates for statistical areas that are larger than census tracts and smaller than public use microdata areas (PUMAs).

3-A FEDERAL AGENCY USES

Federal government agencies have historically used data from the long-form sample for a wide range of purposes. For at least the past two censuses, the Census Bureau and the U.S. Office of Management and Budget (OMB) have required that each item on the census short and long forms be justified as serving a federal agency need. For the long-form sample, each item had to be needed for federal government use for small areas, often as small as census tracts. Uses were classified into three categories: (1) mandated—that is, the use of census data was specified in legislation; (2) required—that is, data were required by legislation and, although the census was not named as the source, it was the only or the historical source of data; and (3) programmatic—that is, the census data were used for agency program planning, implementation, or evaluation or to provide legal evidence. The same general criteria are being applied with the ACS, although congressional oversight committees have indicated that it is not mandatory to pass legislation in order to add a question to the ACS.[1] It should be noted that where laws or regulations specify the use of census long-form-sample estimates, changes in legislation may be required to permit the use of ACS estimates instead.

[1] Personal communication, Lynda T. Carlson, Director, Division of Science Resources Statistics, National Science Foundation.

BOX 3-1
Selected Federal Agency Uses of Census Long-Form-Sample Data

1. The U.S. Department of Justice uses the long-form-sample data on race, Hispanic origin, educational attainment, language spoken at home, how well English is spoken, and citizenship for census tracts and American Indian areas to implement sections of the Voting Rights Act that deal with bilingual voting assistance.
2. The U.S. Equal Employment Opportunity Commission uses the data on occupation, industry, and demographic characteristics for ZIP codes and other geographic areas to analyze statistical evidence in class action charges of employment discrimination.
3. The OMB Statistical and Science Policy Office uses the data on place of work in relation to place of residence, together with population size and density, for counties and places to define metropolitan and micropolitan statistical areas. These areas have many public- and private-sector applications, including use in determining eligibility for some types of federal funding.
4. The U.S. Department of Health and Human Services uses the data on older people, such as marital status, educational attainment, ancestry, disability status, income, year last worked, and housing characteristics, for counties, cities, and census tracts to measure social isolation and housing needs under the Older Americans Act.
5. The U.S. Department of Transportation uses the data on disability, means of transportation to work, and automobile ownership for traffic analysis zones (small areas made up of one or more block groups) to monitor compliance with the Federal Transit Act and the Americans with Disabilities Act.

Types of federal agency uses of the 2000 long-form-sample data and, prospectively, of the ACS data vary widely (Citro, 2000a; National Research Council, 2004b:Ch. 2; National Research Council, 1995:Apps. C, G, H, M). Ten selected long-form-sample uses are summarized in Box 3-1; they give a flavor of the importance of these data to the operation of the federal government. The ACS should be able to serve all of these federal agency uses and more, providing more up-to-date information of higher quality than the long-form sample. Some of the issues that must be considered in using ACS estimates for federal applications are illustrated below in the discussion of two specific uses: formula fund allocation (3-A.1) and determination of income limits for housing assistance programs (3-A.2).

3-A.1 Allocation of Federal Funds

In sheer dollar terms, perhaps the most important use by federal agencies of long-form-sample data is to allocate billions of dollars of federal funds annually to states and localities (National Research Council, 2000b, 2003a; U.S. General Services Administration, 2006). Long-form-sample

6. The U.S. Department of Housing and Urban Development uses the data on rent and utilities, number of bedrooms, plumbing facilities, kitchen facilities, type of heating fuel, and date when the occupant moved into the unit to determine fair market rents for a base year for some metropolitan areas and nonmetropolitan counties. American Housing Survey data and telephone surveys are used to estimate base-year fair market rents for the remaining areas. Fair market rents, updated yearly from the shelter component of local consumer price indexes and telephone surveys, are used to administer rental housing subsidies and to analyze housing costs relative to household income.
7. The Bureau of Labor Statistics uses the data on sex, age, race, Hispanic origin, labor force status, occupation, industry, and class of worker to develop state-level labor force projections, which are used by program planners, policy makers, job training administrators, and career counselors.
8. The Federal Reserve Board uses the data on race, Hispanic origin, and the year a structure was built for census tracts to report on the record of financial institutions in meeting the credit needs of low- to moderate-income neighborhoods under the Home Mortgage Disclosure Act and Community Reinvestment Act.
9. The U.S. Department of Veterans Affairs uses the data on veteran status and other characteristics of veterans for counties and ZIP code areas to assess changes in the veteran population and to allocate resources, such as outreach specialists and employment and training directors.
10. The U.S. Department of Agriculture uses the data on farm acreage and sales to distribute agricultural research and extension funds to states.

data are used in two ways in allocation formulas: directly, in that long-form-sample estimates provide one or more factors in a formula, or indirectly, in that the formula relies on estimates for which long-form-sample data are one input to an estimation process that also uses other data sources.[2] Whether formulas use long-form-sample estimates directly or indirectly has implications for how proactive the responsible program agency needs to be in deciding how best to use ACS estimates in place of long-form-sample estimates.

3-A.1.a. Use of Long-Form-Sample Estimates in Fund Allocation Formulas

Most federal allocation formulas that incorporate long-form-sample data use the long-form-sample estimates directly; see Box 3-2 for seven ex-

[2]Allocation formulas that use long-form-sample estimates (or estimates that incorporate long-form-sample data) may also include other factors that are often based on administrative records, such as per pupil expenditures or taxable resources.

BOX 3-2
Selected Uses of Long-Form-Sample Estimates in Federal Fund Allocation Formulas

1. Special Education Grants to States ($10.6 billion obligated in fiscal 2005): Allocates funds to states for the education of handicapped children in part by a formula that includes long-form-sample estimates of the number of children in the age ranges mandated by the state's program and the number of children in poverty in those age ranges.
2. Head Start ($6.7 billion obligated in fiscal 2005): Allocates funds to states according to long-form-sample estimates of the number of children ages 0–4 living in poor families. Organizations that operate Head Start programs use long-form-sample data as part of their applications to the U.S. Department of Health and Human Services for funding (within the limit of the funds allocated to their state).
3. Community Development Block Grants, Entitlement Grants and State's Program ($4.1 billion authorized in fiscal 2005): Allocates 70 percent of funds to large jurisdictions (metropolitan counties with 200,000 or more people and cities with 50,000 or more people) and 30 percent of funds to the remaining areas of states on the basis of the larger amount computed under two formulas. One formula uses long-form-sample estimates of total population, poverty population, and overcrowded housing units; the other formula uses long-form-sample estimates of total population, poverty population, and housing units built before 1940.
4. Home Investment Partnerships Program ($1.9 billion authorized in fiscal 2005): Allocates funds to states, cities, urban counties, and consortia of local governments by a formula that uses various long-form-sample estimates, such as the estimated number of rental units built before 1950 occupied by poor families.
5. Workforce Investment Act Adult and Youth Activities Programs ($1.9 billion obligated in fiscal 2005): Allocate funds to states, which reallocate most funds to local areas, by formulas that include long-form-sample estimates of unemployment and economic disadvantage for youths and adults.
6. Title V Maternal and Child Health Services Block Grant to the States ($586 million obligated in fiscal 2005): Allocates funds to states as the sum of the state share of funds received for eight antecedent programs as of 1981 plus a share of any funds appropriated above the fiscal year 1983 level according to the state's number of poor children under age 18 estimated from the long-form sample.
7. The New Freedom Program, enacted August 2005 in the Safe, Accountable, Flexible, Transportation Equity Act: A Legacy for Users (SAFETEA-LU, P.L. 109-59): Allocated $78 million in fiscal 2006 for improved transportation services for people with disabilities. Funds are allocated to urbanized areas with 200,000 or more people (60 percent of the funds), and to states for smaller urbanized areas (20 percent) and for nonurbanized areas (20 percent). Within each group, funds are allocated to urbanized areas and states on the basis of the number of people with disabilities.

amples. Long-form-sample estimates enter indirectly into the allocation of funds under Title I of the No Child Left Behind Act (estimated $12.7 billion obligated in fiscal 2005). This program allocates funds to school districts to meet the needs of educationally disadvantaged children by formulas that include estimates of poor school-age children. In the past these estimates were obtained from the most recent census long-form sample; currently, more up-to-date estimates are obtained from statistical models developed by the Census Bureau in its Small Area Income and Poverty Estimates (SAIPE) program.[3]

The SAIPE state- and county-level models include long-form-sample poverty estimates as one input together with more up-to-date information from administrative records to predict school-age poverty from a 3-year average of data from the Current Population Survey Annual Social and Economic Supplement (CPS ASEC). The school district-level model uses the previous census long-form-sample estimates of within-county school district shares of poor school-age children to apply to the updated county model estimates of the number of poor school-age children. The SAIPE program produces annual estimates with a 2-year lag between release and the estimates' income reference year; the lag is due to delays in acquiring administrative records that are required for the modeling.

3-A.1.b Using ACS Estimates in Formulas

Because the 2010 census will not include a long-form sample, policy makers and program managers must develop strategies for introducing ACS estimates into funding program allocation formulas that previously used long-form-sample estimates and decide whether such a change will require legislation or can be handled by regulation. The primary benefits of using ACS estimates are that they will be more timely and up-to-date and probably of higher quality than estimates from the long-form sample, so that the resulting fund allocations will more accurately reflect the distribution of needs among eligible areas.[4] Still, the ACS estimates will have higher sampling error than long-form-sample estimates.

Role of Policy Makers The role that policy makers and program managers play in decisions about the use of ACS estimates in allocation formulas

---

[3]See National Research Council, 2000a; http://www.census.gov/hhes/www/saipe/saipe.html.

[4]This discussion does not address whether the variables in a formula (in the absence of data quality concerns) produce the most equitable fund distributions in light of a program's original goals (see National Research Council, 2003a). The need to replace long-form-sample estimates with ACS estimates could trigger reconsideration of the variables and other features in a formula, but that is outside the panel's charge.

depends at least in part on whether the estimates will enter into a formula indirectly or directly. Indirect uses will require less in-depth consideration by program and policy people because the statistical agency that produces the relevant estimates will presumably tackle the matter. Thus, the Census Bureau SAIPE staff will presumably determine effective ways of including ACS data in their model-based estimates of poor school-age children that are used in the allocation of education funds to school districts under the No Child Left Behind Act.

The Bureau of Economic Analysis (BEA) is already incorporating ACS data into its county-level per capita income estimates, which could be considered for possible use in fund allocation. At present, only the BEA state-level per capita income estimates, which do not require 2000 long-form-sample (or ACS) data, are used in federal fund allocation programs, including the largest program—Medicaid ($193 billion of federal funds obligated in fiscal 2005)—and other programs that use the Medicaid formula.

BEA develops county (and state) per capita income estimates from federal and state administrative records, censuses and surveys, and census-based population estimates (as denominators). Currently, BEA is moving to use the ACS, in place of the 2000 long-form sample, as a source of data on intercounty commuting. This information is needed to convert estimates of per capita income by county of workplace to those by county of residence. The BEA estimates are produced annually for counties about 15 months after the end of a year.[5]

When ACS estimates are to replace long-form-sample estimates directly in a fund allocation formula, then program and policy people must be more involved. Factors in choosing which ACS period estimates to use (1-year, 3-year, or 5-year) include not only the extent of sampling error, but also the desired frequency with which funds are to be reallocated among areas and the types and population sizes of eligible geographic areas. Of course, during the ramp-up period between 2005 and 2010, agencies' choices are constrained by whether the estimates that best serve their needs are available. For example, if 5-year period estimates must be used to obtain an acceptable level of precision, then agencies will need to rely on the long-form-sample estimates until 2010 when ACS 5-year period estimates become available for the period 2005–2009.

Currency, Precision, and Stability Considerations In determining which ACS estimates to use in an allocation formula (assuming they are available for all eligible areas), decision makers should identify key characteristics that the estimates must satisfy. If currency of the information is paramount, so that areas with the greatest present need receive the most funding, then

[5]See http://www.bea.gov/bea/regional/articles.cfm?section=methods.

1-year period ACS estimates will be preferable to 3-year or 5-year period estimates, and 3-year period estimates will be preferable to 5-year period estimates. However, 1-year (or 3-year) period estimates may not be sufficiently precise—that is, may not have low enough sampling error—for fund allocation purposes. If estimates are not precise, then nontrivial changes in funding allocations from year to year may be an artifact of sampling error.[6]

A related consideration is the weight to give to currency for the most equitable allocations versus the practical arguments for moderating the magnitude of year-to-year changes to facilitate program planning and implementation. Many programs moderate fluctuations in program allocations through features of the formula. For example, under a hold-harmless provision, every locality is entitled to receive at least as many dollars as a specified percentage—which could be 100 percent—of its prior-year dollars.

Such legislative provisions have drawbacks, in that their use can delay the responsiveness of the funding formula to changes in need and also create inequitable allocations that are an artifact of sampling error in the estimates. For example, if legislation sets a threshold for eligibility, such as a minimum number of poor school-age children, and an area exceeds that threshold in a particular year because the estimate is greater than the threshold level due to sampling error, it will erroneously receive funding at that time. Moreover, the application of a hold-harmless provision will enable the area to retain funding in subsequent years, even though it was not eligible in the first place. An alternative approach to achieve more stable funding streams, while still responding to changes in need, is to eliminate thresholds and hold-harmless provisions and instead smooth the estimates themselves—for example, by using 3-year period estimates rather than 1-year period estimates for allocations to states (see Zaslavsky and Schirm, 2002). Implementation of this approach could require changes in legislation.

Geographic Area Considerations Yet another consideration in the selection of ACS estimates for fund allocation is the types and population sizes of geographic areas that are eligible for funding. Some formulas apply to a single type of geographic area, such as states, while others include several types of areas, such as states, cities, and counties, and still others have population size thresholds that may vary by type of area.

Consider first a formula allocation program, such as Special Education Grants to States, which uses state-level estimates of all children and poor

---

[6]See Box 2-5 for definitions of sampling error and related terms, such as coefficient of variation and margin of error.

children in specific age ranges, leaving it to state agencies to make further allocations to localities. In this instance, the most straightforward method for taking advantage of the ACS would simply be to substitute up-to-date ACS 1-year period estimates for the long-outdated 2000 census long-form-sample estimates in the formula. The ACS 1-year period estimates should have low sampling error for all 50 states and the District of Columbia. For example, estimates of poor school-age children may have a coefficient of variation of less than 8 percent for the smallest states, with about 600,000 people (see Table 2-7a), while the coefficient of variation of these estimates may be only 1 percent for the largest states, with 20 million people. Moreover, the Special Education Grants Program has minimum funding provisions that would moderate year-to-year fluctuations in allocation amounts from the use of annually updated ACS 1-year period estimates in place of the once-a-decade long-form-sample estimates. Should it be deemed desirable to further smooth funding amounts, the Special Education Grants Program could average 2 years of 1-year period estimates or use 3-year period estimates, which should have very low sampling error for all 50 states and the District of Columbia.

Programs like Community Development Block Grants and Home Investment Partnerships, however, provide funds to different types of governmental units, some of which are smaller in population size than the cutoff of 65,000 people or more for ACS 1-year period estimates. For these programs, it will not be possible to take the simple approach outlined above because ACS 1-year period estimates will not be available for all eligible areas. Moreover, while ACS 3-year period estimates may be available for all eligible areas, they may not be sufficiently precise for some of them. For example, should the needed estimates represent a group as small as poor school-age children, then the 3-year period estimates will not have a reasonably small coefficient of variation until the eligible area has a population of at least 80,000 people (see Table 2-7a).[7]

For such programs as Community Development Block Grants, for which governmental units as small as 50,000 people are eligible for funding, agencies must carefully balance the need for more up-to-date information from using 3-year period estimates against precision requirements that will be better satisfied with 5-year period estimates. For programs for which governmental units must have at least 100,000 people to be eligible for

[7]Table 2-7a should be used only as a very rough guide to expected levels of sampling error for estimates for different size areas from ACS 1-year, 3-year, and 5-year period estimates. The sampling error will differ from that shown in the table for a characteristic that is a different percentage of the population from poor school-age children (as seen in Table 2-8). The sampling error will also depend on the sample size that the ACS achieves in the field for the particular governmental unit.

funding, agencies must trade off the timeliness of 1-year period estimates and the greater precision of 3-year period estimates.

When agencies decide that there is no choice but to use 5-year period estimates from the ACS in a funding formula in order to gain sufficient precision, they should be aware that inequities may result. For example, two areas may have the same 5-year period poverty rate and therefore receive the same allocation, even though one area may have a sharply increasing poverty rate and the other area a sharply decreasing poverty rate over the period. Even in this case, however, the use of ACS 5-year period estimates would represent an improvement over the continued use of the increasingly out-of-date 2000 long-form-sample estimates.

At present, the only federal funding program that makes allocations to areas with fewer than 50,000 people is the No Child Left Behind Act, which allocates funds to school districts, varying in size from a few hundred to several million people (see Table 2-4). The SAIPE estimates that are used for the allocations are more up to date than the direct long-form-sample estimates later in the decade, but they rely on statistical models. The incorporation of ACS data into the SAIPE county and school district models should make it possible to improve their timeliness and precision.

Consistency of Period Estimates In trading off such considerations as currency and precision, in no instance should agencies use in their allocation formulas a mix of different periods of ACS estimates—for example, 1-year (or 2-year) period estimates for larger areas and 3-year or 5-year period estimates for smaller areas—in an attempt to equalize the sampling error across areas. The reason has to do with equity: formulas generally allocate shares of a fixed pie, so that the data used in the allocation should reference the same time period. Otherwise, inequitable outcomes may occur. For example, consider a large county and a medium-sized city, both of which experience rapidly increasing poverty over 5 years. If in a poverty-based formula, 1-year period estimates are used for the large county and 3-year period estimates are used for the medium-sized city, then the county will likely receive more than its fair share of funds over the 5 years compared with the city because the 1-year period estimates will likely exhibit more growth in poverty than the 3-year period estimates.

3-A.2 Determination of Median Incomes for Counties

The U.S. Department of Housing and Urban Development (HUD) obligates $27 billion annually for assisted housing programs in which families that have incomes below specified limits are eligible to live in public housing or receive rent subsidies. The income limits are determined separately for every metropolitan area and nonmetropolitan county as a function of median income. Historically, HUD has used census long-form-sample median family income estimates, adjusted at the national level to agree with estimates from the CPS ASEC for the census income year, as the starting point to develop current fiscal year estimates for each area. To update the long-form-sample estimates, HUD uses the most recent Bureau of Labor Statistics (BLS) wage and salary data for counties, adjusted to match median income estimates for the nine census geographic divisions from the most recent CPS. As a final step, HUD projects the median family income estimates to the middle of the fiscal year for which the agency is setting housing assistance income limits.

The advent of the ACS means that HUD will no longer need to update long-form-sample median family income estimates from data sources, such as BLS person-level wage data, that do not reflect the same concept of total family income. The use of ACS county-level median family income estimates to determine area-specific eligibility for subsidized housing, however, raises at least three important issues: (1) whether achieving comparable levels of precision across areas is preferable to using the same periodicity of ACS estimates (1-year, 3-year, or 5-year) for all areas; (2) the possible effects on the accuracy of ACS income estimates from the moving reference period (respondents are asked about the prior 12 months rather than a consistent prior calendar year); and (3) the possible effects on the accuracy of ACS income estimates from the Census Bureau's procedure for adjusting income amounts for inflation. (See ORC Macro, 2002:162–171, for a fuller discussion of these and other issues.)

3-A.2.a Period Estimates for 1, 3, or 5 Years?

HUD requires median family income estimates each year for all 3,000-plus counties in the United States. One-year period estimates of median family income will probably be reasonably precise for counties with at least 50,000 people, and 3-year period estimates will probably be reasonably precise for counties with at least 20,000 people. (Estimates of median family income are about twice as precise and therefore have only about half the coefficient of variation of estimates of poor school-age children—see Table 2-7a.) However, 1-year period estimates will be available only for counties (and other governmental and statistical areas) with at least 65,000 people, yet three-fourths of counties are smaller than that. Moreover, two-fifths of counties have fewer than 20,000 people so that 5-year period estimates will be the only available source for about 1,300 counties (see Tables 2-4 and 2-5).

A study conducted for HUD by ORC Macro (2002:169) suggested that HUD might want to use 1-year period ACS median family income estimates for counties with 200,000 or more people, 3-year period estimates (when they become available) for counties with 65,000 to 200,000 people, and 5-year period estimates (when they become available) for the remaining three-fourths of counties. This strategy is conservative with regard to sampling error. A reason to be conservative is that HUD is concerned not only with having estimates that are as up-to-date as possible, but also with reducing year-to-year fluctuations in median family income estimates that are due to sampling error.

The previous discussion of using ACS estimates in fund allocation formulas concluded that estimates for different periods should not be used in the same formula because the resulting fund allocations could be inequitable. The HUD use of median income estimates, however, is different in that HUD is not allocating shares of a fixed budget allotment; instead, it is determining an eligibility threshold for an entitlement. Families living in a metropolitan area or a nonmetropolitan county that have incomes below a specified percentage of the median income for that area are entitled to subsidized housing, and the median income levels in other areas are not relevant to this determination. (In practice, entitled families may be put on a waiting list because not enough housing is available.) Given that housing assistance is allocated to individual families on the basis of their incomes as a percentage of the median for their area, it makes sense to use the estimate for each metropolitan area or nonmetropolitan county that has an acceptable level of sampling error and is as up-to-date as possible.

3-A.2.b Moving Reference Periods

Because the ACS is conducted on a continuous monthly basis, the questionnaire items change in their reference period across the year. Many questions (see Table 2-2) refer to the time when the respondent fills out the questionnaire, which could be any date from January to December of a calendar year. Questions on income ask for amounts received in the 12 months prior to when the respondent fills out the questionnaire. Consequently, the ACS 1-year period income estimates will include reference periods that span a full 23 months: for 2005 income estimates, for example, the reference periods range from January–December 2004 for people who responded in January 2005 to December 2004–November 2005 for people who responded in December 2005.

There has been little research on the effects on accuracy of reporting income amounts with a moving reference period of the past 12 months compared with the fixed reference period of the previous calendar year that is used in the long-form sample and the CPS ASEC. A split-sample experiment with mail responses to the ACS questionnaire in October–December 1997 produced the unexpected result of no significant differences in median total income of individuals who were asked to report for the preceding calendar year (January–December 1996) and those who were asked to report for the past 12 months, covering October 1996–September 1997, November 1996–October 1997, and December 1996–November 1997 (Posey and Welniak, 1998). What factors explain this result—for example, whether respondents tend to annualize their current income or to report income for the previous calendar year regardless of the reporting period—are not known. Carefully designed research will be needed to assess the effects of the ACS reference period on income statistics, such as research that compares an external measure of income from administrative records with survey responses for the same individuals.

3-A.2.c *Inflation Adjustments*

For completeness, this section discusses inflation adjustments not only for income, but also for housing amounts. The latter amounts include housing value, monthly contract rent, monthly gross rent (contract rent plus utilities), and monthly selected owners' housing costs (mortgage payments, utilities, taxes, property insurance).[8]

Income To put income amounts that are reported for differing 12-month reference periods on a comparable calendar-year basis, the Census Bureau expresses them in constant dollar terms by using the national consumer price index for urban consumers-research series (CPI-U-RS) for the latest calendar year covered by an estimate.[9] For 1-year period income estimates for 2005, for example, each reported amount on a person record is adjusted by the ratio of the annual average CPI for 2005 divided by the average of the monthly CPIs for the particular 12-month reporting period for that person. For 3-year period estimates for, say, 2005–2007, the incomes for people sampled in 2005 and 2006 (which have already been adjusted to calendar 2005 or 2006 on a 1-year basis) are adjusted to calendar year 2007 by the ratio of the annual average CPI for 2007 divided by the annual average CPI for 2005 or 2006, as the case may be (see Table 3-1 for how this adjustment is carried out).

[8]To create monthly gross rent and selected owners' housing costs, the amounts reported for some costs for either the prior 12 months or as "annual" amounts—see Table 2-2—are converted to monthly amounts.

[9]"The Bureau of Labor Statistics (BLS) has made numerous improvements to the Consumer Price Index (CPI) over the past quarter-century . . . [but] historical price index series are not adjusted to reflect the improvements. Many researchers . . . expressed an interest in having a historical series that was measured consistently over the entire period. Accordingly, the Consumer Price Index research series using current methods (CPI-U-RS) presents an estimate of the CPI for all Urban Consumers (CPI-U) from 1978 to present that incorporates most of the improvements made over that time span into the entire series" (http://www.bls.gov/cpi/cpiurstx.htm).

TABLE 3-1 Hypothetical Inflation Adjustments for Person Income in the ACS

Data: Consumer Price Index for All Urban Consumers (CPI-U), rounded (1983–1984 = 100)

Year	Jan.	Feb.	Mar.	Apr.	May	June	July	Aug.	Sept.	Oct.	Nov.	Dec.	Average
2004	185	186	187	187	189	190	189	190	190	191	191	190	188.9
2005	191	192	193	195	194	195	195	196	199	199	198	197	195.3
2006	198	199	200	201	202	203	204	205	206	207	208	209	203.5
2007	210	211	211	212	213	214	215	216	216	217	218	219	214.5

(1) Adjustment Factors (x) for 2005 ACS 1-Year Period Person Income
(x is applied to prior 12 months' reported income)

January 2005 sample persons (income reported for 01/04–12/04)	$x = [195.3/((185 + \ldots + 190)/12)] = 1.034$
February 2005 sample persons (income reported for 02/04–01/05)	$x = [195.3/((186 + \ldots + 191)/12)] = 1.032$
. . .	. . .
November 2005 sample persons (income reported for 11/04–10/05)	$x = [195.3/((191 + \ldots + 199)/12)] = 1.006$
December 2005 sample persons (income reported for 12/04–11/05)	$x = [195.3/((190 + \ldots + 198)/12)] = 1.003$

(2) Adjustment Factors (x) for 2006 ACS 1-Year Period Person Income

For each monthly sample as in (1) x = [203.5/(average of factors for previous 12 months)]

(3) Adjustment Factors (x) for 2007 ACS 1-Year Period Person Income

For each monthly sample as in (1) x = [214.5/(average of factors for previous 12 months)]

(4) Adjustment Factors for 2005–2007 ACS 3-Year Period Person Income

a. For 2005 sample persons	$x = 214.5/195.3 = 1.09$, x is multiplied by the adjusted 2005 income (1)
b. For 2006 sample persons	$x = 214.5/203.5 = 1.05$, x is multiplied by the adjusted 2006 income (2)
c. For 2007 sample persons	$x = 214.5/214.5 = 1.00$, x is multiplied by the adjusted 2007 income (3)

(5) 2005–2007 ACS 3-Year Period Person Income Estimates for All Persons

Calculated as (4.a + 4.b + 4.c)/3

SOURCE: See http://www.bls.gov for monthly CPIs through February 2006; other months are hypothetical.

This adjustment expresses all of the reported income amounts for a given period (1 year, 3 year, or 5 year) in a comparable manner with regard to purchasing power as of the most recent calendar year in the period. Such an adjustment should not be confused with a current-year estimate. For example, an inflation-adjusted 5-year period median income estimate covering years 2005–2009 is not an estimate of median income for the latest year (2009); instead, it is an estimate of the median of all of the reported income amounts over the 5 years expressed in 2009 dollars.

It is possible that for many applications users may prefer an inflation adjustment to the most recent calendar year to no adjustment at all. For some applications, users may find that an inflation adjustment to the latest year is not adequate. For example, users frequently wish to compare ACS income estimates with those from other household surveys. Yet a 1-year period income estimate from, say, the 2005 ACS that expresses income amounts in constant 2005 dollars for reference periods spanning January 2004 through November 2005 is not comparable to an estimate from a survey, such as the 2006 CPS ASEC, that collects all income amounts for the same fixed reference period of calendar 2005. The reasons are that prices are not income, and incomes may grow faster (or slower) than prices.

Turek, Denmead, and James (2005) illustrate the problems of using price change as a proxy for income change when comparing survey estimates. For 1998—a period of strong economic growth—they estimated that the Census Bureau's inflation adjustment would make up only 22 percent of the difference between average person total income from a simulated 1998 ACS sample compared with average person total income reported for calendar year 1998. The simulations used the Survey of Income and Program Participation, which collects income on a 1-month or 4-month basis over a multiyear period. The analysis compared income amounts reported by people for the 12 months preceding each month in 1998 unadjusted for inflation (average $17,304 person total income), the same income amounts adjusted for inflation to calendar 1998 (average $17,447), and income amounts reported by the same people for all 12 months of 1998 (average ($17,945). Presumably, the difference between the second and third figures occurs because, on average, people received pay raises or returns on assets between their income reporting period and the end of the calendar year that exceeded the rate of inflation (for example, a big raise in June 1998 for an individual who reported income for June 1997–May 1998).

Many applications, such as HUD's use of county-level median income to determine eligibility for housing assistance programs, require current-year estimates. The ACS inflation-adjusted period estimates will not be optimal for such applications, given that they represent averages over the period expressed in dollars for the latest year in the period instead of estimates for the latest year. The inability of the inflation adjustments to represent latest-year income amounts is likely to be much more pronounced for the ACS 3-year and 5-year period estimates compared with the 1-year period estimates. Yet for the county-level estimates of median income required by HUD, only one-fourth of counties will have 1-year period estimates available, and over 40 percent of counties will have only 5-year period estimates available.

Even when users find inflation-adjusted period income estimates to be reasonably satisfactory for an application, they may prefer adjustments that reflect variations in price changes, such as the use of different price indexes for different geographic areas. However, only limited data are available for this purpose (see Section 4-D.3).

Finally, in the special case of poverty estimates, the Census Bureau's method for determining poverty status for families and their members does not require adjusting income amounts for inflation. This situation arises because the Census Bureau compares the income of a family (or unrelated individual) for a 12-month reporting period, *not* adjusted for inflation, to 12-month nominal dollar thresholds by family size and type for that same period. These thresholds are derived from a base-year threshold (1982) using the national CPI, as is done in the official poverty measure. The only difference from the official measure, which uses calendar-year thresholds, is that the threshold for each family is the average of the CPI-adjusted monthly thresholds for that family's 12-month income reporting period. For a 5-year period estimate, then, the poverty rate is the average rate of everyone in the sample over the 5 years.

Housing For housing amounts, such as value, rent, utilities, property taxes, and others, the Census Bureau makes no inflation adjustments for the 1-year period estimates. When, however, the 1-year period estimates for housing amounts are cumulated over 3 or 5 years, the Census Bureau adjusts them for inflation by using the ratio of the annual average CPI value for the latest year of the period to the annual average CPI value for the year for which the amounts were reported.

The issues that can affect uses of the inflation-adjusted income amounts can also affect the inflation-adjusted housing amounts. The ACS 3-year and 5-year period estimates for rent, housing value, utilities, and other housing amounts expressed in dollars for the latest year are not the same as estimates for the latest year. Moreover, increases (or decreases) in housing amounts often differ across areas and by item—for example, housing values in recent years have increased much more than many other items in the national CPI and have increased much more in some areas than others.

Section 4-D.3 discusses several issues involved in adjusting ACS period estimates of income and housing amounts for inflation. A key question that needs to be resolved by discussion among users and the Census Bureau is the purpose of the adjustment. Assuming that users largely prefer an inflation adjustment to the latest year of the period, then the question becomes one of the specific method(s) to use. Another issue is how to assist users who require current-year estimates rather than averages expressed in latest-year dollars.

3-A.2.d Alternative Sources of Median Income Estimates

Abstracting from the previous discussion, there are at least three possible approaches for HUD (and other users) to obtain median family income estimates for the previous calendar year that are reasonably precise for all counties:

1. As suggested by ORC Macro (2002:162–171), HUD could ask the Census Bureau to produce 1-year period ACS median family income estimates for combinations of small counties to accompany the estimates that are published for larger counties. (If PUMA combinations of counties are suitable, then HUD could use the 1-year period estimates that will be regularly produced for PUMAs.)
2. HUD could plan to use the SAIPE model-based median household income estimates for all counties once the model is modified to incorporate information from the ACS. A drawback of the SAIPE estimates is the 2-year lag between release and the calendar year reference period of the estimates. Also, at present the SAIPE estimates represent a 3-year average, but this may change if the ACS is used as the dependent variable in the model equations in place of the CPS ASEC. An advantage of model-based estimates is that they exhibit less variability in precision across areas than direct estimates (see Bell, 2006, for comparisons for states).
3. HUD could decide to use ACS 3-year or 5-year period estimates for counties and ask the Census Bureau to develop an alternative method for adjusting income responses in the ACS to reflect HUD's need for current-year estimates. For example, appropriate year-to-year ratios of the BLS wage data for counties could be applied to the ACS 3-year or 5-year household income estimates, *not* adjusted for inflation, to produce current-year median income estimates.

3-B STATE AGENCY USES

State governments have many uses of census long-form-sample data for program planning, implementation, and evaluation that are similar to those of federal government agencies (see Section A above). They also have many uses that are similar to those of local government agencies (see Section 3-C below). Because of these similarities, the only major use by states that we explore in detail here is allocation of state funds to localities in Section 3-B.1. Strategies for using ACS data instate fund allocations are considered in Section 3-B.2.

It is worth noting that many state uses are to respond to requirements of the federal government. For example, HUD requires states and localities to have a Comprehensive Housing Affordability Strategy. This plan includes an assessment of the housing needs of families residing in a jurisdiction that is developed, in part, from long-form-sample data on demographic and housing unit characteristics for individual census tracts in the area. Such applications in the ACS context will require use of the 5-year period estimates for census tracts, which will likely need to be aggregated into larger areas to obtain sufficient precision.

3-B.1 Allocating State Funds to Localities

Under many federal fund allocation programs, states are responsible for distributing most or all of their funds to localities by using long-form-sample data. Many states also allocate their own funds by means of formulas to local jurisdictions, such as counties and school districts (see examples in National Research Council, 2000b:Table 2-1). The most used sources of data for state funding formulas are estimates from the previous long-form sample and state administrative records, such as school lunch data and income tax records.

The problems with long-form-sample estimates, as noted throughout this report, include that they are not timely, that they become increasingly out of date over a decade, and that they suffer from high levels of item nonresponse because long-form data collection takes a back seat to completing the basic census count. The long-form-sample estimates also have large sampling errors for small areas.

Administrative records have problems as well. They may not correspond that closely to the target population for a program—for example, school lunch data, which are often used in state formulas to target funds to school districts with poor children, may not closely track the poverty population because children in families with incomes as high as 185 percent of the poverty threshold are eligible for reduced-price lunches. In addition, program participation may be affected by such factors as outreach activities that operate more strongly in some areas than others. To the extent that this is true, the use of administrative data on school lunch or food stamp participants as a proxy for the poverty population may not give consistent estimates across areas (see National Research Council, 2000a:App. D).

3-B.2 Strategies for Using ACS Data in State Fund Allocations

States should consider the use of ACS estimates in place of the data sources they currently use for allocating state funds to localities. The same considerations apply as discussed for federal fund allocation, such as the value placed on having the most up-to-date estimates in contrast to the stability of funding streams and the types and population sizes of eligible geographic areas. It is most likely that states would need to use ACS 5-year period estimates to allocate funds to local areas given that 1-year and even 3-year period estimates are not available or not sufficiently precise for many jurisdictions.

There may be instances in which a state believes it is important that fund allocations (or another application) reflect data that are as current as possible and when reasonably precise 1-year (or 3-year) period estimates are available for many but not all eligible jurisdictions. Should a state find itself in this situation, it could consider using a simple procedure to update the 5-year period estimates for jurisdictions for which they are the only reasonably precise estimates available (see Section B.3 below). The intent would be to put the 5-year period estimates on a comparable basis with 1-year (or 3-year) period estimates and not have to discard the more up-to-date estimates for those jurisdictions for which they are available and sufficiently precise for their intended use. Federal agencies may also be able to use this procedure for some applications.

For other applications, the goal may be currency of estimates, but there may be reason to believe that a simple updating procedure will not give good results because its underlying assumptions about change over time among areas are unrealistic. In such instances, a more advanced form of small-area estimation is called for. Such estimation requires additional data from administrative records or other sources, similarly to the way that the Census Bureau's SAIPE program uses food stamp and federal income tax data to generate updated county estimates of poor school-age children.

Before deciding to use any type of updating procedure, simple or complex, it is essential to carefully examine the procedure's underlying assumptions. It may be that less current estimates are preferable to more current estimates produced with an unrealistic updating procedure.

3-B.3 Example of a Simple Updating Procedure

Table 3-2 provides an example of a simple procedure to produce current county-level estimates of poor school-age children for possible use in allocating state funds to counties. The state in this example plans to use ACS 1-year period estimates for as many counties as practicable and to adjust ACS 3-year or 5-year period estimates for the remaining counties to

TABLE 3-2 Example of Simple Method to Update ACS 5-Year Period Estimates for 2010–2014 to Latest Year (2014), Four Small Counties (A, B, C, D) in State *X*, Using Data for Two Public Use Microdata Areas (PUMAs)

	PUMA	County		PUMA	County	
	1	A	B	2	C	D
Total Population (20% are school-age children)	100,000	50,000	50,000	100,000	50,000	50,000
Estimated Number of Poor School-Age Children						
1. 5-year period ACS estimate, 2010–2014	4,000	1,500	2,500	2,000	1,000	1,000
2. 1-year period ACS estimate, 2014	5,000	(not available)		2,100	(not available)	
Change in School-Age Poverty						
3. Ratio of 2014 PUMA estimate to 2010–2014 PUMA estimate (line 2/line 1)	1.25	(not applicable)		1.05	(not applicable)	
Estimated Number of Poor School-Age Children, 2014						
4. For PUMAs: ACS 1-year period estimate (line 2) For counties: Simple method, using county ACS 5-year Period estimate and PUMA change ratio (line 1 × line 3)	5,000	1,875	3,125	2,100	1,050	1,050

How well does the simple method to update a 5-year average estimate of poor school-age children to the latest year work?

- Assume that the actual number of poor school-age children for the four counties in 2014 is 2,100 for County A, 2,900 for County B, 800 for County C, and 1,300 for County D.
- For Counties A and B in PUMA 1, which both experienced an increase in school-age poor children from the average 5-year estimate to the latest year (1,500 to 2,100 and 2,500 to 2,900, respectively), the simple updating method makes their 5-year period estimates more current.
- For Counties C and D in PUMA 2, the simple method is less satisfactory. Because County C bucked the overall trend and had a decrease in school-age poor children (from 1,000 to 800), the PUMA 2 change ratio between the 2014 estimate and the 2010–2014 estimate is very small. Consequently, the simple updating method does not capture either the substantial decrease in school-age poor children in County C or the substantial increase in school-age poor children (1,000 to 1,300) in County D.

NOTE: See text on the need to understand and evaluate the assumptions that underlie any modeling procedure, even the simplest, before using a particular procedure to update ACS 5-year (or 3-year) period estimates to 1-year period estimates. The method illustrated assumes that the numbers of poor school-age children grew at the same rate for each county in a PUMA, or, alternatively, that each county's share of poor school-age children in a PUMA remained the same over time.

represent the latest year of the period. Say that the state has 1 million total people resident in eight counties: four counties are large, each with 200,000 people, and four counties are small, each with 50,000 people. The four smaller counties make up two PUMAs. (In actuality, two-thirds of counties are smaller than this.)

The state could use 1-year period ACS estimates of poor school-age children for the four large counties directly in the allocations. Also available would be 1-year period estimates for the two PUMAs, individually and combined, which could be used to adjust 5-year period estimates for the four smaller counties to the same 1-year period. The simple updating procedure would apply the ratio for the PUMAs of the 1-year and the 5-year period estimates of poor school-age children to the 5-year period estimates for each county component.

Using separate ratios for the two PUMAs (as in Table 3-2) would better capture differences among the smaller counties than would using a single ratio for the two PUMAs combined, but the combined ratio would be more precise than the two separate ratios. Even using separate ratios, the updated estimates for the counties in PUMA 2 are not as realistic as those for the counties in PUMA 2 because one county in PUMA 2 experienced a decrease in school-age poverty and not an increase as in the other three counties.

The simple procedure works best when it only has to be used—and, therefore, its assumptions only have to be invoked—for a small fraction of the total number of jurisdictions. Because only about half a dozen states have ACS 1-year (or even 3-year) period estimates available for most counties, the procedure may not be widely useful when the goal is to adjust 5-year period estimates for smaller counties to the latest year.

The Census Bureau's SAIPE program currently uses this type of simple procedure to produce updated estimates of poor school-age children for school districts within counties. In that application, good administrative data are available with which to update the county estimates, but the updating procedure for school districts has to assume that the within-county proportions of poor school-age children for school districts are the same for the estimation year as they are for the previous long-form-sample year. Work is under way that shows promise of improving the currency of SAIPE school district estimates of poor school-age children by using IRS personal income tax data coded to the block level (Maples, 2004). The ACS estimates for school districts may also be helpful in the SAIPE school district-level model.

3-C LOCAL GOVERNMENT USES

Local governments—counties, cities, towns, townships, school districts, and areas governed by Alaska Native or American Indian tribes—will likely be major users of the ACS, particularly local governments with sizeable populations. To date, local governments have been limited to once-every-10-year updates of socioeconomic characteristics for their area from the decennial long-form sample. The Census Bureau provides updated estimates of total population throughout the decade for counties and places (the updates include age, sex, and race/ethnicity detail for counties), as well as updated estimates of school-age poverty for counties and school districts. In addition, many local governments have their own sources of data from administrative records and, in some cases, their own surveys. However, most jurisdictions rely heavily on the detailed socioeconomic information in the long-form sample for a myriad of applications involving program planning, allocation of resources, location of service facilities (for example, health clinics, police stations, schools), preparation of supporting material to accompany requests for state and federal aid, and understanding of important trends for their jurisdiction in terms of economic growth or decline, changing age, race, and ethnic composition of the population, and the like.

Illustrative applications in which ACS estimates are used in place of long-form-sample estimates are discussed below for large cities (3-C.1), small jurisdictions in a rural state (3-C.2), and jurisdictions with large seasonal populations (3-C.3). These examples highlight some of the important considerations that local governments need to take into account when they begin to work with ACS estimates. They also illustrate that large areas will benefit greatly from the ACS, while areas with fewer than 50,000 people will confront a mixed situation: on the positive side, the ACS estimates will be more current than the long-form sample estimates; on the negative side, the ACS estimates will be imprecise for estimates of many population groups—more imprecise than the long-form-sample estimates.

3-C.1 Large City Applications of the ACS

This section considers strategies for large cities to work with multiple ACS estimates (1-year, 3-year, 5-year) and analyze change over time. It also provides a case study that illustrates how useful ACS estimates can be for subcity-area analyses.

3-C.1.a Working with Multiple Estimates

Large cities, considered as those with at least 250,000 people (for which 1-year period ACS estimates for small population groups should meet common standards of precision—see Table 2-7a), can benefit from the full set of ACS 1-year, 3-year, and 5-year period estimates. (Such cities are referred to as BIG CITY throughout the text and examples.) The challenge is how to make the most effective use of the various period estimates to understand citywide trends and, at the same time, assess varying neighborhood conditions that are important for program planning and implementation.

Both 1-year and 3-year period estimates will be available not only for BIG CITY as a whole, but also for its PUMAs, which are defined to have at least 100,000 people. Five-year period estimates will be available for BIG CITY, its PUMAs, and small neighborhood areas—namely, census tracts, which average about 4,000 people, and block groups, which average about 1,500 people. (Cities—in contrast to counties—do not contain separate independent governments, such as towns and school districts, so there are no subcity areas with populations between 20,000 and 100,000 for which 3-year estimates will be provided under current plans.) The 5-year period estimates for census tracts and block groups will be extremely imprecise for many population groups of interest because these areas are so small in population size.[10] Hence, users must combine groups of tracts or block groups into larger areas—such as health service areas, school attendance areas, planning districts, and the like—for which 5-year period estimates will be reasonably precise.

Given that 5-year period estimates must be used for subcity areas, there is the issue of which set of estimates to use for BIG CITY as a whole for comparative analysis. In the presence of economic growth (or decline), in-migration (or out-migration) of various population groups, and other social and economic changes, a city's 5-year period estimates may differ appreciably from its 3-year period estimates, and even more so from its 1-year period estimates. Moreover, some neighborhoods may lag behind or be ahead of the overall city trend (see Table 3-3 for an example).

Which estimate to use for BIG CITY will depend on the application, but many users will want to minimize the confusion caused by using estimates for different periods in any given analysis. One strategy is to use the 1-year period estimates for public and media consumption regarding citywide trends (see Section 3-F). The 5-year period estimates for BIG CITY and user-defined subcity areas would be reserved for detailed analyses that are released at a later time and used primarily by the city's own staff for planning and related purposes.

Another strategy is to request special tabulations from the Census Bureau of 1-year or, more likely, 3-year period estimates for user-defined subcity areas that meet the Census Bureau's population thresholds of at least 65,000 people for 1-year period estimates and at least 20,000 people for 3-year period estimates. Cities should give early attention to their possible need for such custom estimates and work with the Census Bureau

---

[10]Research on sampling error by the Census Bureau (Starsinic, 2005) found that ACS estimates for census tracts exhibit much more error compared with long-form-sample estimates than is the case for ACS county estimates compared with long-form-sample estimates. A likely explanation is that census tract estimates, in contrast to county-level estimates, are not adjusted to housing unit or population controls.

TABLE 3-3 School-Age Poverty Rates for BIG CITY/COUNTY and Three Subareas, Illustrative ACS 1-Year, 3-Year, and 5-Year Period Estimates for 2010–2014

	BIG CITY/COUNTY 250,000 people	Area A* 100,000	B* 65,000	C* 85,000
	ACS 1-Year Period Estimates of Percent Poor School-Age Children			
2010	15.0	10.0	22.0	13.0
2011	16.0	11.0	25.0	12.0
2012	17.0	12.0	28.0	11.0
2013	18.0	12.0	31.0	11.0
2014	20.0	16.0	35.0	9.0
	ACS 3-Year Period Estimates of Percent Poor School-Age Children			
2010–2012	16.0	11.0	25.0	12.0
2011–2013	17.0	11.7	28.0	11.3
2012–2014	18.3	13.3	31.3	10.3
	ACS 5-Year Period Estimates of Percent Poor School-Age Children			
2010–2014	17.2	12.2	28.2	11.2

How do the 1-year, 3-year, and 5-year period estimates compare with each other?

- School-age poverty *increased* in BIG CITY/COUNTY, so the 5-year period estimate (2010–2014) of 17.2 percent is lower than the latest 3-year period estimate (2012–2014) of 18.3 percent, which, in turn, is lower than the latest 1-year period estimate (2014) of 20 percent.
- The same pattern is evident for Areas A and B.
- School-age poverty *decreased* for Area C, so the 5-year period estimate (2010–2014) of 11.2 percent is higher than the latest 3-year period estimate (2012–2014) of 10.3 percent, which, in turn, is higher than the latest 1-year period estimate (2014) of 9 percent.

How do the 3-year and 5-year period estimates compare with continuing to use a 2010 census estimate (if 2010 included a long-form sample and provided estimates equal to those shown for the ACS for 2010)?

- The latest 5-year period estimate more accurately depicts current school-age poverty than would continuing to use a 2010 census estimate.
- The latest 3-year period estimate even more accurately depicts current school-age poverty than would continuing to use a 2010 census estimate.

**Availability constrains the choice of estimates*:

- In BIG CITY, 1-year and 3-year period estimates will only be available for the city as a whole and for PUMAs with at least 100,000 people, so the 1-year and 3-year period estimates shown for Areas B and C will not be available.
- In BIG COUNTY, 1-year and 3-year period estimates will be available for the county as whole, PUMAS, and any places or towns with 65,000 or more people; in addition, 3-year period estimates will be available for governmental jurisdictions with at least 20,000 people, but large sampling errors will limit their usefulness for comparisons among areas and over time.

to develop specifications for the subcity areas and the table content. The subcity areas must be large enough in population size and the table content must not be too detailed if 1-year, or even 3-year, period estimates are to be reasonably precise. (See Section 4-D.4 for a recommendation that the Census Bureau consider producing 3-year and even 1-year period estimates for areas smaller than PUMAs in large cities.)

3-C.1.b Analyzing Change over Time

In addition to comparative analyses among subcity areas, users will likely want to analyze trends over time for BIG CITY as a whole and for its subareas. The sampling errors for estimates of differences are always larger than the sampling errors for the individual estimates that are being compared. Consequently, users should anticipate that estimates of year-to-year differences will often be very imprecise and should take care to avoid playing up differences that may appear important in policy terms but are, in fact, within the margin of error. In addition, for analyses of year-to-year differences that must use 3-year or 5-year period estimates and not 1-year estimates, there is the problem of how to interpret the results. Yet an investment in learning how to work with multiple years of ACS estimates, which may require seeking statistical advice, should benefit users who want to exploit the continuous availability of updated information for time trend analysis.

The following text highlights selected aspects of using the ACS to measure change over time. Chapter 6 has a technical discussion of measuring change with ACS period estimates and the implications of alternative approaches for the precision and usefulness of the resulting estimates.

Using 1-Year Period Estimates to Estimate Change for BIG CITY as a Whole Consider two consecutive 1-year period ACS estimates of poor school-age children for BIG CITY (which is assumed to have 50,000 school-age children in a total population of 250,000)—for example, 17 percent poor school-age children in 2010 and 19 percent poor school-age children in 2011. An increase of this magnitude for the nation would be a significant change, both statistically and substantively—over 1 million more children would be poor, and the estimate of change would be very precise. However, the increase for BIG CITY in this example is only 1,000 more poor children, and the estimate of change would not be precise: the 90 percent margin of error for the estimated 2 percentage point increase in school-age poverty would likely be about plus or minus 4.6 percentage points compared with about plus or minus 3.2 percentage points for each year's individual estimate of percentage poor school-age children.[11] This means that, on the basis of two successive 1-year period ACS estimates for BIG CITY, the school-age poverty rate may have increased by as much as 6.6 percentage points or *decreased* by as much as 2 percentage points, or it may have stayed the same. Users cannot conclude whether a change has occurred because the estimates are not precise enough to indicate what has happened.

Only if BIG CITY experiences a large real change is the estimate of the difference between two successive 1-year period ACS estimates likely to be statistically significant. Yet BIG CITY will benefit greatly once a time series of 1-year period ACS estimates is available, because the patterns of yearly change will be informative regarding the existence (or not) of a trend in such characteristics as the percentage of poor school-age children. Consider an example in which BIG CITY had an estimated 15 percent school-age poverty rate in both 2000 and 2010, but poverty increased from 15 to 22 percent in 2005 and then decreased to 15 percent in 2010. Two consecutive long-form-sample estimates for BIG CITY, while quite precise, would completely miss the intercensal dynamics of school-age poverty, whereas a time series of 1-year ACS estimates for the city could track the intercensal trends, even though the year-to-year estimates of change were not precise.

Table 3-4 illustrates changes in school-age poverty rates for BIG CITY (population 250,000) and VERY BIG CITY (population 1 million) over the period 2010 to 2014. The year-to-year differences in school-age poverty rates for BIG CITY (Part A, line 2) are not statistically significant, even though the example purposefully accelerates the increase in school-age poverty over the time period (from a 1 percentage point difference between 2010 and 2011 to a 3 percentage point difference between 2013 and 2014). There is only one significant year-to-year difference for VERY BIG CITY (Part B, line 2), which is the 3 percentage point difference between 2013 and 2014.

As 1-year period estimates accumulate, however, the differences from the first year—2010—are significant by 2014 for BIG CITY (Part A, line 3) and by 2013 for VERY BIG CITY (Part B, line 3). The reason is that the size of the differences between the estimation year and 2010 increases over time (from 1 percentage point between 2010 and 2001 to 2 percentage points between 2010 and 2012, 4 percentage points between 2010 and 2013, and 7 percentage points between 2010 and 2014). It could also be

[11]When two estimates are approximately independent, as is the case for two ACS 1-year period estimates (for which, the samples do not overlap), the standard error of an estimate of change is the square root of the sum of the squared standard errors for the two individual estimates. In the example in the text, the standard errors for each year of about 1.91 (2010) and 1.99 (2011) are squared, summed, and the square root taken to give a standard error of the estimate of change of about 2.76. Times 1.65, the 90 percent margin of error is plus or minus 4.55.

TABLE 3-4 Analyzing Trends Over Time for School-Age Poverty Rates, Illustrative ACS 1-Year Period Estimates, 2010–2014, BIG CITY and VERY BIG CITY

A. BIG CITY (250,000 people)	2010	2011	2012	2013	2014
(1) Percent poor school-age children	15.0	16.0	17.0	19.0	22.0
90% MOE	±2.99	±3.07	±3.15	±3.28	±3.47
(2) Difference from prior year	—	1.0	1.0	2.0	3.0
90% MOE	—	±4.29	±4.39	±4.55	±4.78
(3) Difference from 2010	—	1.0	2.0	4.0	7.0
90% MOE	—	±4.29	±4.34	±4.44	±4.58*

B. VERY BIG CITY (1,000,000 people)	2010	2011	2012	2013	2014
(1) Percent poor school-age children	15.0	16.0	17.0	19.0	22.0
90% MOE	±1.49	±1.53	±1.57	±1.64	±1.73
(2) Difference from prior year	—	1.0	1.0	2.0	3.0
90% MOE	—	±2.14	±2.20	±2.27	±2.39*
(3) Difference from 2010	—	1.0	2.0	4.0	7.0
90% MOE	—	±2.14	±2.17	±2.22*	±2.29*

* = statistically significant at the 90% confidence level.

NOTE: MOE = margin of error (refer to Box 2-5).

What can the user conclude about changes in school-age poverty rates over time?

- The year-to-year differences (line 2) for BIG CITY are *not* statistically significant, even though the example purposefully accelerates the increases in school-age poverty compared with Table 3-3; the only significant 1-year difference for VERY BIG CITY is the 3 percentage point increase in school-age poverty between 2013 and 2014.
- As 1-year estimates accumulate beginning in 2010, however, the differences from *2010* (line 3) are significant by 2014 for BIG CITY and by 2013 for VERY BIG CITY, as the size of the difference increases (from 1 percentage point between 2010 and 2011 to 7 percentage points between 2010 and 2014).

possible to use time-series modeling to improve the statistical power of the analysis (that is, the power to detect statistically significant differences) by taking the entire series into account.

Using 5-Year Period Estimates to Estimate Change for Smaller Cities or Subareas of Large Cities Now consider cities and subcity areas for which there are not precise 1-year or 3-year period estimates for population groups as small as poor school-age children. In this situation, analyses of change must use 5-year period estimates, but comparisons of successive 5-year period estimates will not have the precision that one might assume from the additional sample.

Part A of Table 3-5 compares pairs of successive 5-year period estimates for the rates of school-age poverty in SMALL CITY or BIG CITY SUBAREA (population 50,000, including 10,000 school-age children). Each pair of estimates is 1 year apart. For example, an estimate for 2010–2014 is compared with the corresponding estimate for 2011–2015, and so on through the comparison of an estimate for 2014–2018 with the corresponding estimate for 2015–2019. For simplicity it is assumed that the population size remains constant across the decade.

The individual 5-year period estimates are constructed by assuming that the underlying 1-year period estimates increase each year by 1.2 percentage points from 2010 (15 percent school-age poverty rate) to 2019 (25.8 percent school-age poverty rate). Consequently, the estimated difference between each pair of 5-year period estimates that are 1 year apart is also 1.2 percentage points, with an estimated 90 percent margin of error of about 2.1 percentage points. Consequently, none of these differences is statistically significant for SMALL CITY or BIG CITY SUBAREA: school-age poverty could have increased by more than 3 percent or it could have decreased by as much as 1 percent (1.2 ± about 2.1).

The reason that none of the differences is statistically significant is that each pair of 5-year period estimates being compared shares 4 of 6 years in common. For example, in the comparison between the 2010–2014 estimate and the 2011–2015 estimate, the years 2011, 2012, 2013, and 2014 are shared in common. The only new data in the comparison are for the first and the sixth years—2010 in the 2010–2014 estimate and 2015 in the 2011–2015 estimate.

Statistically, the comparisons between adjacent pairs of 5-year period estimates are the equivalent of taking one-fifth of the 5-year difference between year 1 and year 6 as if one had available the 1-year period estimates for those 2 years (assuming that the population size remains the same over the period—see Chapter 6 for further detail). Thus, in Table 3-5, for the comparison between the estimates for 2010–2014 and 2011–2015, one-fifth of the difference between an assumed 15 percent poor school-age children in 2010 and an assumed 21 percent poor school-age children in 2015 is 1.2 percent (6 percent divided by 5).

Such a comparison will have a large sampling error for an area with only 50,000 people and 10,000 school-age children, which can be seen by considering the sampling errors for the assumed underlying 1-year period estimates of school-age poverty. For example, the assumed estimate of 15 percent poor school-age children in 2010 will have a coefficient of varia-

TABLE 3-5 Analyzing Trends Over Time for School-Age Poverty Rates, Illustrative ACS 5-Year Period Estimates, SMALL CITY or Subarea of BIG CITY with 50,000 People and 10,000 School-Age Children, 2010–2019

A. Estimating Year-to-Year Change by Comparing Overlapping Pairs of 5-Year Period Estimates, Assuming an Underlying Linear Upward Trend in School-Age Poverty

	Percent Poor School-Age Children		Difference from Prior Period	
	Estimate	90% MOE	Estimate	90% MOE
2010–2014	17.4	±3.2	—	—
2011–2015	18.6	±3.3	1.2	±2.0
2012–2016	19.8	±3.3	1.2	±2.1
2013–2017	21.0	±3.4	1.2	±2.1
2014–2018	22.2	±3.5	1.2	±2.2
2015–2019	23.4	±3.5	1.2	±2.2

NOTES: MOE = margin of error. The formula for calculating standard errors for estimates of change has been adjusted in the case of overlapping pairs of estimates to take account of the data shared in common; see Table 6-4.

- To create BIG CITY subareas, the user must aggregate 5-year period estimates for census tracts.
- The above 5-year period estimates are assumed to reflect 1-year period estimates of school-age poverty as follows:

2010	15.0%	2015	21.0%
2011	16.2	2016	22.2
2012	17.4	2017	23.4
2013	18.6	2018	24.6
2014	19.8	2019	25.8

What can the user learn from this example (Part A, which compares pairs of overlapping 5-year period estimates, assuming a linear upward trend in school-age poverty)?

- None of the differences between adjacent overlapping pairs of 5-year period estimates for SMALL CITY or a subarea of BIG CITY (each with 50,000 people) is statistically significant.
- The reason is the substantial overlap between adjacent pairs of 5-year period estimates—they share 4 of 6 years in common (for example, 2011, 2012, 2013, and 2014 in the comparison between the 2010–2014 estimate and the 2011–2015 estimate)—in which the only new data are for 2010 and 2015.
- Assuming no change in the size or demographic composition of the population over time, the differences between adjacent pairs of 5-year period estimates, with 4 of 6 years overlapping, are the equivalent of computing one-fifth of the change between year 1 and year 6 as if one had 1-year period estimates for those two years (see text; see also Section 6-C for the mathematical proof).
- One-fifth of the change between years 1 and 6 is likely to be a small number—it is only 1.2 percent in the data shown above (for example, one-fifth of the difference between 15 percent in 2010 and 21 percent in 2015). Consequently, the sampling error relative to the size of the estimate will be large for an area as small as 50,000 people with only 10,000 school-age children (see text).

TABLE 3-5 Continued

B. Estimating Change by Comparing Pairs of 5-Year Period Estimates That Overlap Less and Less, Assuming an Underlying Linear Upward Trend in School-Age Poverty

	Percent Poor School-Age Children		Difference from Prior Period	
	Estimate	90% MOE	Estimate	90% MOE
(i) 2 years apart (3 years in common)				
2010–2014	17.4	±3.2	—	—
2012–2016	19.8	±3.3	2.4	±2.9
(ii) 3 years apart (2 years in common)				
2010–2014	17.4	±3.2	—	—
2013–2017	21.0	±3.4	3.6	±3.6
(iii) 4 years apart (1 year in common)				
2010–2014	17.4	±3.2	—	—
2014–2018	22.2	±3.5	4.8	±4.3*
(iv) 5 years apart (0 years in common)				
2010–2014	17.4	±3.2	—	—
2015–2019	23.4	±3.5	6.0	±4.7*

NOTES: * = statistically significant at the 90 percent confidence level. The above 5-year period estimates are assumed to reflect the same 1-year period estimates used in Part A above.

What can the user learn from this example (Part B, which compares pairs of 5-year period estimates that overlap less and less, assuming a linear upward trend in school-age poverty)?

- The differences between 5-year estimates that are 2 years apart and 3 years apart are not significant, but the differences between 5-year estimates that are 4 years and 5 years apart are significant.
- The reason is the decreasing extent of overlap between pairs of 5-year period estimates, which adds more new data to the comparison, thereby increasing the precision of the estimated difference (see discussion in the text).

C. Estimating Change by Comparing Pairs of 5-Year Period Estimates That Overlap Less and Less, Assuming a Jump in School-Age Poverty in 2015

	Percent Poor School-Age Children		Difference from Prior Period	
	Estimate	90% MOE	Estimate	90% MOE
(o) 1 year apart (4 years in common)				
2010–2014	17.0	±3.1	—	—
2011–2015	18.2	±3.2	1.2	±2.0
(i) 2 years apart (3 years in common)				
2010–2014	17.0	±3.1	—	—
2012–2016	19.4	±3.3	2.4	±2.9

continued

TABLE 3-5 Continued

	Percent Poor School-Age Children		Difference from Prior Period	
	Estimate	90% MOE	Estimate	90% MOE
(ii) 3 years apart (2 years in common)				
2010–2014	17.0	±3.1	—	—
2013–2017	20.6	±3.4	3.6	±3.6
(iii) 4 years apart (1 year in common)				
2010–2014	17.0	±3.1	—	—
2014–2018	21.8	±3.5	4.8	±4.3*
(iv) 5 years apart (0 years in common)				
2010–2014	17.0	±3.1	—	—
2015–2019	23.0	±3.5	6.0	±4.7*

NOTES: * = statistically significant at the 90 percent confidence level. The above 5-year period estimates are assumed to reflect 1-year period estimates of school-age poverty estimates as follows:

2010	17%	2015	23%
2011	17	2016	23
2012	17	2017	23
2013	17	2018	23
2014	17	2019	23

What can the user learn from this example (Part C, which compares pairs of 5-year period estimates that overlap less and less, assuming an upward jump in school-age poverty in 2015)?

- The estimated differences between 5-year period estimates that are 1 year apart are the ***same*** as in Part A above; the estimated differences that are 2, 3, 4, or 5 years apart are the ***same*** as in Part B above.
- The user will need to use auxiliary knowledge, which may include 1-year or 3-year period estimates for larger geographic areas, to distinguish the nonlinear underlying pattern of school-age poverty in Part C from the linear underlying pattern in Parts A and B.

tion of about 27 percent and a 90 percent margin of error of ±6.7 percent, meaning that the 90 percent confidence interval ranges from 8.3 to 21.7 percent (see Tables 2-7a and 2-7b). Consequently, for a difference between the 2010 estimate and the corresponding estimate for 2015 to be statistically significant, that difference must be very large.

Part B of Table 3-5 compares pairs of 5-year period estimates for school-age poverty in SMALL CITY or BIG CITY SUBAREA that overlap less and less, in which the underlying trend is also a steady increase of 1.2 percent in the percentage poor school-age children from one year to the next. The differences between 5-year estimates that have 3 years' overlap

(i) or 2 years' overlap (ii) are not significant, but the differences between 5-year estimates that have only 1 year's overlap (iii) or no overlap (iv) are significant. The reason is that the decreasing extent of overlap between pairs of 5-year period estimates adds more new data to the comparison, thereby increasing the precision of the estimated difference.

Imagine that one had available estimates of the difference between each pair of years that are 5 years apart. Then, from Part B of Table 3-5:

(i) The comparison between two 5-year period estimates that overlap by 3 years (instead of 4 years as in Part A) is the equivalent, statistically, of taking two-fifths of the average 5-year difference between the ***two*** pairs of years that are not shared in common. For example, in comparing the 5-year period estimates for 2010–2014 and 2012–2016, years 1 and 6 (2010, 2015) and years 2 and 7 (2011, 2016) provide new data not shared in common. The difference in the assumed school-age poverty rates between each of these pairs of years is 6 percent, and two-fifths of the average difference (6 percent) is 2.4 percent, which is the difference shown between the two 5-year period estimates for 2010–2014 and 2012–2016. This difference is more precise than when the overlap between pairs of 5-year period estimates is 4 years and only one pair of years is not shared in common, but not precise enough for statistical significance.

(ii) The comparison between two 5-year period estimates that overlap by 2 years (instead of 3 or 4 years) is the equivalent, statistically, of taking three-fifths of the average 5-year difference between the ***three*** pairs of years that are not shared in common: for example, years 1 and 6 (2010, 2015), years 2 and 7 (2011, 2016), and years 3 and 8 (2012, 2017). The estimated difference between the two 5-year period estimates, which works out to 3.6 percent, still does not attain statistical significance in this example.

(iii) The comparison between two 5-year period estimates that overlap by 1 year (instead of 2, 3, or 4 years) is the equivalent, statistically, of taking four-fifths of the average 5-year difference between the ***four*** pairs of years that are not shared in common: years 1 and 6 (2010, 2015), years 2 and 7 (2011, 2016), years 3 and 8 (2012, 2017), and years 4 and 9 (2013, 2018). The estimated difference between the two 5-year period estimates, which works out to 4.8 percent, is statistically significant.

(iv) Finally, the comparison between two 5-year period estimates that do not overlap at all is the equivalent, statistically, of taking the average 5-year difference between all ***five*** pairs of years that are not shared in common. The estimated difference between the two

5-year period estimates, which works out to 6.0 percent, is also statistically significant.

The drawback of making comparisons with 5-year period estimates that overlap very little or not at all is that the user must wait for the second set of estimates to become available for the analysis. SMALL CITY will have 3-year period estimates available, and the wait for a second, nonoverlapping 3-year period estimate would not be as long as for a second, nonoverlapping 5-year period estimate. However, unless SMALL CITY experienced a very large increase in school-age poverty over a 6-year period, the comparison of 3-year period estimates that were 3 years apart and did not overlap (for example, 2010–2012 and 2013–2015) would not likely yield a significant result, so the user may need to turn to comparisons of 5-year period estimates.

The example of comparing 5-year period estimates in Table 3-5, Parts A and B, is simplistic because it assumes that the total number of school-age children does not change over the period. Also, the example projects a constant linear increase of 1.2 percentage points each year in school-age poverty from 15.0 percent in 2010 to 25.8 percent in 2019. Of course, poverty (and other characteristics) may change at varying rates and in different directions, and the user will not know the underlying dynamics of year-to-year change in 5-year (or 3-year) period estimates.

Part C of Table 3-5 provides an example that produces the same estimates of differences between 5-year period estimates as in Parts A and B but with a distinctly different underlying trend in the data: in this example, school-age poverty is assumed to be static at 17 percent for the years 2010–2014 when it jumps to 23 percent in 2015 (perhaps because a large employer left town) and remains at that level for the years 2016–2019. The interpretation of the differences between pairs of 5-year period estimates is the same as in Parts A and B—namely, that each difference is a fraction (one-fifth, two-fifths, three-fifths, four-fifths, or five-fifths) of the average difference between the pairs of years that are not shared in common. The user must use other information, however, to differentiate between the linear upward trend in poverty in Parts A and B and the jump in poverty in Part C. Examining 1-year or 3-year period estimates for larger geographic areas, such as counties or PUMAs, may help assess the underlying dynamics of change for SMALL CITY or BIG CITY SUBAREA.

3-C.1.c Case Study

The following case study of a rezoning initiative in a large city illustrates the potential usefulness of up-to-date ACS estimates. It starts by describing how data for a housing planning policy for a neighborhood would be obtained before the advent of the ACS and then indicates how ACS data could be used to better inform the policy makers.

Background Area *X* is a neighborhood in BIG CITY that for decades has housed primarily working-class Polish, Hispanic, and Orthodox Jewish families. In recent years, increasing numbers of artists, young professionals, and students priced out of the upscale portion of BIG CITY have moved into the area. In response to strong demand for housing in Area *X*, BIG CITY's planning department initiated an effort to allow new housing development on underused waterfront land abutting the area. As part of an environmental review before the rezoning could take effect, city planners had to determine whether the introduction of new housing could displace the existing residential population through rising rents. Since the proposal would allow the development of luxury apartments in a neighborhood that consisted mostly of modest worker housing, it seemed likely that existing residents could be priced out of the market by newcomers. However, it also seemed likely that residents who had recently moved to the area already had many of the socioeconomic characteristics expected of residents in the proposed new housing—in short, that demographic change was already well under way and indirect displacement was already occurring in certain parts of the study area.

Data and Analysis Needs In order to determine specifically which populations were potentially vulnerable to displacement, neighborhood-level analysis of socioeconomic and housing data was necessary. In 2000, Area *X* included 33,000 occupied housing units and 80,000 residents. However, the 2000 census data did not capture the rapid social and economic change that had occurred in the area in recent years, and no post-2000 data were available to evaluate trends. In an attempt to validate anecdotal evidence of change, BIG CITY's planners supplemented the 2000 census data with other evidence of socioeconomic change, including newspaper articles, new housing construction permits, interviews with brokers about rising rents, surveys of illegal loft conversions, and documented cases of new capital investment and economic activity. BIG CITY did not have an ongoing household survey (as some cities have undertaken at some times), and it did not have the time or resources to conduct one.

Having current and historical statistical data on income, occupation, rents, housing value, and other items to assess changes in the characteristics of area residents would be invaluable for a task that has significant implications for policy and program development. Once the ACS is fully implemented and there is no longer a need to wait 10 years for a new long-form sample, then BIG CITY's planners would be in a much better position to undertake an analysis of this nature. The example below assumes that the year is 2016.

Strategies for Using the ACS An initial strategy for BIG CITY's planners to consider, assuming that Area *X* is contained within a single PUMA and hence represents a large proportion of the PUMA's population, is to use the ACS 3-year period estimates for the PUMA as a proxy to track population growth and changes in socioeconomic composition of Area *X*. In fall 2016, 3-year period estimates for the PUMA could be compared for, say, 2007–2009, 2010–2012, and 2013–2015. While 1-year period estimates would also be available for the PUMA, they might not be sufficiently precise for the purpose—see Table 2-7a. A variant of this strategy would be to average two years of 1-year period estimates for the PUMA and compare the 2-year averages for, say, 2006–2007, 2008–2009, 2010–2011, 2012–2013, and 2014–2015.

The analysts would need to consider three potential problems that could affect the results. First, the head count and age, race, and sex composition of the population would likely differ for the PUMA before and after the 2010 census because of inaccuracies in the pre-2010 population controls (see Section G). Second, because the PUMA in this example is a subcity area and not a county, the sampling error of its estimates would likely be higher than if it had benefited from PUMA-level population controls rather than the county-level controls used in the ACS. Third, the PUMA in this example is somewhat larger than Area *X*, and it is possible that the PUMA population outside Area *X* differs from the Area *X* population in ways that could affect the results.

A second strategy would be for the planners to use the ACS 5-year period estimates for an aggregation of the census tracts or block groups making up Area *X*. The combined 5-year period estimates could then be compared for, say, 2006–2010 and 2011–2015. Again, corrections to the population controls from the 2010 census could distort the precensus and postcensus comparisons.

A combined strategy could make good use of all of the available data. In such a strategy, comparing the 5-year period estimates for the census tracts making up Area *X* to the 5-year period estimates for the larger PUMA could help assess the validity of using 3-year (or 2- or 1-year) period estimates for the entire PUMA as a proxy for Area *X*. The advantage of being able to use 2-year or 3-year period estimates is that they will better capture trends than the 5-year period estimates that average the data over a longer time span.

Whichever strategy the planners ultimately select, the availability of ACS estimates would be a vast improvement over the current situation in which indirect or partial measures of change had to suffice. The ACS data would permit BIG CITY to make a much more informed assessment of the extent of displacement of current residents that was already occurring and would likely occur with the rezoning. The results of the analysis would inform policy makers, lawmakers, and advocacy groups about neighborhood change, ultimately affecting which policies would be supported and where limited resources would be spent.

3-C.2 Small Jurisdiction Applications of the ACS

Generally, smaller counties, cities, and other governmental and statistical areas will not benefit as much from the ACS as larger areas, if only because larger areas will have more sets of estimates published for them (1-year, 3-year, and 5-year period estimates for areas with at least 65,000 people, and 3-year and 5-year period estimates for areas with at least 20,000 people). In some states, sizeable proportions of the population live in small counties, cities, towns, and school districts that will have only 5-year period estimates from the ACS. In 2000, for example, the percentages of people living in counties with fewer than 25,000 residents exceeded 20 percent in 7 states: Alaska (22 percent), Arkansas (27 percent), Idaho (25 percent), Montana (34 percent), North Dakota (47 percent), South Dakota (57 percent), and Wyoming (31 percent) (from the 2002 Census of Governments, U.S. Census Bureau, 2002a:Table 6).

Five-year period estimates for areas this small will be subject to large levels of sampling error (refer back to Tables 2-7a, 2-7b, and 2-7c), although the oversampling of housing units in very small areas will help their precision somewhat. Consider the 15 percent of people in North Dakota and 24 percent of people in South Dakota who live in cities with fewer than 1,000 residents. Over a 5-year period, these areas will be sampled initially at rates of 1 in 3 housing units (if they have between 500 and 1,000 residents) or 1 in 2 housing units (if they have fewer than 500 residents), compared with the average ACS initial sampling rate of 1 in 9 housing units (refer back to Table 2-3, Part A). This oversampling will reduce the sampling error of estimates for these areas by about 40-50 percent compared with the sampling error of estimates for areas with a 1 in 9 sampling rate (assuming that the areas have the same combined mail and computer-assisted telephone interviewing [CATI] response rates and therefore the same computer-assisted personal interviewing [CAPI] subsampling rates).

Oversampling also benefits many larger areas that contain very small cities, townships, or school districts. Selecting just one of many such examples, in 2000, Iowa County, Wisconsin, had 22,780 residents living in 11 cities and 14 towns (U.S. Census Bureau, 2002a:Table 16). Careful examination of the population size of each subcounty jurisdiction would be required to determine the effect of oversampling, but it seems likely that the entire county would be sampled at 5-year cumulative rates of 1 in 2 or 1 in 3 households instead of at an average rate of 1 in 9 or less. The effect would be to reduce the sampling error for estimates of the entire county to the point that they could meet common standards of acceptable precision.[12]

Even with oversampling, however, the 5-year period estimates for very small governmental units will fall far short of common standards of precision for many population groups of interest. For example, the 90 percent confidence interval for an estimate of 15 percent poor school-age children for an oversampled area of 1,500 people, based on the assumptions in Table 2-7c, would likely range from 7 to 23 percent, which is not very informative about the extent of school-age poverty. By contrast, the 90 percent confidence interval for an estimate of 15 percent poor school-age children for an area of 50,000 people would likely range from 12 to 18 percent, which is a considerable improvement.

It is important to remember that the 2000 long-form-sample estimates were also subject to considerable sampling error for small areas. However, they were somewhat more precise than the corresponding estimates from the ACS cumulated over 5 years.

The precision of the 5-year period ACS estimates can be improved by aggregating small areas into larger units. Indeed, this is the recommended strategy for large jurisdictions—namely, to aggregate census tracts and block groups into larger subcity or subcounty areas for such purposes as planning the location of governmental service sites and services. A strategy of aggregation is not as suitable for small governmental jurisdictions, given that each typically provides its own services and is interested in estimates for its jurisdiction alone.

Small jurisdictions could ask the Census Bureau to provide estimates for, say, 8-, or 10-year periods that are more precise than the 5-year period estimates. The drawback of this approach is that lengthening the period of the estimates averages underlying patterns of variation in social and economic phenomena over longer periods and does not produce large gains in precision. For the case of a town of 1,500 people, the 90 percent confidence interval for an estimate of 15 percent poor school-age children would be reduced from 7 to 23 percent for the 5-year period estimate to 8.7 to 21.3 percent for an 8-year period estimate and to 9.3–20.7 percent for a 10-year period estimate (under the assumptions used in Table 2-7c). By comparison, the 90 percent confidence interval for the same estimate from the 2000 long-form sample would be 9.7 to 20.3 percent.

To produce reasonably precise estimates for small population groups in

[12]The beneficial effects on sampling error for county estimates that result from oversampling subcounty areas are not as great when the subcounty areas are sampled at varying rates, such as 1 in 2, 1 in 3, 1 in 6, and 1 in 9.

small jurisdictions would require a significant expansion of the ACS. Many such groups are of interest to users, including not only poor school-age children, as discussed in this report, but also ethnic and language minorities, veterans, and people with disabilities. Increasing the final 5-year ACS sample size (after subsampling for CAPI follow-up) to equal the originally proposed size (which was double the current size—see Section 1-B.3) would certainly help. However, acceptable precision for small groups could still often require aggregating estimates over 8 to 10 years.

Of course, ACS estimates for larger population groups will be more precise than those for small groups, and the 5-year period estimates for some large groups in small jurisdictions may reach acceptable precision, particularly if the jurisdiction's housing units are oversampled. For example, a 5-year period estimate of 15 percent *total* poor people in an oversampled jurisdiction of 1,500 people will have a 90 percent confidence interval of 11.4 to 18.6 percent, which is much narrower than the interval of 7.0 to 23.0 percent for poor school-age children.

Small jurisdictions may be able to use the levels and trends in the more precise 5-year period estimates for similar but larger jurisdictions to improve understanding of what is occurring for their jurisdiction. Moreover, small jurisdictions, just as large jurisdictions, will benefit from the fact that ACS multiyear period estimates never become as outdated as the long-form-sample estimates do before they are replaced by estimates from the next census.

3-C.3 Special Case of Seasonal Populations

Some jurisdictions in the United States have large, seasonal fluctuations in population. Examples include many college towns, the west and east coasts of Florida, parts of Arizona, the northern parts of Wisconsin, Minnesota, and Michigan, and the Atlantic beaches. Because of the continuous sampling and data collection for the ACS and its use of a 2-month residence rule instead of the "usual residence" rule of the decennial census, the ACS estimates for an area with seasonal fluctuations in population will likely differ from the long-form-sample estimates for the same area.

Table 3-6 works through a simplified example for a hypothetical county in Florida. This county is assumed to have a year-round population of 100,000, of whom 20,000 (20 percent) are poor, and a winter (December-March) population of 300,000, of whom 35,000 are poor (11.7 percent, averaging the 20 percent year-round poverty rate with a rate of 7.5 percent for the richer, part-time residents). Over the entire year, *on average*, there were 166,667 people in the county, of whom 25,000 were poor (15 percent poverty rate, averaging the year-round poverty population for 8 months and the winter poverty population for 4 months).

TABLE 3-6 Hypothetical County in Florida with Winter Influx of Residents

County Characteristics, January–December 2010							
	Assumed Distribution				Measured by the ACS (Before Controls)		
	Year-Round Pop.		Seasonal Population		Total Population		
	Total	Poor	Total	Poor	Total	Poor	(%)
January	100,000	20,000	200,000	15,000	300,000	35,000	11.7
February	100,000	20,000	200,000	15,000	300,000	35,000	11.7
March	100,000	20,000	200,000	15,000	300,000	35,000	11.7
April	100,000	20,000	—	—	100,000	20,000	20.0
May	100,000	20,000	—	—	100,000	20.000	20.0
June	100,000	20,000	—	—	100,000	20,000	20.0
July	100,000	20,000	—	—	100,000	20,000	20.0
August	100,000	20,000	—	—	100,000	20,000	20.0
September	100,000	20,000	—	—	100,000	20,000	20.0
October	100,000	20,000	—	—	100,000	20,000	20.0
November	100,000	20,000	—	—	100,000	20,000	20.0
December	100,000	20,000	200,000	15,000	300,000	35,000	11.7
12-month average	100,000 (20.0% poor)	20,000	66,667 (7.5% poor)	5,000	166,667	25,000	15.0

NOTE: For ease of presentation, the example unrealistically assumes zero year-round population growth over the year and that all seasonal residents arrive December 1 and leave March 31.

Population Controls, 2010			
	Total	Poor	(%)
April 1, 2010 census estimate	100,000	N.A.	N.A.
July 1, 2010 population estimate	100,000	N.A.	N.A.

NOTE: The July 1, 2010, population estimate updates the 2010 census estimate with administrative records. For ease of presentation, the example unrealistically assumes zero population growth from April to July; actual growth might be a fraction of 1 percent.

Hypothetical Estimates of Total Population and Number and Percent Poor, 2010			
	Total	Poor	(%)
Census long-form-sample estimate, 2010 (based on March–June data with April population control)	100,000	20,000	20.0
ACS 1-year period estimate, 2010 (controlled) (based on 12-month average data with July population control)	100,000	15,000	15.0
ACS 1-year period estimate, 2010 (not controlled) (based on 12-month average data, no control applied)	167,000	25,000	15.0

Assuming the winter population has left the area entirely by March 31, the Census Bureau would estimate the county's April 1 population at 100,000 and its July 1 population at about the same number. Therefore, in this very stylized example, both the ACS and a long-form sample conducted in the same year would provide a total population figure for the county of about 100,000 (since the ACS weighting procedure adjusts the ACS sample to conform to the July 1 county population estimates), but the composition of the population would differ between the two surveys. The long-form sample would provide an estimate of 20,000 poor people (20 percent poverty rate for the year-round population). The ACS would provide a 1-year period estimate of 15,000 poor people (15 percent average for the year-round and seasonal populations combined). Note that the *percentage* of people in poverty from the ACS estimate reflects the average composition of the population over the year; however, the *number* of poor people is lower than both the long-form-sample estimate and a 12-month average of the ACS that is not constrained to the July population control.

This example is exaggerated, but it does point up the differences between the long-form sample and the ACS for areas that experience significant seasonal fluctuations of population and for which the socioeconomic characteristics of the seasonal and year-round populations differ appreciably. In these instances, the long-form sample provides the numbers and characteristics of the population as of April 1. The ACS provides comparable population numbers by age, race/ethnicity, and sex based on July 1 postcensal estimates, even though the total population, as well as demographic groups (for example, young and older people), may change during the year. For socioeconomic characteristics, the ACS provides percentages that reflect the average experience of the area over the year; however, the percentages are applied to the July 1 population figures so that the numbers are neither the same as the long-form-sample estimate nor the same as an average estimate from the ACS that is not controlled to the census-based population estimates (see further discussion in Section 4-A.5).

For most areas, this problem will not be significant because seasonal increases (or decreases) in population are a small percentage of the year-round population, or the characteristics of seasonal and year-round residents do not differ appreciably. In areas for which users believe that seasonal differences may be significant, they may wish to make a case to the Census Bureau of the need for tabulations of their population at different times of the year (see Section 7-D.2).

3-D TRANSPORTATION PLANNING USES

Transportation planners are devoting considerable effort to understanding the ACS, determining how to work with the data, and identifying concerns to raise with the Census Bureau about the data products. Their efforts in this regard go back to the beginning of the ACS (see, for example, Bureau of Transportation Statistics, 1996). With funding from federal transportation agencies, a committee of the National Research Council's Transportation Research Board organized a conference on "Census Data for Transportation Planning—Preparing for the Future" in May 2005. The conference covered a wide range of topics and issues regarding the opportunities and challenges presented by the advent of the ACS (see http://trb.org/conferences/censusdata/Program.pdf).

The transportation community's interest in the ACS is explained by the central role that the long-form-sample data have historically played in transportation applications ranging from nationwide program planning and evaluation to local analysis of commuting patterns. Questions on place of work, means of transportation to work, length of commute, and vehicle ownership have been included on the long-form questionnaire for three or more decades, as have questions about disabilities that make it difficult for people to work or to go outside their homes (see Citro, 2000b).

The U.S. Department of Transportation has worked closely with the Census Bureau and with state transportation departments and metropolitan planning organizations to improve the quality of the data on place of work (by, for example, encouraging large employers to inform workers of the addresses to report for particular workplaces) and to develop special tabulations for transportation users. The Census Transportation Planning Package (CTPP) has been produced from censuses beginning in 1970 and includes tabulations of households and workers by place of residence, workers by place of work, and flows between place of residence and place of work for each traffic analysis zone (TAZ). There are a large number of such zones, designated by states and regional transportation agencies, each comprised of one or more blocks, block groups, or census tracts within metropolitan areas.[13]

Regional and metropolitan transportation planning organizations are also heavy users of the long-form-sample PUMS 5 percent sample files, which in 2000 provided records for 14 million individuals, with geographic identification by state and PUMA (areas of about 100,000 population). The long-form-sample PUMS files are the basis of sophisticated transportation activity modeling systems that contain synthetic population models for a base year and, say, 20 years into the future. The population models are calibrated to control totals for the base year and future years on total households, households by income level, and other characteristics that are estimated by the regional organization at the county or TAZ level. The models are then used to predict activity and travel patterns at the person, household, or trip level.

---

[13]See http://www.trbcensus.com/ctpp.html; National Research Council (1995:App. G).

3-D.1 Using the ACS 1-Year PUMS Files

Transportation planners are concerned that the ACS yearly PUMS product will contain only about 3 million person records. This reduction from 14 million persons means that the sampling error of estimates from the ACS 1-year PUMS will be much larger than those of estimates from the 2000 long-form-sample 5 percent PUMS, and estimates from the long-form-sample 5 percent PUMS are already subject to about 1.8 times more sampling error than estimates from the full long-form sample.

As an example, for a PUMA with 50,000 workers, an estimate from the 2000 long-form sample 5 percent PUMS that 15 percent of workers carpooled to get to work would have a 90 percent margin of error of approximately plus or minus 1.6 percent—1.83 times the margin of error of about ±0.9 percent for the full long-form sample (see the fourth row in Table 2-7b). This margin of error equates to a coefficient of variation of 6.5 percent. However, a corresponding estimate from the ACS 1-year PUMS would have a 90 percent margin of error of at least plus or minus 3.6 percent based simply on the difference in the number of records. This margin of error equates to a coefficient of variation of 14.5 percent, which does not meet accepted standards for precision. Moreover, the weights in the ACS PUMS will be more variable than those in the 2000 long-form-sample PUMS due to the subsampling for CAPI follow-up in the ACS. Consequently, estimates from the ACS PUMS will likely be even less precise compared with estimates from the 2000 long-form-sample PUMS than indicated above.[14]

A possible solution for the smaller size of the ACS PUMS is to combine two or more PUMS. While transportation modelers will not likely want to fully analyze each new PUMS release because of the time and resources that would require, the availability of an annual PUMS will make it possible to periodically check and recalibrate their models. Similarly, the availability of updated ACS 5-year period estimates will make it possible to reestimate control totals for the models at the county and TAZ levels more often than once a decade.

[14]The current scheme for selecting the ACS PUMS files draws an equal-probability systematic sample of all ACS housing unit records and their household members in each state, with the records sorted by several characteristics (see the 2005 PUMS accuracy statement at http://factfinder.census.gov/home/en/acs_pums_2005.html). A different selection scheme would retain a higher proportion of the CAPI cases so as to equalize the weights of CAPI and non-CAPI cases, yielding a PUMS that would produce more precise estimates than the current PUMS. This scheme could be extended toward equalizing the weights of all sampled housing unit records within PUMAs.

3-D.2 Using the ACS TAZ Data

A concern in using the ACS 5-year period estimates for traffic analysis zones is that the 60-month averages that underlie the estimates may obscure short-term changes in commuting patterns that occur in response to marked changes in the local economy or the transportation infrastructure. To address this concern, transportation planners in large cities and metropolitan areas can examine 3-year or 1-year period estimates for the area as a whole, for PUMAs, and, in some cases for smaller cities and towns. Analyses of these estimates can provide an overall sense of changes in commuting modes and times to work that can inform assessments of the usefulness of the 5-year period estimates.

Precision is also a very serious concern for 5-year period TAZ estimates. Statistical mapping techniques may help transportation planners extract useful information from the estimates in some instances. For example, by geographically displaying such variables as mode of transit to work, where workers live, and where workers work on maps of transportation routes, places of employment, and other local features, planners may see patterns that suggest how to combine TAZ estimates to produce meaningful larger areas that have more precise estimates. (Such statistical mapping techniques may help users in other fields extract value from ACS 5-year period estimates for census tracts and block groups.)

The usefulness of 5-year period TAZ estimates also depends importantly on two other factors: (1) procedures that the Census Bureau uses for imputing missing responses and (2) decisions it makes regarding the data that can be provided while protecting confidentiality. Regarding imputation, the Census Bureau needs to engineer its data processing so that imputations for missing responses to commuting items can be made at the outset at the block level. In the long-form-sample processing, imputations for these items were made initially at the city level and only subsequently, in the CTPP, carried out at the block level.

Regarding confidentiality, the Census Bureau needs to consider carefully the added confidentiality protection afforded by 5-year averages compared with point-in-time estimates. The added protection results from the fact that many people change one or more characteristics of interest over a 5-year period, such as place of work, occupation, place of residence, commuting mode, etc. Consequently, the risk of reidentification of a specific individual in 5-year aggregations is reduced. Taking account of this added protection should enable the Census Bureau to release sufficient information on commuting (and other topics) to be useful at the level of traffic analysis zones, block groups, and census tracts (see Section 4-D.1).

3-D.3 Conclusion on Using the ACS for Transportation Planning

While transportation planners face significant challenges in using ACS data for applications for which they have previously used the long-form sample, the frequency of release of updated ACS estimates offers benefits to them. The 1-year and 3-year period estimates can help transportation planners track overall trends in commuting patterns and other aspects of household transportation and alert them to the need for special surveys or other data collections to update their models. The 5-year estimates can provide intercensal checks on local-area transportation patterns that would not be possible with the decennial long-form sample, although estimates for traffic analysis zones will often need to be combined to attain an acceptable level of precision. The ACS PUMS can be used in a variety of ways, and it is issued more frequently than the long-form-sample PUMS.

3-E ACADEMIC RESEARCH USES

Researchers in universities, colleges, research institutes, and other settings have made extensive use of long-form-sample data to understand key social processes, such as migration flows, changes in marriage patterns and family living arrangements, and the social and economic effects of the aging of the population. They have also used long-form-sample data to develop insights on such important topics as trends in educational attainment, magnitudes and effects of immigration from abroad, and concentrations of people in poverty.

Some research applications have used summary files of detailed tabular data for small areas, such as Summary Files 3 and 4 from the 2000 census. For example, summary files have supported analyses of migration flows among regions, states, counties, and places and of concentrated populations of the poor, minorities, and immigrants from different countries. Summary information on neighborhood characteristics has been appended to the records of respondents to such ongoing research surveys as the Panel Study of Income Dynamics. This additional information has permitted rich contextual analyses of family social and economic dynamics.

Other research applications have used the PUMS files, which have been constructed for most censuses back to 1850 (Ruggles, 2000). PUMS files permit detailed, multivariate analyses on such topics as the interactions among disability, educational attainment, labor force attachment, and income and the characteristics that distinguish people who migrate long distances from those who migrate shorter distances or not at all.

3-E.1 Using Summary Files for Research

At present, no equivalent of Summary File 3 or 4 from the 2000 long-form sample exists for the ACS. This lack is a drawback for the research community because summary files permit ready analysis of detailed information across multiple geographic areas, population groups, and subject areas. In contrast, the tables that are available online for the ACS can only be displayed one at a time for a specified type of geographic area within a larger unit—for example, a table of age by sex for one or all towns in a particular county. The detailed tables and single-year and multiyear profiles are also available as spreadsheets through the ACS FTP site, and in that format the data can be manipulated (for example, calculating percentages or adding or subtracting categories), but the spreadsheet contents are limited to a specified geographic area, such as a county or township. An ACS Download Center provides access to up to 50 tables for a geographic summary level, such as all states or all counties. None of these data products are as useful for research purposes as a summary file in the same format as the decennial census summary files.

The Census Bureau recently began work to specify and implement an ACS equivalent of Summary File 3 from the 2000 long-form sample. This is a welcome development, not only for the research community, but also for many other users who require the ability to easily manipulate large amounts of data for multiple areas and population groups. The initial prototype 2005 ACS summary file has just been released and contains all of the detailed tables for every geographic area with 65,000 or more people; eventually, the ACS summary files will be released annually for each year's 1-year, 3-year, and 5-year period estimates. Users have been invited by the Census Bureau to comment on the prototype summary file.[15]

Researchers who work with the new product will need to be cognizant of the larger sampling errors of the ACS tables compared with the 2000 long-form-sample tables and develop strategies for effective use of the ACS. Such strategies include combining data for census tracts and block groups into larger areas, collapsing data categories, and combining ACS summary files for nonoverlapping periods. The advantage of the ACS will be that researchers will not need to wait for 10 years to track trends in migration flows and other social, demographic, and economic phenomena.

3-E.2 Using PUMS Files for Research

Many researchers will turn to the ACS 1-year PUMS files for their analyses. The availability of PUMS files year after year will afford much

[15]See http://www.census.gov/acs/www/Special/Alerts/Alert44.htm#News2, ACS Alert 44, December 28, 2006.

flexibility to researchers. For example, in analyzing economic change, they can plan to use two (or more) PUMS files that coincide with different stages of a recession and subsequent recovery. Such an analysis was not possible with the once-every-10-years long-form-sample PUMS. The identification of PUMAs of about 100,000 population on each year's ACS PUMS file also affords flexibility for analysis.

A drawback of the ACS 1-year PUMS files, as noted above for transportation, is the larger margins of errors compared with the 2000 long-form-sample 5 percent PUMS file. Many research uses of the PUMS data can benefit from combining two or more ACS PUMS files to increase the sample size for the analysis and thereby increase the precision of estimates. Researchers may be able to develop custom PUMS files for particular applications—such as a merged file of two or more 1-year PUMS for analyzing economic returns to education—that can be shared with other researchers.

Researchers will also need to grapple with the different reference periods for different respondents in the ACS PUMS files and develop appropriate analytical strategies. For income amounts for the previous 12 months, the Census Bureau will provide the reported amount, not adjusted for inflation. It will also provide a single inflation factor, which will adjust the values, on average, to July dollars for the latest year covered in a PUMS file (for example, 2005 for the 2005 PUMS file, which contains income reference periods that span January–December 2004 through December 2004–November 2005). A single inflation factor is used because the ACS PUMS files do not indicate the month of interview in order to protect confidentiality. Reconsideration of this decision and inclusion of the month (or season) of interview in the PUMS records and in selected summary tables would greatly increase the analytical value of the files (see Section 4-D.1).

3-F MEDIA AND GENERAL PUBLIC USES

This section discusses using ACS profiles and rankings, which will be appealing products for occasional users and the media (refer back to Box 2-2). It also discusses comparisons of ACS estimates with other data sources, which can confuse users when differences between the ACS and the other data sources are not understood.

Journalists who frequently use statistical information to track local, regional, and national trends will use not only the ACS profiles and rankings, but also more detailed tables. They, like other involved federal, state, and local data users, will need support from the Census Bureau to understand how to properly apply the data (see Section 7-A).

3-F.1 Using ACS Profiles and Rankings

ACS products that are likely to be of broad general interest include single-year profiles, providing key 1-year period estimates for governmental and statistical areas with at least 65,000 people; multiyear profiles, providing the same key 1-year period estimates for the current year and four prior years for areas with at least 65,000 people; and single-year ranking tables and charts comparing states and large cities on selected 1-year period estimates.[16] These products will be timely and easy to reference. They will be the starting point for press releases from government officials and media articles describing what has occurred in a city, county, metropolitan area, or state since the year before and in comparison with other areas.

For trend analysis using multiyear profiles, public officials and the media must take care to avoid making too much of year-to-year differences that are within the margin of error (see Section 3-C.1.b). Just as the media have educated the public about the margin of error in public opinion polls, so should they take on the responsibility to educate readers about the margin of error from ACS estimates in profiles and other data products. The Census Bureau will provide margins of error for estimates in single-year profiles. In multiyear profiles, it will indicate estimates for each year that are statistically significantly different from the estimates for the current year.[17]

Similarly, for comparisons across areas using 1-year period estimates, public officials, the media, and readers must learn that, in most cases, the difference between, say, the city with the highest school-age poverty rate and the city with the next highest rate is not necessarily indicative of a real difference or even of the real ordering. In fact, the estimates for 5 or 10 of the cities with the highest rates may be not be statistically different, so that it is appropriate to say only that City A falls into the top, middle, or bottom group of cities rather than to assign it an individual rank-order number. Moreover, when the subsequent year's period estimates are released, and City A has moved, for example, from number 1 to number 2, 3, 4, or 5 in school-age poverty, the reader should not conclude that school-age poverty has necessarily declined in City A relative to the other cities on the basis of one year's difference in rankings.

Although sampling error affects such uses of the ACS data as trend analysis and comparative rankings, the regularly updated ACS estimates will be more helpful to users than the once-a-decade estimates from the

[16]Multiyear profiles will be published for geographic areas defined according to the latest known boundaries for all years shown.

[17]The Census Bureau provides 90 percent margins of error; for agreement with standard statistical practice, it should provide 95 percent margins of error instead (refer back to Box 2-5).

census long-form sample. From the example above, it is useful information to know what group of cities—top, middle, or bottom—a particular city is part of and to know, year to year, whether that city has remained about the same in relative ranking or, in contrast, has experienced a major change.

3-F.2 Comparisons with Other Data Sources

Often, estimates will be available not only from the ACS, but also from another data source, and the public, policy officials, and the media will want to know the reasons if the ACS and the other source do not agree. In fact, it is likely that differences will occur between estimates from two data sources because of differences in concepts and definitions, data collection procedures, and other aspects of the two sources.

In addition, users who want to compare 2005 ACS estimates for governmental or statistical areas with 65,000 to 250,000 people with estimates from an earlier period must use a different source—namely, the 2000 long-form sample—as their point of comparison. (The 2005 ACS estimates for areas with 250,000 or more people can be compared with estimates from the Census 2000 Supplementary Survey (C2SS) or any of the 2001–2004 ACS test surveys.) As described in Chapter 2, there are important differences between the long-form sample and the 2005 ACS, involving sample size, population covered, data collection mode, population controls, and others, so that assessment of changes between 2000 and 2005 must be made with great care. In the future, the yearly releases of ACS data will make it possible to assess trends using just the ACS, but the 2000 long-form sample will remain an important comparison source for small areas for some time to come.

With regard to comparisons between the ACS and another source for the same time period, an object lesson is afforded by experience in comparing state estimates of poverty from the CPS ASEC and the ACS supplementary surveys. National estimates from the CPS ASEC are the official poverty estimates for statistical use according to OMB Directive 14. To respond to user needs, the Census Bureau began publishing poverty estimates for states in 1990 from the CPS by averaging 2 and 3 years' worth of estimates to improve precision. The Census Bureau has also published state poverty estimates from the C2SS and the 2001–2004 ACS test surveys and, now, the 2005 ACS.

Comparisons of trends from the CPS ASEC state poverty estimates averaged over 2 years with those from the C2SS and the ACS 2001–2004 test surveys revealed instances in which the two data sources did not agree on the poverty rate or the direction of change (increase or decrease in poverty). There are many reasons that may explain these differences:

- Sampling error: A difference between the CPS ASEC and the ACS may not be significant because it falls within the margin of error. Each year's CPS ASEC sample is about 77,000 housing units, compared with about 600,000 housing units for the C2SS and each of the 2001–2004 ACS test surveys and 2 million housing units for the full ACS (all figures are for responding units).
- Population coverage: The CPS ASEC covers the civilian noninstitutional population, while the C2SS, the ACS test surveys, and the 2005 ACS cover the civilian and military household population. The 2006 ACS covers cover virtually the entire population, including civilian and military residents of households and institutional and noninstitutional group quarters (refer back to Table 2-1).
- Residence rules: The CPS ASEC employs a usual residence rule, while the ACS employs a 2-month residence rule.
- Reference periods: The CPS ASEC reference period for household composition—which is used to determine poverty thresholds—is February, March, or April (these are the months of interview each spring). The CPS ASEC reference period for income is the previous calendar year, which centers on July 1. The ACS reference period for household composition is the month of interview, which extends from January through December. Its reference period for income is the previous 12 months (extending from January of the preceding calendar year to November of the current calendar year) with adjustments made for inflation.
- Mode of data collection: The CPS ASEC is conducted in person using CAPI for sample cases having their first interview and by telephone using CATI for sample cases having their second (or later) interview. The ACS is a mail survey with CATI and CAPI follow-up.
- Imputation and weighting procedures: The CPS ASEC procedures for imputing an amount for unreported income are carried out on a national basis, whereas the ACS imputation procedures are carried out state by state, thereby capturing state differences in income patterns. The CPS ASEC population controls are applied for demographic population groups at the national level, and there are no housing unit controls, whereas the ACS population (and housing unit) controls are applied for counties or groups of small counties.
- Question content: The CPS ASEC includes questions on 50 different sources of income; the ACS asks the standard long-form-sample questions, which include 8 sources of income. Past research has shown that asking more detailed questions elicits more complete reporting of income; however, the 2005 CPS ASEC (2004 income)

and the 2004 ACS supplementary survey produced about the same level of total income (Nelson, 2006). There were differences by source (the CPS estimated more wages and less self-employment income compared with the ACS), but the aggregates were very close.

Research is needed to understand the contributions of each of the above factors to differences between the CPS ASEC and the ACS. For users, now that the ACS is in full production with a vastly larger sample size than the CPS ASEC, it seems reasonable that they look to the ACS estimates for states and substate areas. However, users who want to analyze income by source and examine the correlates of income for population groups at the national level should stay with the CPS ASEC, which not only is the source of official income statistics, but also contains a wealth of variables to use in analysis.

3-G WHAT HAPPENS IN A DECENNIAL YEAR?

An important element of the ACS design is to control each year's estimates at the level of the county (or group of small counties) by total housing and by total population, categorized by age, sex, race, and Hispanic origin. The population control totals for each year are produced by updating the previous decennial census totals with administrative records on births, deaths, and net migration. The housing unit control totals for each year are produced by updating the previous census totals with housing permit records (see Chapter 5). The use of control totals is important to reduce sampling error in the estimates and to adjust the ACS estimates for possible undercoverage of housing and of the population, which may be particularly pronounced for some demographic groups.

The problem with the population control totals is that they become increasingly prone to error as each year passes from the previous census. While birth and death records are very accurate, there is considerable uncertainty about the quality of estimates of net migration, both net immigration from abroad, including illegal immigration, and net migration flows among counties. In the 2000 census, the estimate of the total U.S. population updated from the 1990 census was 1.8 million people *fewer* than the 2000 census count of 281.4 million people, and there were significant errors also in estimating the population of subnational areas. The national underestimate, which was particularly large for people ages 18–29 and for minorities, was attributed to an underestimation of illegal immigrants during the economic boom of the last half of the 1990s (National Research Council, 2004b:Table 5.1).

The postcensal housing unit controls are also subject to error, given that housing permit data do not necessarily correspond with actual housing units constructed and occupied and given the problems of estimating demolitions and conversions to nonresidential use. The errors in the housing unit controls may well vary across geographical areas and may also cumulate over time.

While one cannot be sure what the magnitude or direction of the errors will be in the 2000 census-based estimates of the population and housing for 2010, one can be reasonably sure that there will be discrepancies between the estimates and the 2010 census counts and, furthermore, that the discrepancies will be greater for many counties and combinations of counties that are the basis for the ACS weighting controls. For the ACS, this means that there will be a discontinuity in many areas in totals for important demographic groups between 1-year period estimates for years preceding a census and for years including and following a census. This discontinuity will also exist for 3-year and 5-year period estimates between those that completely antedate a census year and those that include and follow a census year (for example, when comparing 5-year period estimates for 2005–2009 and 2010–2014). For 3-year and 5-year period estimates that span a census year (for example, a 5-year period estimate for 2008–2012), the Census Bureau plans to use an average of controls in which the population estimates for precensus years are adjusted to be consistent with the census counts.

One might consider that ACS estimates of percentages, as opposed to levels, would not be affected by the problem of differences in precensus and postcensus population controls. This will be the case, however, only if the discrepancies between the two sets of controls are relatively uniform by demographic category. If the discrepancies differ by category, which is likely, then the percentages will be affected as well (see Table 3-7 for an example).

There is no universal solution for the problem that will result from discrepancies between precensus and postcensus population controls. Users must address the situation for their applications and areas of interest, given that the problem will be more significant for some areas and population groups than others. The Census Bureau can help users in this regard by producing concurrent series of estimates that are based on precensus and postcensus controls. For example, the Census Bureau could produce two series of 1-year period estimates for, say, 2008–2010, in which the first series would use the 2000 census-based controls (the official series for those years), while the second series would backcast the 2010 census-based controls.

TABLE 3-7 Hypothetical Effect of Decennial Census on ACS 1-Year Period Estimates, BIG CITY, 2008–2012

	Population Control		Estimated Number of Poor People (Equal to the Percent Poor Times the Control)		
Year	Non-Hispanic	Hispanic	Non-Hispanic (10% Poor)	Hispanic (20% Poor)	Total
ACS, Control Based on 2000 Census					
2008	200,000	50,000	20,000	10,000	30,000
2009	200,000	52,000	20,000	10,400	30,400
2010	200,000	54,000	20,000	10,800	30,800
2010 Census	200,000	64,000	20,000	12,800	32,800
ACS, Control Based on 2010 Census					
2011	200,000	67,000	20,000	13,400	33,400
2012	200,000	70,000	20,000	14,000	34,000

NOTES:

- The population controls for Hispanics and non-Hispanics (and for age, sex, and race groups) are implemented for the ACS by estimation area (county or group of small counties; this example assumes that BIG CITY is its own county). The controls are developed from the previous census updated with administrative records on births, deaths, and net migration.
- For ease of presentation, the example unrealistically assumes constant poverty rates from the ACS for the Hispanic population (20%) and the non-Hispanic population (10%) and that the city experienced no growth in the non-Hispanic population.
- For ease of presentation, the example unrealistically assumes that all of the error in the population controls prior to 2010 pertained to the Hispanic population.

What happened?

- The 2010 census gave the same non-Hispanic count (200,000) as the 2000-based population controls updated to 2010. But the 2010 census gave a different Hispanic count from the updated controls—64,000 instead of 54,000. Consequently, the controls were revised going forward from 2010.

What does the example tell the user?

- The number of Hispanics and non-Hispanics is determined by the population controls, while the percentage of poor people in each group is determined by the ACS. Hence, the number of poor people in each group is the product of the control and the estimated ACS percentage.
- The 2010 census results indicate that the Hispanic population and, consequently, the poverty population grew faster prior to 2010 than previously estimated, so that the ACS estimates of the number of poor Hispanics and total poor were too low for the years 2008–2010.
- Users will not know until the 2020 census is taken the extent of error that may occur in the 2010 census-based population estimates that are used as controls for the ACS in the period 2011–2020.

3-H PREPARING TO USE THE ACS

Users can ease the transition from using long-form-sample estimates to using ACS estimates for their applications by becoming knowledgeable about general guidelines for effective use of the ACS (3-H.1) and by taking concrete steps in advance to prepare for the time when 1-year, 3-year, and 5-year period estimates will be released each year (3-H.2).

3-H.1 General Guidelines for ACS Use

Abstracting from the specific applications discussed above, this section provides the panel's basic general guidelines for appropriate use of ACS estimates.

a. Always examine margins of error before drawing conclusions from a set of estimates. Users should follow this practice for the long-form sample, the ACS, and any other survey on which they rely. More specifically:

- When using ACS data to estimate a number or percentage for a single area or population group, such as a city or county, the ACS period estimate chosen—1 year, 3 years, or 5 years—should satisfy the precision requirements appropriate for the purpose for which it is being used (refer back to Table 2-7a).
- Five-year period estimates will not be precise for estimates of small population groups (for example, poor school-age children, poor elderly people, minorities, high school dropouts) for areas with fewer than about 50,000 people, which includes most counties, cities, towns, townships, and school districts, as well as every census tract and block group (refer back to Table 2-7a). Consequently, users should work with 5-year period estimates for small areas only with great care.
- When it is unduly burdensome to examine numerous individual margins of error—as, for example, when working with a large number of 5-year period estimates for multiple geographic areas—users should at least examine some of the individual error margins to check that the estimates are of adequate precision for their purpose.

b. Review the available information about nonsampling errors for estimates of interest and use this information in interpreting findings from the ACS.

- Research on nonsampling errors that may systematically bias survey estimates upwards or downwards is difficult to conduct, and the available information is rarely complete or definitive about the magnitude of the biases. Hence, users are rarely in a position to adjust estimates of interest to correct for nonsampling errors. Nonetheless, users should acknowledge known nonsampling errors in their uses of the ACS data.
- As examples of possible biases in the ACS, a comparison of the C2SS with the 2000 long-form sample found significantly lower estimates of median income in the C2SS than in the long-form sample, while comparisons of the C2SS and the 2001–2003 ACS test surveys with the CPS consistently found significantly lower estimates of unemployment in the ACS surveys than in the CPS (see Section 2-B.2.e). Further research is needed to determine which survey is more accurate.

c. Carefully consider the pros and cons of alternative strategies for extracting value from ACS 5-year period estimates for very small areas, such as aggregating small-area estimates into estimates for larger, user-defined areas.

- Large cities and counties should use ACS 5-year period estimates for census tracts and block groups as building blocks to define larger areas that are meaningful for analysis and for which 5-year period estimates are sufficiently precise. For example, a city might aggregate census tracts into several planning areas, or it might use combinations of block groups that do not necessarily respect census tract boundaries. Statistical mapping techniques may help identify which tracts and block groups would be most useful to combine into subareas for analyzing such phenomena as commuting patterns. For user-defined subareas, a city might ask the Census Bureau to develop 3-year period estimates for large population groups to obtain more information on trends.
- Small governmental units may not be able to satisfy their data needs by aggregating 5-year period estimates into larger areas. However, with due care they may be able to work with 5-year period estimates for large population groups in their jurisdiction and 5-year period estimates for smaller groups for a larger area, such as their county, to assess changes in the composition of their own area. Small governmental units might also ask the Census Bureau to develop ACS estimates for their area for periods longer than 5 years.

- In addition to the basic strategies outlined above, in the future it may be possible to extract more value from the ACS 5-year period estimates by linking them with administrative records and other sources of local-area information (see Section 7-D.1).

d. When using ACS data to estimate shares of some total, compare estimates among areas or population groups, or assess trends over time, use ACS estimates that pertain to the same time period (1-year, 3-year, or 5-year) for all geographic areas or population groups that are being compared. Do not use a mixture of different period estimates.

- For example, when determining the share of federal or state program funds that is to be allocated to each county in a state, the ACS estimates that are used will most likely need to be 5-year period estimates. The reason is that 1-year and 3-year period estimates are available for only about one-fourth of counties (refer back to Table 2-5), and it is not equitable to use a mixture of 1-year, 3-year, and 5-year period estimates to determine each county's share of funds.
- An exception to the need to use 5-year period estimates for fund allocations to counties is when a state has only a small number of counties that lack 1-year (or 3-year) period estimates. In this case, it may be appropriate to update the 5-year period estimates for the smaller counties by using information for larger areas, so that the equivalent of 1-year (or 3-year) period estimates can be used for all counties in the state. A simple procedure for accomplishing the updating is described in Section B.2 above; the use of this or another procedure depends on the reasonableness of the underlying assumptions.
- As a matter of good practice, differences that are observed in comparing areas or population groups or in assessing trends over time should be evaluated not only for statistical significance, but also for substantive importance—that is, whether the differences are large enough to matter for policy, planning, or research purposes.

e. When analyzing trends over time for an area or population group, use ACS 1-year period estimates whenever they are available and sufficiently precise for the purpose of interest and be cognizant of changes in geographic area boundaries that may affect comparability. Keep in mind that the sampling error for the estimate of the difference between pairs of 1-year period estimates will be larger than the sampling error of either estimate.

f. If only 3-year or 5-year period estimates are available or sufficiently precise, use them with care for analyzing trends over time for an area or population group. In general, avoid analyses of changes over time that are based on overlapping period estimates (for example, 5-year period estimates for 2010–2014 and 2011–2015).

- It is not straightforward to interpret the meaning of differences that are observed between pairs of 3-year or 5-year period estimates: an observed difference may reflect a gradual change over the period, or it may reflect another pattern of change, such as stability in a characteristic followed by a sudden increase or decrease. Examining 1-year (or 3-year) period estimates for a larger area may help determine the appropriate interpretation of differences that are observed between pairs of 5-year period estimates for smaller areas within the larger area.
- The less that pairs of 5-year (or 3-year) period estimates overlap in time, the more precise will be an estimate of differences between them—for example, a difference observed between estimates for 2010–2014 and 2015–2019 will be more precise than a difference observed between estimates for 2010–2014 and 2011–2015. Indeed, to obtain an acceptable level of precision for analysis of population groups, it will generally be necessary to wait for a second, nonoverlapping estimate to become available to compare to an earlier estimate.

g. Take advantage of the availability of 1-year and 3-year period estimates for PUMAs, which include about 100,000 people, to assist with analyses for smaller areas.

- As one example, 5-year period estimates for small areas (census tracts in a city, towns in a county, small counties in a state) could be updated by adjusting their 5-year period estimates to the latest 1-year (or 3-year) period estimates for the applicable PUMA, as in Section B.2 above. Such adjustments need to be performed with care.
- As another example, 1-year period estimates for large cities or counties could be compared with the PUMA estimates for the rest of the state (or the rest of the county in the case of a large city within a very large county).

h. Take care to label ACS estimates, including those for 1 year, 3 years, and 5 years, as **period** ***estimates.***

- ACS 3-year and 5-year period estimates do not refer to a particular year, such as the end year or the middle year. They are *period* estimates—averages of characteristics over a 36-month or 60-month period—and should be labeled as such. Otherwise, readers may draw an incorrect inference—for example, assuming that a 5-year period estimate of 15 percent poverty is the rate for the end year, when the end-year rate could be considerably higher or lower.
- ACS 1-year period estimates are also an average over 12 months (except for the special estimates released in June 2006 for January–August 2005 and September–December 2005 for areas affected by Hurricanes Katrina and Rita).

i. Use ACS 3- and 5-year period estimates for income, housing value, and housing costs with care. To compensate for the differing time periods for which dollar amounts are collected, those amounts are adjusted to a common calendar year by the change in the national CPI. This inflation adjustment expresses all of the reported dollar amounts in a comparable manner with regard to purchasing power as of the most recent calendar year in the period. However, the resulting estimates should not be interpreted as current-year estimates.

j. Use care in comparing ACS estimates with estimates from other data sources, including the 2000 long-form sample and other surveys, and be cognizant of the differences that could affect the comparisons. Such differences may include population coverage, sample size and design, reference periods, residence rules, and interview modes.

3-H.2 Suggestions for Users During the Ramp-up Period

In the next few years, users who plan to make extensive use of the ACS will have an opportunity to prepare for the full range of 1-year, 3-year, and 5-year data products that will be available beginning in 2010. It is important for users—including federal, state, and local agencies, and private organizations—not to stint attention or resources in order to make the most of this opportunity, so that they are well prepared to work with the full richness of the ACS data by 2010.

We outline below some of the steps that the technical staff of an agency can take to ensure that their agency is well prepared to work with the ACS data (see also the recommendations in ORC Macro, 2002:Ch. 10).

a. Steps to prepare agency management:

- Schedule briefings with agency program managers to acquaint them with the ACS and how it will replace and improve on the once-a-decade long-form sample.
- Make the case for sufficient resources from management to support planning for effective use of the various ACS products, which should save resources in the long run by minimizing inappropriate or ineffective use of the data.
- As methods are developed to work with the ACS data for key applications, keep agency management informed of solutions.
- Apprise management of the need to take such actions as seeking legislative authority or modification of regulations to permit ACS data to be used in place of long-form-sample data for particular applications.

b. Steps to determine which data and methods to use for particular applications:

- Make use of information from the Census Bureau about the likely sampling error for different size areas to determine the most useful ACS estimates for the agency's application(s). For example, if a city will have 1-year period estimates provided but their sampling error will be high, then the city may want to rely on the 3-year period estimates for planning and program applications.
- Use the detailed 1-year period estimates that first became available in summer and fall 2006 to help develop the most useful profiles and other products to generate from the detailed 3-year and 5-year period estimates when they become available.
- Use the training data sets released by the Census Bureau in spring 2007 of 1-year, 3-year, and 5-year period estimates fo 34 ACS test site counties for the years 1999–2005. Use of these data can provide valuable experience in working with multiple sets of estimates prior to the availability of the full set of 3-year and 5-year period estimates.
- Determine the most appropriate geographic aggregations of 5-year period estimates for subareas of, say, a city or county. For example, the technical staff might divide a city of 500,000 into 10 service areas of approximately 50,000 population and aggregate 5-year period estimates for census tracts and block groups accordingly. Similarly, a large county might aggregate 5-year period estimates for townships into subcounty regions.
- If resources permit, commission a detailed, comprehensive analysis of the alternatives for using various ACS data products for key

applications, similar to the study commissioned by HUD (ORC Macro, 2002).

- Consider how often to analyze updated ACS estimates in light of the agency's needs and resources. For example, it is unlikely that updates of small-area data analyses can be conducted more than twice a decade, nor may it be effective or efficient to do so.
- Determine if any special tabulations will be needed from the Census Bureau, develop detailed specifications for them, and discuss feasibility and costs with the Census Bureau well in advance. For example, areas with large seasonal populations may want to request special tabulations.
- Determine whether and how additional data sources may be helpful in some applications of ACS data. For example, a state might want to use administrative records information in conjunction with ACS estimates in fund allocation formulas.
- Determine if model-based or composite estimates that are developed from the ACS and other data sources by statistical agencies could support particular applications, thereby saving on the program agency's technical resources.
- Request that the Census Bureau inform users of helpful guides that are developed by State Data Centers and other organizations and individuals to assist users—for example, the recent publication, *American Community Survey Data for Community Planning* (Taeuber, 2006).

c. Steps to work with public officials, the media, and other constituents:

- Develop templates for appropriate interpretative language to use in press releases and talking points about each summer's issuance of the latest ACS estimates from the Census Bureau. Given that the media and public officials will inevitably want to compare trends across time and levels across areas using the most recent estimates regardless of their precision, the agency technical staff should develop suitably cautionary language to include in statements by public officials and in speaking with the media.
- Develop standard formats for tables to provide to constituent groups (for example, neighborhood advisory commissions or council members in a city or county). Be sure to include appropriate explanatory material about sampling error and other aspects of the data.

d. Steps to look toward the future:

- Keep up to date with Census Bureau information on the ACS web site, such as users' guides and design and methodology reports, and take advantage of training opportunities afforded by the Census Bureau, state data centers, and other organizations.
- Feed back questions, concerns, and data needs to the state data centers and to the Census Bureau. On one hand, be cognizant that the Census Bureau has a heavy workload in collecting, processing, and disseminating the continuous ACS, but, on the other hand, remind the Census Bureau that the ACS must be an evolving data system that responds to user needs.
- Liaise with other users with similar interests to develop and evaluate strategies for effective use of ACS data products, and put forward coordinated requests for new and improved data products, training materials, and other support from the Census Bureau.
- If there is a need for new or modified questions, work with the Census Bureau and stakeholders to determine what is the minimum set of changes that would serve the purpose. The Census Bureau has a protocol for the testing that must be undertaken before a new question can be added to the ACS (see Section 7-C.2).
- Similarly, work with the Census Bureau and stakeholders to adjust geographic boundaries for census tracts and block groups in ways that reflect population change but minimize discontinuities in local geographic boundaries over time. If, for example, most changes to these small areas involve splitting them to reflect population growth (or, alternatively, combining them to reflect population decline), then it will be easier to use successive 5-year period estimates.
- Participate in forums in which users share their experiences with analysis and presentation techniques that make effective use of the ACS data for a range of applications.

In conclusion, the ACS will offer not only significant challenges to data users, but also significant benefits. Having more timely and up-to-date information that is likely of higher quality will benefit all applications that previously used the long-form-sample estimates. In the future, there will be opportunities for new uses of the ACS that would never be possible with the long-form sample. Users should take steps during the ramp-up period to prepare for the ACS, anticipate problems, and work together and with the Census Bureau on solutions.

PART II

Technical Issues

4

Sample Design and Survey Operations

The panel is impressed by the extent of research and development that the Census Bureau has devoted to the design and operation of the American Community Survey (ACS) throughout 10 years of testing and partial implementation. Given that the ACS has just been fully implemented and given its complex design, it will continue to require a high level of research, evaluation, and experimentation that can not only inform users and ACS managers, but also lead, as appropriate, to modifications that increase the quality and usefulness of the data and the efficiency of the survey operations. Such research needs to systematically evaluate various aspects of the survey in the context of full implementation and also to address unforeseen problems that may arise in data collection and processing.

The sections of this chapter address the following specific aspects of the ACS sample design and operations that, in the panel's judgment, require continued research, evaluation, and experimentation by the Census Bureau:

- Sampling operations for housing units, including initial sampling using the Master Address File (MAF) as the sampling frame and subsampling for nonresponding housing units;
- Data collection for housing units, including mode of data collection and residence rules;
- Sampling and data collection for group quarters; and
- Data preparation, including confidentiality protection, collapsing of tables for large sampling errors, inflation adjustments, tabula-

tion specifications with respect to the universe and geographic areas for which various estimates are provided, and data quality review.

In each section, descriptive information precedes a discussion of issues and the panel's assessment. Weighting procedures are discussed tangentially; for a detailed discussion, see Chapter 5, which reviews the construction and interpretation of the ACS estimates for 12 months (1-year period estimates), and Chapter 6, which reviews the construction and interpretation of the ACS estimates for 36 months (3-year period estimates) and 60 months (5-year period estimates). A report from the U.S. Government Accountability Office (2004) discusses some of the same ACS issues as this report, including residence rules, methods for deriving independent population and housing controls, inflation adjustments for dollar amounts, and understanding the ACS multiyear period estimates.

Chapters 4, 5, and 6 necessarily emphasize aspects of the ACS that appear to be or may be problematic and hence require continued research and evaluation. Readers should keep in mind the substantial benefits of the ACS in comparison with the 2000 long-form sample that are spelled out in Chapters 2 and 3. These benefits include timeliness, frequency of updating, improved data quality in terms of completeness of response, and consistency of measurement with the long-form sample for most items.

4-A SAMPLING OPERATIONS FOR HOUSING UNITS

This section briefly describes the development of the initial ACS sample of housing units from the MAF (4-A.1), sampling rates for the initial sample (4-A.2), and subsampling rates for nonresponse follow-up (4-A.3). It then outlines the panel's concerns and recommendations for the MAF (4-A.4) and the ACS sample size and design (4-A.5).

4-A.1 Developing the Initial Sample

The initial ACS sample of housing unit addresses in the 50 states and the District of Columbia for 2005 and subsequent years consists of approximately 250,000 housing units per month and approximately 3 million housing unit addresses for the year (about 2.3 percent of 129.5 million housing units on the MAF in 2005).[1] The initial sample—that is, the sample before subsampling for nonresponse follow-up by computer-assisted per-

[1]Refer back to Box 2-1 for a brief description of sampling and other procedures in the Puerto Rico Community Survey; for further information about the housing unit sampling procedures in the United States and Puerto Rico, see Asiala (2004, 2005); Hefter (2005a); U.S. Census Bureau (2006:Ch. 4).

sonal interviewing (CAPI)—is selected using systematic sampling from the MAF so that each monthly sample is spread throughout the United States in an unclustered way. The initial sampling occurs in two phases (see Box 4-1): subdivision of the MAF into yearly segments (first phase) and selection of addresses for the ACS sample for each data collection year from the applicable segment (second phase).

The first phase is designed to allocate housing unit addresses on the MAF to five equal segments, each of which is assigned to year t_1 through year t_5 of specified 5-year periods, which are 2005–2009, 2010–2014, 2015–2019, and so on. The addresses in each segment are eligible to be selected for the ACS sample only for the years to which they are assigned (for example, 2005, 2010, 2015, and so on for t_1 addresses; 2006, 2011, 2016, and so on for t_2 addresses). This segmentation procedure ensures that no address will be included in the ACS sample more than once every 5 years.

The first-phase segmentation of MAF addresses proceeds on a continuous basis in two waves each August and January. The process began in August 2004 when all of the housing unit addresses on the MAF at that time were assigned to equal segments for years 2005–2009. Then, each January and August, newly added addresses are assigned equally to one of the five segments for the period then in progress: for example, new addresses identified in January 2006 were assigned equally to years 2005–2009. In August 2009, all addresses assigned to segments for 2005–2009 that still exist as housing unit addresses on the MAF at that time will be reassigned to the same segments for 2010–2014, and the process of assigning newly added addresses to segments for these 5 years will proceed each January and August until August 2014, when the process will begin anew. The assignment to the five yearly segments is carried out using systematic sampling of addresses, which are arranged in each county by sampling rate stratum (see below) and geographical location.

The second-phase sampling is designed to select ACS sample addresses from a given year's first-phase segment to meet specified sampling rates that are chosen to improve the precision of estimates for small governmental units. The second-phase sampling proceeds in two stages, corresponding to the stages of first-phase sampling. In August of year $t - 1$, a main sample is selected from the segment of MAF addresses assigned to year *t*; then, in January of year *t*, a supplemental sample is selected from newly added MAF addresses assigned to year *t*'s segment. The main sample addresses are assigned equally to the 12 months of year *t* for data collection, while the supplemental sample addresses are assigned equally to months April–December of year *t* for data collection.

BOX 4-1
Developing the Initial ACS Sample, Phases One and Two, Area *X* with 20,000 Housing Units (50,000 People)

Phase One:
Allocate Master Address File (MAF) Housing Unit Addresses to Five Segments

August 2004:
1. MAF addresses for Area *X*: 20,000 housing unit addresses
   1a. Divide into 5 equal segments: 4,000 addresses each
      Segment 1: 2005
      Segment 2: 2006
      Segment 3: 2007
      Segment 4: 2008
      Segment 5: 2009

January 2005:
   2. Newly added MAF addresses for Area *X*: 500 housing unit addresses
   2a. Divide into 5 equal segments as above: 100 addresses each

August 2005:
   3. Newly added MAF addresses for Area *X*: 625 housing unit addresses
   3a. Divide into 5 equal segments as above: 125 addresses each

January 2006:
   4. Newly added MAF addresses for Area *X*: 250 housing unit addresses
   4a. Divide into 5 equal segments as above: 50 addresses each

August 2006:
   5. Newly added MAF addresses for Area *X*: 100 housing unit addresses
   5a. Divide into 5 equal segments as above: 20 addresses each

January 2007:
   6. Newly added MAF addresses for Area *X*: 100 housing unit addresses
   6a. Divide into 5 equal segments as above: 20 addresses each

And so on . . . *until*:

August 2009:
MAF addresses for Area *X*, including all additions, demolitions, and ineligible units:
   Divide into 5 equal segments
      Segment 1: 2010 (addresses previously assigned to 2005)
      Segment 2: 2011 (addresses previously assigned to 2006)

4-A.2 Initial Sampling Rates

Initial sampling of housing unit addresses from the applicable segment for a data collection year each August and January (prior to nonresponse follow-up subsampling) uses one of five different sampling rates for the addresses within each geographic block (an area of, on average, about 15–20 housing units). The five sampling rates pertain to five strata containing the following kinds of blocks:

Segment 3: 2012 (addresses previously assigned to 2007)
Segment 4: 2013 (addresses previously assigned to 2008)
Segment 5: 2014 (addresses previously assigned to 2009)

January 2010:

Newly added MAF addresses for Area *X*

Divide into 5 equal segments as above, and so on . . .

MAF Addresses in Segments 1-3 After Phase One, August 2004, 2005, 2006

Assumes no demolitions or ineligible units.

7. Segment 1 - August 2004 (line 1a):	4,000 addresses
8. Segment 2 - August 2005 (lines 1a + 2a + 3a):	4,225 addresses
9. Segment 3 - August 2006 (lines 1a + 2a + 3a + 4a + 5a):	4,295 addresses

Phase Two: Select Housing Unit Addresses for Each Year's ACS Sample from Applicable Segment

Assume sampling rate is 11.5 percent (2.3 percent times 5—see Section A.2).

2005 ACS:	
August 2004: Draw main sample from Segment 1 (line 7)	460 units
January 2005: Draw supplemental sample from Segment 1 (line 2a)	12 units
TOTAL ACS sample for 2005	472 units
2006 ACS:	
August 2005: Draw main sample from Segment 2 (line 8)	486 units
January 2006: Draw supplemental sample from Segment 2 (line 4a)	6 units
TOTAL ACS sample for 2006	492 units
2007 ACS:	
August 2006: Draw main sample from Segment 3 (line 9)	494 units
January 2007: Draw supplemental sample from Segment 3 (line 6a)	2 units
TOTAL ACS sample for 2007	496 units

And so on . . .

NOTE: Phase Two initial sampling rates will be reduced as the size of the MAF grows to maintain the overall ACS initial sample size of about 3 million housing unit addresses.

1. Blocks in the smallest governmental units that are eligible for oversampling (refer back to Tables 2-3 and 2-4)—defined as eligible governments with an estimated fewer than 200 occupied housing units;
2. Blocks in smaller governmental units—defined as eligible governments with an estimated 200 to fewer than 800 occupied housing units;

3. Blocks in small governmental units—defined as eligible governments with an estimated 800 to fewer than 1,200 occupied housing units;
4. Blocks in large census tracts in large governmental units—defined as census tracts with an estimated more than 2,000 occupied housing units; and
5. All other blocks.

The designation of initial sampling rates is based on estimates of occupied rather than total housing units because blocks in governmental units or census tracts with large numbers of seasonally vacant housing units would be undersampled if total housing units were the criterion. The estimates of occupied housing units are obtained from the current MAF address count times an estimate from the 2000 census of the proportion of occupied housing units for blocks in the governmental unit or census tract. These estimated proportions will presumably be updated at each census.[2]

Initial sampling rates are calculated for each of the five strata that will produce approximately equal precision for estimates of a given characteristic for small governmental units and large census tracts outside these units and keep the overall initial ACS sample at about 3 million housing unit addresses each year. A budget constraint necessitates that the initial sampling rate be reduced for some census tracts in order to pay for a higher level of CAPI nonresponse follow-up in tracts with lower-than-average response by mail and computer-assisted telephone interviewing (CATI). For this purpose, the initial sampling rate is reduced by 8 percent for census tracts in strata 4 and 5 (see above) for which at least 75 percent of addresses are mailable and it is projected that at least 60 percent of responses will be obtained by mail or CATI.

For 2005, the initial (and reduced initial) overall sampling rates for the five strata were as follows:

1. blocks in the smallest governmental units eligible for oversampling: 10 percent;
2. blocks in smaller governmental units: 6.9 percent;
3. blocks in small governmental units: 3.5 percent;
4a. blocks in large census tracts in which sample reduction not made: 1.7 percent;

[2]In Alaska Native and American Indian areas, blocks are assigned to a stratum by applying the estimated percentage of the population that is Alaska Native and American Indian to the estimate of occupied units for the block; the purpose of this procedure is to boost the sample in Alaska Native and American Indian areas.

4b. blocks in large census tracts in which sample reduction made as above: 1.6 percent;
5a. all other blocks in tracts in which sample reduction not made: 2.3 percent; and
5b. all other blocks in tracts in which sample reduction made as above: 2.1 percent.

The second-phase sampling of housing unit addresses for a data collection year uses the above sampling rates multiplied by 5 to allow for the fact that only one-fifth of the addresses in the MAF are included in the first-phase segment for that year. For example, 50 percent of the addresses in the first-phase segment for blocks in category 1 above will be sampled (10 percent multiplied by 5), as will 34.5 percent for blocks in category 2, 17.5 percent for blocks in category 3, and so on. In years after 2005, the 2005 initial sampling rates will be reduced as necessary to maintain an overall initial sample size of about 3 million housing unit addresses. The exception is that no reduction will be made in the sampling rate for stratum 1.

4-A.3 Subsampling for CAPI Follow-up

Even though response to the ACS, like the decennial census, is mandatory, the Census Bureau has never expected that as high a proportion of housing units sampled in the ACS would return their questionnaires by mail as occurs in the publicity-rich environment of the census. In order to reduce costs, the Census Bureau planned from the beginning to use CAPI to collect data for a subsample of nonresponding ACS sampled housing units instead of all of them. Before drawing the subsample, the Census Bureau planned to try to collect data by telephone using CATI for as many as possible of the sampled housing units not responding by mail.

The Census Bureau specified three CAPI subsampling rates to apply to housing unit addresses that were mailed a questionnaire but did not respond by mail or CATI (see Hefter, 2005a, for details):

1. addresses in census tracts with predicted levels of mail and CATI responses between 0 and 35 percent: 50.0 percent (1 in 2);
2. addresses in census tracts with predicted levels of mail and CATI responses between 36 and 51 percent: 40.0 percent (2 in 5); and
3. addresses in other census tracts: 33.3 percent (1 in 3).

In addition, two-thirds (66.7 percent) of nonmailable addresses and addresses in remote Alaska are followed up in the CAPI operation.

The higher (lower) rates are used to subsample nonresponding housing units in census tracts with predicted lower (higher) levels of mail and

CATI response in order to roughly equalize the precision of the estimates for areas with differing levels of predicted mail and CATI response rates. The predicted levels for the 2005 subsampling operation were developed from mail response rate information from the Census 2000 Supplementary Survey and the 2001–2003 ACS test surveys when available and otherwise from a model that included data from the 2000 census; ACS mail and CATI response rate information will be used for all areas in the future.

4-A.4 MAF Concerns and Recommendations

The MAF plays a critical role as the sampling frame for the ACS. It is the Census Bureau's inventory of known residential addresses (housing units and group quarters) and selected nonresidential units in the United States and Puerto Rico. It contains mailing and location address information and other attribute information about each address. It also contains geographic codes, such as county and place codes, obtained by linking to the Census Bureau's Topologically Integrated Geographic Encoding and Referencing (TIGER) database.

For purposes of sampling housing unit addresses for the ACS, the following types of housing unit records are currently included in the ACS version of the MAF (see U.S. Census Bureau, 2006:Ch. 3):

- housing units in existence in the 2000 census and those added from the postcensus program to resolve challenges by localities to their population counts (the count question resolution program);
- new housing units added from semiannual updates of the U.S. Postal Service's (USPS) Delivery Sequence File (DSF), along with housing units that were deleted in the 2000 census but continue to appear on the DSF;[3]
- new housing units added from ongoing listings of addresses in areas of new construction that are conducted for the Census Bureau's other household surveys; and
- new housing units added from the Community Address Updating System (CAUS), which annually lists addresses in about 20,000 blocks, out of a total of 750,000 largely rural blocks, where use of the DSF does not provide adequate coverage.

Corrections to housing unit addresses are obtained from all of the above updating programs and from ACS interviewers.

Because the ACS is a continuous monthly survey nationwide, it is es-

[3]To the extent that demolished housing units are not systematically deleted from the DSF, then the retention of housing units that remain on the DSF but were deleted from the 2000 census MAF may result in unnecessary follow-up costs in areas with heavy demolitions.

sential that its sampling frame—the MAF—be as complete and accurate as possible and that it be updated on a continuous basis in all areas of the country. The panel is concerned about the quality of the MAF updating, not only in areas with city-style addresses (house number and street name—see Section A.4.a below), but also in rural areas (see Section A.4.b; see also National Research Council, 2004a, which raises many of the same points).

4-A.4.a The MAF in Urban Areas

The MAF updating for city-style-address areas between censuses depends almost entirely on the USPS DSF, for which the Census Bureau receives updated versions every 6 months. The DSF is a mail delivery file and is not meant to be a complete address list. Research conducted prior to the 2000 census indicated that the DSF is deficient as a source for the MAF in urban areas in at least three respects (see U.S. General Accounting Office, 1998:17-18):

1. The DSF misses many addresses in new construction areas, where it takes time to establish separate mailboxes and mailing addresses.
2. Portions of the DSF are not updated at the same rate all around the country.
3. The DSF often does not clearly identify addresses in small multi-unit structures—in many of these units, mail may be delivered to a central hall or desk and not to the individual apartments.

These deficiencies in the DSF led to a decision by the Census Bureau for the 2000 census to conduct a complete canvass of all 8.2 million blocks in 1999 in order to bolster the completeness of the 2000 Decennial Master Address File. Previously, the Census Bureau had planned to conduct a complete canvass only in rural areas and to spot-check addresses in urban areas.

For the 2010 census, the Panel on Research on Future Census Methods (National Research Council, 2004a) recommended partnerships with state, local, and tribal governments to collect address list and geographic information throughout the decade in order to reduce the need for block canvassing in 2009, but such partnerships were not developed. Instead, the Census Bureau plans to repeat the very costly complete block canvass operation in 2009. It also plans to conduct a Local Update of Census Addresses (LUCA) program in 2008, in which local governments are given the opportunity to review and update the residential address listings for their jurisdiction, similar to a LUCA program conducted just prior to the 2000 census.[4]

[4]The 2000 LUCA program experienced scheduling and communication problems, and participation was spotty across the country (National Research Council, 2004b:145).

While the block canvass in 2009, supplemented by a LUCA Program, should improve the MAF not only for the 2010 census, but also for the ACS, the improvements will be made at a point in time rather than continuously over the period leading up to 2010. The consequence is that the ACS sample for years prior to the block canvass will to some extent underrepresent growing areas of the country.

The ACS may also not accurately represent residences in small multiunit structures (those with 2–9 apartments). Evidence from the 2000 census indicates that the problem of missed or erroneously identified addresses in these types of structures persisted in the 2000 MAF even after the block canvass and LUCA programs. At present there is no research in progress to investigate the problems of addresses in multiunit structures or duplicate addresses, even though research on the accuracy of the 2000 MAF shows that over 2 million duplicate housing unit addresses may not have been weeded out and, at the same time, over 2 million housing unit addresses may have been missed (National Research Council, 2004b:140-141).

Finally, research is needed on questionnaires that are returned by postal carriers because they are "undeliverable as addressed"—about 12 percent of mailed-out ACS questionnaires in 2005 (U.S. Census Bureau, 2006:7-7). These questionnaires are not missed in the ACS sample because they are included in the workload for CAPI follow-up, but they indicate problems in the MAF that need investigation.

Independent housing unit control totals for 1,951 estimation areas (counties and groups of small counties) are used to adjust the weights of ACS housing unit responses with the intent to reduce net coverage error (see Section 5-C for how these controls are developed using the previous census and local information on building permits). The application of the controls will increase (decrease) the housing unit weights in areas for which the unadjusted ACS estimates fall short of (exceed) the controls. While the effectiveness of these controls requires more research, their use may help identify and adjust for possible housing unit coverage errors in the ACS. However, their use will not adjust for coverage errors for specific kinds of housing—for example, the same weight adjustment is made to single-family homes, small multiunit apartments, and large building apartments in an estimation area. Moreover, because the census-based housing unit controls are subject to error, there will likely be inconsistencies between ACS estimates of housing units for 2010 and the 2010 census results.

4-A.4.b The MAF in Rural Areas

In rural areas, the CAUS was developed because of the difficulty of using the DSF to identify addresses that should be added to the MAF. Many DSF addresses in rural areas are rural route or post office box number addresses that do not indicate street name and house number. Consequently, when the Census Bureau receives an updated DSF that has more addresses than are currently listed on the MAF for a rural area, it is not easy to determine which addresses are new. (Research is under way to determine if there are effective ways to use the DSF for address updates in rural areas.)

The identification of CAUS counties for listing is based on an algorithm that considers the address characteristics of existing MAF records for the county, changes in postcensal housing unit estimates for the county, and changes in the DSF tallies for the county. A second stage takes the counties identified for some potential CAUS listings and identifies blocks that would be expected to yield the most new units. Two ACS sources are used to identify CAUS-eligible blocks. One is blocks with addresses in which ACS fieldwork returned numerous outcomes such as "unable to locate" or "address nonexistent." The other is blocks with a high percentage of addresses that were unmailable from the ACS mailout operation because they lacked a house number/street name/ZIP code address. A third source is from field representatives who identify blocks needing updating while they are in the field completing other block listing assignments. The number of selected blocks from those ranked highly by the algorithm is dictated by budget and operational constraints.

Dean and Peterson (2005) conducted the first evaluation of CAUS. They examined the CAUS listings completed between September 2003 and August 2004 to evaluate the targeting of blocks for CAUS work, review the quality of the CAUS listings, and find out if other Census Bureau operations would have captured the address updates or if CAUS was the only means to collect the information. The study found that CAUS was successful in adding addresses to the MAF that would not have been added by other means, but the study was limited in scope and did not address the issue of addresses that are missed because of constraints on the CAUS operation. No further evaluation has been conducted of CAUS.

4-A.4.c Recommendations for MAF Research and Development

Recommendation 4-1: Given the centrality of the MAF to the ACS, the Census Bureau should ensure that adequate resources are provided to attain the highest possible completeness and accuracy of MAF address information on a continuous basis.

Recommendation 4-2: The Census Bureau should plan now for programs to follow the 2010 census to ensure that the MAF is updated on a continuous basis more completely than is being done prior to 2010. These programs should include not only the current updates from the DSF and the CAUS but also such initiatives as continuing local review, the use of ACS field interviewers to investigate address problems, and the use of address information from the Census Bureau's e-StARS database of linked administrative records.

Recommendation 4-3: The Census Bureau should support a continuing research program on the quality of the MAF and the cost-effectiveness of the various operations that are designed to update the MAF. This program should include periodic field checks on MAF addresses, comparisons with housing unit estimates for specific areas, comparisons with the e-StARS database, and comparisons with the results of the 2009 complete block canvass that will be used to prepare the 2010 census MAF. The program should also include studies of methods to improve the listing of small multiunit addresses in urban areas, characteristics of duplicate housing units, and characteristics of undeliverable mail addresses. In addition, the program should examine the effectiveness of the CAUS and explore ways to improve its performance.

The e-StARS database referenced in recommendations 4-2 and 4-3 is the Census Bureau's electronic Statistical Administrative Records System (StARS). This database consists of addresses and other linked information for households and people from a number of federal and state administrative records, including Social Security, unemployment compensation, Medicare, and others. Addresses are geocoded to small geographic areas. The Census Bureau is using the e-StARS database for wide-ranging research on such topics as ways in which administrative records could improve census operations (Resnick and Obenski, 2006). The Bureau is also experimenting with the use of e-StARS to reduce the variance of ACS estimates for subcounty areas (Fay, 2006; see Section 6-B).

4-A.5 Sample Design Concerns

The ACS sample design is complex and differs from the point-in-time design of the decennial census long-form sample. The panel is concerned that users understand the differences, including the role of housing and population controls. The panel is also concerned about the consequences of smaller sample sizes (compared with the long-form sample) and variable sampling rates (including CAPI subsampling) for the sampling errors of ACS estimates.

4-A.5.a Long-Form and ACS Sampling Frames

The 2000 census long-form sample was a systematic sample of housing units from the DMAF, using variable sampling rates to provide more precise estimates for small governmental jurisdictions. A number of steps were taken to achieve high coverage for the DMAF and to remove duplicate and nonexistent addresses. Although the final DMAF had some remaining errors of omission and erroneous inclusion (as noted above), there is evidence that, on balance, it included virtually all of the housing stock at the time of the census in spring 2000 (National Research Council, 2004b:140-141). In turn, the long-form sample represented the characteristics of the housing stock as of spring 2000 and the people living in that housing as of spring-summer 2000 (allowing that not all enumerations were completed until August and that there were undoubtedly errors in reporting where people lived as of Census Day, April 1). Recall that long-form-sample questions refer to the time of the census or the prior calendar year and that long-form-sample population and housing estimates are adjusted to agree with Census Day control totals from the complete count census of both short and long forms for small weighting areas.

The ACS representation of housing and people in any one year differs in at least three respects from the point-in-time long-form-sample representation for the census year. These differences involve: (1) the composition of the housing sample each year; (2) the composition of the sample of people in the sampled housing stock, and (3) the housing unit and population estimates that are used as ACS controls.

1. The ACS sample of housing units in a given year t represents addresses recorded on the MAF as of January of year t; there is no provision to add newly constructed or newly identified units until the following year.[5] Given that the ACS collects data continuously throughout the year, measures of the *characteristics* of the January housing stock from the ACS (for example, occupancy status, owner/renter, housing value) represent averages over the entire year, which may not be the same as how these characteristics would be measured in January. For housing units newly added to the MAF between August of year $t - 1$ and January of year t, they are included in the year t sample for only 9 months of the year (April-December), so their characteristics (for example, vacancy status) will represent averages over the last 9 months rather than all 12 months of the year.
2. The ACS sample of household members in year t represents people who lived in the January MAF housing stock at some time in the

[5]Each year's ACS estimates actually pertain to responses *obtained* during that year, some of which may come from housing units that were included in the sample for November or December of the preceding year but did not provide data until January or February. This does not materially affect the discussion in the text.

year according to the residence rules for the ACS, which uses a 2-month rule and not the census "usual residence" rule (see Section B.2 below). The measures of household member *characteristics* in year *t* from the ACS represent averages over the entire year. Reflecting the continuous data collection in the ACS, the questionnaire items (population and housing) refer to the time during the year when respondents fill out the survey form, or, for some items, to the previous 12 months.

3. The housing units and people for whom the ACS collects data in year *t* are weighted to agree with independently derived estimates of total housing units and population by age, race, sex, and ethnicity as of July 1 for each of 1,951 estimation areas (described above). The application of July 1 controls may make it appear that the ACS representation of housing and population is not so different from the long-form sample, which represents April, but this is not the case. As explained below, the July 1 controls, which derive from the previous census, are not consistent with the underlying ACS data.

4-A.5.b ACS Housing and Population Controls

For 1-year period estimates from the ACS for housing, the application of a single control for total housing units—assuming it is accurate—will capture growth (or decline) in the housing stock in an estimation area that occurred between January and June. However, it will not capture changes in the composition of the housing stock—for example, in single-unit versus multiunit dwellings—due to growth (or decline) between January and June, nor will it capture changes in the housing stock between July and December. Moreover, the application of housing unit controls for estimation areas will not capture differences in housing growth (or decline) among smaller areas within an estimation area, such as cities for which independent housing unit estimates are available but are not currently used as ACS controls. Finally, the ACS estimates of characteristics of the January housing stock will be averages over the year and not point-in-time estimates for July 1 or any other time during the year. (See Section 5-C for further discussion of the housing unit controls.)

For 1-year period estimates from the ACS for people living in the January housing stock, the application of the population controls will adjust a few more dimensions than just total population to a July 1 reference date—namely, sex, age (13 categories), and race and ethnicity (6 categories), although in practice some collapsing of the cross-classification of these dimensions is common. Yet the population adjustments will have all of the problems of the housing unit adjustments enumerated above. In addition, in areas with seasonal fluctuations in population, the use of July 1 controls derived from the previous census can distort the ACS average estimates of the numbers of various types of people in the area over the entire year (refer back to Table 3-6 for an example; see Section 5-D for further discussion of the population controls).

For many—and perhaps most—areas of the country, the somewhat different representation of population and housing in the ACS 1-year period estimates compared with the decennial long-form sample will not be a significant problem. For some areas, however, the differences may be more pronounced. In fast-growing areas, the restriction of the sample to the January MAF housing stock, even when weighted to represent the housing stock as of July, may cause the ACS estimates to lag the situation on the ground. This could happen, not only for total housing, but also for some housing characteristics if new construction differs markedly from older housing stock. In areas with large seasonal fluctuations in population, as was just noted, the application of census-based July 1 population controls to data that were collected throughout the year may result in estimates of household member characteristics that represent neither a point in time nor an average number.

4-A.5.c Smaller Initial Sample Size and CAPI Subsampling

Budget constraints limit the size of the sample initially selected for the ACS to 3 million housing units per year, cumulating to 15 million housing units over 5 years. Even if data were collected for the full initial sample, the total 5-year ACS initial sample size is smaller than the 18 million housing units that received the 2000 census long-form questionnaire (16.4 million housing units with usable data were included on the final edited data file). The 5-year ACS initial sample size is smaller yet than the expected sample of about 21.7 million housing units that would result if the average 1-in-6 long-form sampling rate were applied to the 130 million MAF housing unit addresses in 2005. Moreover, the initial ACS sample is reduced by 8 percent in census tracts outside oversampled jurisdictions that are expected to have high mail and telephone response rates.

In addition, unlike the long-form sample design, the ACS design subsamples housing units that do not respond by mail or telephone for follow-up with CAPI. The CAPI subsampling uses three different rates in order to approximately equalize the precision of estimates for areas with higher and lower mail/CATI response rates. The effect of the CAPI subsampling and the 8 percent reduction in the initial sample in census tracts expected to have high mail and telephone response is to reduce the size of the final sample to about 2.1 million housing units per year nationwide, or about 10.5 million housing units cumulated over 5 years. This reduced sample size is only 70 percent of the initial full sample and less than half of the likely size of a long-form sample in 2010.

The use of CAPI subsampling increases the sampling error of estimates from the ACS in two ways compared with a design that follows up all nonrespondents. First, as noted above, the nominal sample size after subsampling (the number of mail, CATI, and CAPI responses) is reduced to about 70 percent of the initial size. Second, the effective sample size (the size that determines the precision of the estimates) is further reduced from the nominal size. The reason is the variation in sampling rates due to the subsampling, which equates to variation in the weights assigned to respondents. This variation in weights leads to a loss of precision of estimates compared with estimates from an equally weighted sample of the same size. (See Box 4-2 for a simple illustration.) The benefits of the CAPI subsampling are cost savings from reducing the number of expensive CAPI interviews and the size of the CAPI interviewing staff.

4-A.5.d Variable Initial Sampling Rates

Similar to the long-form sample design, the ACS sample design specifies a limited set of variable initial sampling rates that are introduced in order to make the estimates for small governmental jurisdictions about as precise as the estimates for census tracts in larger jurisdictions. Yet many of the estimates for small areas will not meet commonly accepted statistical standards given the overall size limit of the ACS sample. Small areas must be aggregated into larger geographic areas to obtain reasonably precise estimates from the ACS for them, particularly for small population groups (refer back to Table 2-7a). Such aggregation makes sense for block groups and census tracts for some forms of analysis in larger counties and cities, but it is not likely to be suitable for analyses of small governmental units.

Moreover, the use of a small set of discrete initial sampling rates for different-sized governmental units, when combined with features of governmental organization in the United States, has at least three adverse consequences for the sampling errors of ACS estimates for some jurisdictions. (These problems also affected the sampling errors of long-form-sample estimates.) First, because of the variety of governmental units among and within states, there will likely be some anomalous situations. For example, some states have many small school districts, places, and functioning townships, while other states are principally organized into counties and larger cities. States of the first type will have larger samples, proportionate to their population, than states of the second type.

Second, the use of a small number of discrete initial sampling rates means that areas that differ little in population size may have markedly different sampling errors because they fall into different sampling rate categories. For example, the standard errors of estimates for a governmental

BOX 4-2
Illustration of the Effect of CAPI Subsampling on Precision of ACS Estimates

(1) MAF universe for area (housing units)	4,350
(2) ACS 1-year sample (2.3 percent annual sampling rate—see Table 2-3a)	500
(3) Mail or CATI respondents (52 percent)	260
(4) Remaining sample (line 2 – line 3)	240
(5) CAPI subsample (one-third of line 4)	80
(6) Realized number of sample cases (line 3 + line 5—assume there are no final nonrespondents, vacant units, or ineligible units)	340
(7) Effective sample size for estimation (line 6 reduced by the loss of precision due to unequal weights—the 80 CAPI housing units are assigned a weight 3 times as large as the 260 mail and CATI units)	255
(8) Standard errors for estimates based on an effective sample size of 255 housing units compared with an equally weighted sample of 340 units	15% larger
(9) Standard errors for estimates based on an effective sample size of 255 housing units compared with a long-form-sample size of 725 units	69% larger

NOTES: The effective sample size for the unclustered ACS design (line 7) is given by $(\Sigma w_i)^2/\Sigma w_i^2$, where w_i is the weight of the respondent—see, for example, Kish (1992).The calculation of differences in standard errors when compared to the long-form sample (line 9) does not take account of weighting factors, such as housing unit and population controls, that are intended to reduce sampling error in both the ACS and the long-form sample. Including them would further favor the long-form sample because the census-based controls used for the long-form sample are applied at a local area level. Moreover, inaccuracies in the estimated controls used for the ACS can lead to bias in ACS estimates, a feature that is not reflected in the sampling errors (see Chapter 5).

unit with 801 households that is initially sampled at an annual rate of 3.5 percent will be about 40 percent larger than that for a governmental unit with 799 households that is initially sampled at an annual rate of 6.9 percent (assuming equal CAPI subsampling rates). The use of discrete CAPI subsampling rates also has this effect.

Third, governmental units may be restructured in ways that have implications for their sampling rate categories and the sampling error of their estimates. For example, a county with 1,000 occupied housing units and five equal-size school districts that are then consolidated into a single countywide district could move from an initial annual sampling rate of 10 percent (because of the constituent districts) to one of 3.5 percent, with a 69 percent increase in standard errors for county and subcounty estimates (assuming equal CAPI subsampling rates).

There are alternative approaches that could be considered. For example, if the initial sampling rate were a smoother function of the measure of governmental unit size, there would be no jurisdictions of very similar size with markedly different sampling rates (see Kalton et al., 1998:19). It is also possible that making school districts ineligible for oversampling could reduce the number of anomalous situations, such as states with disproportionately larger samples. School districts were first made eligible for oversampling in the 2000 census because of the need for more precise estimates for allocation of federal elementary and secondary education funds (see Section 3-A). This need persists, but the costs of oversampling school districts may outweigh the benefits. Oversampling school districts contributes to anomalous situations, as noted above. In addition, school districts frequently change boundaries, and in the ACS context, such changes could contribute to abrupt changes in sampling rates when districts combine or split up.

Regardless of the approach used to oversample small jurisdictions, one result is that many larger jurisdictions, such as counties and cities, contain blocks with very different sampling rates. For example, a county with a large city surrounded by small townships may have initial sampling rates that vary on an annual basis from as much as 10 percent (for the smallest government units) to as little as 1.6 percent (for large census tracts predicted to have at least 60 percent mail/CATI response). After subsampling for CAPI follow-up, the final sampling rates may vary on an annual basis from as much as 7.0 percent (for the smallest government units predicted to have, say, 50 percent mail/CATI response) to as little as 1.2 percent (for large census tracts predicted to have, say, 60 percent mail/CATI response). Accumulated over 5 years, the final sampling rates after CAPI subsampling may vary from 35 to 6 percent—a 6-to-1 ratio; in contrast, the 2000 long-form sampling rates varied from 50 to 12.5 percent—a 4-to-1 ratio. (This discussion ignores the effects of other weight adjustments, such as population and housing unit controls.)

The wider variation in final sampling rates will increase the sampling error of ACS estimates relative to long-form-sample estimates for geographic areas and population groups that incorporate varying sampling rates—either from the initial sampling, the CAPI subsampling, or both sources. This increase in sampling error will be in addition to the increase from the smaller overall initial sample size of the ACS and the use of CAPI subsampling.

4-A.5.e Recommendations to Review Sample Size and Design

The above discussion of sampling errors did not have the advantage of actual data from the full ACS. Now that the first year of data collection for the full ACS has been completed, the Census Bureau can begin to estimate the expected 5-year sampling errors for small governmental units and census tracts in larger jurisdictions, investigate disparities in sample allocation among states that differ in governmental organization, and determine the extent of other anomalous situations, such as jurisdictions with similar populations that fall into disparate sampling rate categories. Using that information, the Census Bureau should review the sample design decisions that led to the initial sample sizes and effective sample sizes after CAPI subsampling and consider alternatives that might reduce anomalies and make the allocation of the sample as equitable as possible. A review should be conducted of such alternatives as making the CAPI subsampling rates a smoother function of mail and CATI response rates and informing the choice of subsampling rates by the theoretical results on optimum subsampling rates for initial nonrespondents developed by Hansen and Hurwitz (1946).

Yet whatever the particulars of the sample design, given the available budget, the bottom line is that the sampling error of ACS estimates for small governmental jurisdictions will be larger, often substantially so, than the corresponding long-form-sample estimates. The same conclusion applies to ACS estimates for census tracts in larger jurisdictions, although these estimates can much more readily be combined into larger areas for analytical purposes.

The panel thinks that it is critically important to maintain and, if possible, increase the overall size of the ACS sample. A goal could be to increase the final ACS 5-year sample size (after subsampling for CAPI follow-up) to at least the number of housing units in the 2000 long-form sample, which was 16.4 million. This increase would provide a sample about 55 percent larger than the current ACS. To attain this larger final sample size would require an initial 5-year ACS sample size of about 23.5 million housing unit addresses instead of the current 15 million.[6]

Even with an increase in the ACS sample size of the magnitude just outlined, many small-area estimates, particularly for small population groups,

[6]The originally planned initial ACS sample size over 5 years was 30 million housing units, which would have generated a final sample size of about 19 million housing units (see Section 1-B.3).

would still not meet commonly accepted standards of precision or even the levels of precision of the long-form sample. They would, however, be 25 percent more precise than comparable estimates from the current ACS.

Recognizing fiscal constraints, increases in the ACS sample size would likely have to be made on an incremental basis. Eliminating the institutional group quarters population from the ACS, as suggested in Section 4-C below, could permit a small increase in the household sample size within the current budget. It is also possible that making school districts ineligible for oversampling would permit some redistribution of the sample to other types of small governmental units.

The panel urges the Census Bureau to work closely with the user community to identify and assess the merits of alternative sample sizes and designs for the ACS. It is unlikely that any single design will be optimal for all users, so that trade-offs and compromises will be necessary, as is true of the current design.

Recommendation 4-4: The Census Bureau should identify potential ways to increase the precision of ACS estimates for small geographic areas, particularly small governmental jurisdictions, through reallocation of the sample and through increases in the overall sample size. Cost savings should be sought to support such increases, although increases that could significantly improve the precision of estimates will require additional funding from Congress. Sample reallocation should also be considered to minimize anomalies across areas (for example, jurisdictions with very similar populations that fall into different sampling rate categories).

4-B DATA COLLECTION FOR HOUSING UNITS

4-B.1 Mode of Collection

The ACS, like many surveys, uses a mixed-mode data collection design in order to maximize response while containing costs. The ACS uses three modes of data collection:

1. mailout-mailback, assisted by an advance letter, postcard reminder, and second questionnaire mailed to nonrespondents;
2. CATI from three telephone call centers to try to reach mail nonrespondents (the telephone is also used to follow up mail respondents for whom edit checks indicate a problem with the coverage of household members or failure to answer a minimum number of items); and

3. CAPI of a subsample of mail/CATI nonrespondents. CAPI interviewers, who operate from the Census Bureau's 12 regional offices, may first attempt to complete an interview by telephone, but approximately 80 percent of CAPI cases require a personal visit to the sample address.

The 2000 long-form sample, in contrast, used two modes of data collection—mailout-mailback (assisted by an advance letter and reminder postcard) and personal paper-and-pencil interviewing.

There is evidence from comparisons of the 2000 long-form sample with the Census 2000 Supplementary Survey (C2SS), which used ACS procedures, that the professional, fully trained ACS CATI and CAPI interviewers, assisted by the built-in computer edits and questionnaire routing of the CATI and CAPI instruments, obtained more complete data than the minimally trained, temporary census enumerators (see Section 2-B.2). The CATI and CAPI interviews were even more complete for most items than the ACS mailout-mailback responses (National Research Council, 2004b: Table 7.5).

Yet the panel has two related concerns with the three different data collection modes in the ACS. One concern is that mode effects may bias responses for the same item in different ways. A second concern is that mode effects may vary among population groups and geographic areas because of differences in their response patterns by mode.

4-B.1.a Mode Effects on Questionnaire Items

Survey literature documents that responses for the same item obtained in different ways—writing on a paper questionnaire, typing on an Internet questionnaire, responding over the telephone, responding in person—often have different properties (see, for example, de Leeuw, 2005; Dillman, 2000: Ch. 6). Some of these differences may be due to respondent-interviewer effects that are not present for mail or Internet reports; other differences may be due to different presentations of the items in the various modes—for example, providing marital status categories on a mail questionnaire but asking an open-ended "What is your marital status?" question in a telephone interview.

Only limited research has been conducted to date of mode effects in the ACS. Some mode differences were found in the Census Bureau studies that compared the C2SS and the 2000 long-form-sample responses for various questionnaire items, such as disability and race and ethnicity (see Section 2-B.2; see also Stern and Brault, 2005, which reports on the response effects of changing the placement of disability questions on the 2003 ACS mail questionnaire).

The panel recommends that research on mode effects on item reporting in the ACS be conducted using appropriate experimental designs. Even though it is difficult to design an experiment that can estimate the pure mode effect on reporting because of the confounding mode effect on unit nonresponse (see Biemer and Lyberg, 2003), some work is possible and should be done, given the centrality of multiple reporting modes to the ACS. For example, a sample of mail respondents could be reinterviewed by CATI or CAPI to compare the two sets of responses, or a subsample of mail nonrespondents for which telephone numbers are available could be sent to CAPI instead of CATI interviewing and their responses compared with responses obtained by CATI.

4-B.1.b Differences in Response Mode for Population Groups

Census Bureau research has shown that households responding by mail in the decennial census differ from households requiring follow-up. Households that respond by mail are more likely to own their own homes and be headed by an older person; they are less likely to be headed by a nonwhite or Hispanic person (National Research Council, 2004b:101-102). Analysis of mail response rates for the C2SS, based on housing units in census tracts with 75 percent or more people reporting a specific race or ethnicity, found marked differences in mode of response by the race and ethnic composition of the tract—see Table 4-1.

TABLE 4-1 Weighted Distribution of Respondents by Mode for Census Tracts with Concentrations of Race and Ethnicity Groups, Census 2000 Supplementary Survey

Population Group (housing units) (weighted)	Response Mode (percent)			Total Response
	Mail	CATI	CAPI	
Predominantly white census tracts	60.5	7.4	28.1	96.0
Predominantly Asian census tracts	58.6	4.1	32.5	95.2
Predominantly black census tracts	34.9	8.9	48.6	92.4
Predominantly Hispanic census tracts	34.2	8.3	53.3	95.8
Predominantly American Indian and Alaska Native census tracts	16.6	2.6	69.9	89.1
Total housing units	56.2	7.3	31.9	95.4

NOTES: The distributions represent the percentages of housing units that responded by mail, CATI, and CAPI (with CAPI responses weighted to account for subsampling) among the estimated number of housing units that were eligible to be interviewed (excluding nonresidential addresses). The distributions shown apply to housing unit responses in census tracts in which 75 percent or more of the population reported a specific race or ethnicity.

SOURCE: U.S. Census Bureau (2002b:Tables 2, 3, 4).

It is likely that differences in response modes characterize other groups as well. For example, non-English-speaking households may be less likely to respond by mail or CATI and more likely to respond by CAPI compared with English-speaking households. There may also be important geographic area differences in response modes.

Overall, the use of mailout-mailback, CATI, and CAPI interviewing results in high housing unit response rates to the ACS. Thus, in the C2SS (see Table 4-1), the overall weighted response rate was 95.4 percent, including 56.2 percent mail response, 7.3 percent CATI response, and 31.9 percent CAPI response (applying weights to CAPI respondents to account for the subsampling). The 2005 ACS overall weighted response rate was even higher (97 percent), although, based on data from January to March 2005, the distribution of responses by data collection mode has changed. Thus, only about 51 percent of the eligible sample in January-March 2005 responded by mail, while 9 percent were interviewed by telephone and 38 percent were CAPI interviews, with 2 percent nonresponse (U.S. Census Bureau, 2006:Figure 7-2).

Differences in response patterns (the mix of the three modes) among population groups and geographic areas—and changes in response patterns over time—may result in different levels and directions of response biases among groups and areas. Whether such effects are important and for which characteristics remains to be established by research.

4-B.1.c *Recommendation for Mode Effects Research*

> **Recommendation 4-5:** The Census Bureau should conduct experimental research on the effects of the different data collection modes used in the ACS—mailout-mailback, CATI, and CAPI—on ACS estimates and, when possible, on response errors for questionnaire items. In addition, the Census Bureau should assess how different patterns of responding by mail, CATI, and CAPI among population groups and geographic areas affect comparisons of ACS estimates and inform data users of consequential differences.

4-B.2 Residence Rules

4-B.2.a *Two-Month Rule*

The decennial census employs a *usual place of residence* concept; in the 2000 census, this meant that a person was to be counted at the place where he or she lived or stayed most of the time. Most other household surveys also use a similar concept. In contrast, because of its continuous design in which data collection occurs throughout the year, the ACS changed to a *current residence* concept, as is more common in polls and other person-based surveys. Specifically, the ACS residence concept is based on a "2-month rule:" people who live for more than 2 months at a sample address are assumed to be residents of that unit. The rule is intended to be prospective as well as retrospective—that is, people who have lived in a unit for more than 2 months at the time of the ACS interview and people who have just moved into the unit and expect to stay there for more than 2 months are considered residents of the unit.

The Census Bureau has identified three exceptions to this general concept (U.S. Census Bureau, 2006:6-2, 6-3): (1) children younger than college age who are away at boarding schools or summer camps are to be considered residents of their parents' or caregivers' homes; (2) children who live under joint custody agreements and move often between the residences of their parents are to be considered current residents of the sample unit at which they are staying when contact is made; and (3) commuter workers who stay in a residence close to their work and return regularly to a family residence are to be considered residents of the family residence and not the work residence. In addition, people staying at a unit at the time of the interview who have no other place to stay are to be considered residents of the unit.

While the 2-month rule generally seems reasonable, it is not clear why 2 months was chosen and not another value (for example, 1 month or 3 months). Some of the exceptions to the 2-month rule, particularly for commuter workers, also do not have a clear conceptual basis. In addition, while the 2-month rule acknowledges that not everyone stays in the same "usual residence" all the time (for example, people with summer and winter homes, commuter workers), it does not address other kinds of situations in which people have multiple residences. Examples include people with weekday and weekend residences, people who live and travel throughout the year in recreational vehicles, and people who move among the residences of several relatives or friends.

The 2-month residence rule is applied at the time the data collection takes place. For example, if no mail return comes back from a sampled address and there is no success with CATI, but the address is included in CAPI in the third month of data collection, respondents are asked about residence under the 2-month rule at the time of the interview. Thus, the reference period is a function of the time of interview rather than a fixed time interval related to the month of mailout.

The CATI and CAPI computerized instruments may include questions to enable the ACS residence rules to be applied as intended. However, the mail questionnaire does not clearly or fully explain these rules, as shown in Box 4-3. The accompanying guide for respondents does not provide further instruction (see U.S. Census Bureau, 2006:App. B, which reprints the mail

BOX 4-3
Residence Rule Guidance on the ACS Mail Questionnaire

Page 1:

Asks respondent to provide the number of people who "are living or staying at this address."

Page 2, left-hand margin:

Asks respondent to "READ THESE INSTRUCTIONS FIRST."

- LIST everyone who is living or staying here for more than 2 months.
- LIST anyone else staying here who does not have another usual place to stay.
- DO NOT LIST anyone who is living somewhere else for more than 2 months, such as a college student living away.

If this place is a vacation home or a temporary residence where no one in this household stays for more than 2 months, do not list any names in the List of Residents.

IF YOU ARE NOT SURE WHOM TO LIST, CALL [number].

questionnaire and instruction booklet; see also National Research Council, 2006, Section 8-C, for a more detailed discussion). To date, no research has been carried out to estimate the extent to which mail respondents follow the intended ACS residence concept.

4-B.2.b Recommendation on Residence Rules Research

A separate Committee on National Statistics panel was charged to conduct a comprehensive review of the residence rules for the decennial census. In its report (National Research Council, 2006), the panel comments on the ACS residence rules, noting the lack of a clearly articulated basis for the rules (including the exceptions to the 2-month rule noted above) and the lack of clear instructions on the mail questionnaire on how to apply the rules. The report cites literature on relevant respondent behavior, such as the tendency to ignore instructional material, which can lead respondents to misapply residence rules even if they are clearly specified. The report recommends research leading to the addition of questions on the census about other places where people live to assist the Census Bureau to determine usual place of residence. The report also recommends research leading to the inclusion of a question on usual place of residence in the ACS in addition to residence according to the 2-month rule, in order to make it possible to relate census and ACS results (National Research Council, 2006:Recs. 8-3, 8-4).

We support research with the ACS's experimental methods panel (see Section 7-C.2) to assess the extent to which respondents give different answers to the decennial census usual residence rule and the ACS 2-month residence rule and the extent to which they follow the specific ACS rules, such as the rule to count boarding school students at home. The inclusion of questions on other residences at which respondents spend time would facilitate the determination of respondents' usual residence and 2-month residence to use in analyzing the experimental results.[7] Such research might in the future lead to improvements in the way in which the 2-month rule is explained to respondents, as well as possibly to a decision to modify the 2-month rule in some respect.

In addition, we support research on the effects of the residence rules, assuming they are applied as intended, on estimates for different geographic areas and population groups. For example, the application of the 2-month residence rule should provide a basis for identifying seasonal fluctuations in population in ways that would not be possible with a usual residence rule. A possibly confounding effect could occur from the 3-month data collection window for each month's sample that is part of the ACS design. What is the effect, for example, on estimates of occupied versus vacant housing units when a seasonal resident does not respond by mail or CATI and has left the area by the time of the CAPI interview? Questions such as these should be addressed through appropriate research, including experimentation.

> **Recommendation 4-6:** The Census Bureau should conduct experiments to determine the extent to which ACS respondents give different answers to the decennial census usual residence rule and the ACS 2-month residence rule and the extent to which they apply the specific ACS residence rules (for example, reporting commuter workers at the family residence, applying the 2-month rule prospectively). To help clarify residence according to the census and ACS concepts, the experimental questionnaire should ask about other residences at which respondents spend time. The Census Bureau should assess the implications of the experimental results for ACS population estimates for different geographic areas and population groups. Depending on the results, the Census Bureau should consider appropriate changes

[7]The ACS questionnaire currently asks three relevant questions (see Table 2-2): whether any household members live at the address year round, number of months members live here, and main reason members stay at the address, but these questions are slated to be eliminated in 2008. Moreover, the questionnaire does not ask for information on other residences.

in the ACS questionnaire instructions on residence or in the residence rules themselves.

4-C GROUP QUARTERS SAMPLING AND DATA COLLECTION

About 3 percent of the U.S. population resides in group quarters and not in households. A group quarters (GQ) is defined as a place where people stay that is normally owned or managed by an entity or organization providing housing (and often other services) for the residents. People living in GQs are normally not related to one another. Group quarters include not only institutions, such as prisons and nursing homes, but also noninstitutional group quarters, such as college dormitories, military quarters, and group homes of various kinds (see listing in Table 2-1). Housing unit addresses at which large numbers of (mostly) unrelated people live used to be classified as GQs, but in 2000, these units were classified as households. Similarly, they are included in the ACS household population. Boarding schools and summer camps for children below college age are not included in the ACS GQ universe because of the Census Bureau's rule that children at these facilities are to be reported at their parental or caregivers' residences (see Section B.2.a above). Data collection procedures for GQs in the ACS were tested in 1999 and 2001 in the 36 test counties and revised as appropriate. GQs were not included in the C2SS or the 2001–2004 ACS test surveys, nor were they included in the 2005 ACS because of budget constraints. They were included in the 2006 ACS and are included in the 2007 ACS. Some GQ types are out of scope for the ACS for privacy reasons or because monthly data collection would be too difficult and costly: domestic violence shelters, soup kitchens, mobile food vans, targeted nonsheltered outdoor locations, natural disaster shelters, and quarters for crews of maritime vessels.

This section describes the development of the MAF for GQs (4-C.1), sampling of GQs and residents within them (4-C.2), data collection for GQs (4-C.3), and the panel's concerns and recommendations about GQs (4-C.4, 4-C.5). For details about MAF development, sampling, and data collection procedures for GQs in the ACS, see U.S. Census Bureau (2006:3-7 to 3-8; 4-8 to 4-10; Ch. 8).

4-C.1 Group Quarters and the MAF

For the 2000 census, the Census Bureau originally constructed separate MAFs for GQs and housing units using somewhat different procedures. In the 1990s, the Census Bureau developed an inventory of GQs from various sources. It did not check the GQ inventory against the housing unit MAF until late 1999; these checks identified problems of duplicate GQ and housing unit enumerations, as well as erroneous geographic coding of some GQs (for example, college dormitories that were assigned the geographic location of the university administrative headquarters). Additional geographic coding problems were identified by localities after release of the census counts. After August 2000, when the GQ enumeration was completed, addresses for GQs were added to the MAF, with a flag to indicate that an address was a GQ.

The GQ MAF for the 2006 ACS was constructed by merging an updated GQ inventory file, extracts from the final 2000 MAF, and a file of GQs that were closed on April 1, 2000, but may be open at other times of the year. The Census Bureau also obtained a file of federal prisons and detention centers from the U.S. Bureau of Prisons and a file of military bases and vessels from the Department of Defense. In addition, the Census Bureau conducted Internet research to identify new state prisons and state prisons that had closed. On an ongoing basis, information on new GQs and updated address information for existing GQs is collected by CAUS and the current demographic surveys listing operations.

There has been no formal evaluation of the GQ MAF for the quality of the GQ addresses or for the completeness of the list of GQs. In 2009, there will be a validation of GQ addresses in preparation for the 2010 census. It is likely that the ACS GQ population based on the current GQ MAF will differ in some respects from the 2010 census GQ population.

4-C.2 Sample Design for Group Quarters

The sampling for GQs is different from the sampling for housing units (see Hefter, 2005b). All GQ samples are selected in the main sampling phase in August preceding the data collection year. Two strata are created to sample GQs: the first stratum includes small GQs estimated to have 15 or fewer people as well as GQs listed as closed on Census Day, 2000; the second stratum includes larger GQs estimated to have more than 15 people.

For the small GQ stratum, a two-phase sample of GQs is selected, similar to how the housing unit sample is obtained. The first-phase sampling began in August 2005, when all small GQs were assigned to one of five 20 percent segments or subuniverses. One of these subuniverses is the 2006 first-stage sample, and the rest are assigned to 2007–2010. The 2006 subuniverse will not be eligible for sampling again until 2011. In August 2006, all small GQs that were new since the previous year were assigned equally to the five existing subuniverses, as will be done each August through 2009. In August 2010, the plan is to reassign the GQs in the 2006–2010 subuniverses to subuniverses for 2011–2015 and to assign new GQs likewise.

The second-phase sample of small GQs is designed to yield a 2.5 percent systematic sample of such GQs within each state, sorted by GQ type and geography. To achieve this sampling rate for each year, 12.5 percent (1 in 8) of GQs in the appropriate one-fifth subuniverse are selected. Every person in the selected GQs is eligible to be interviewed, although, if there turn out to be more than 15 people residing in the GQ, a field subsampling operation is implemented to reduce the sample to 10 people.

For the larger GQ stratum, there is no assignment to subuniverses. Instead, all of these GQs in a state are sorted by type and geography, and a measure of size is calculated, which equals the estimated number of residents divided by 10. These groups of 10 constitute the first-stage unit of sample selection: a 2.5 percent (1 in 40) systematic sample of groups is selected each year. GQs with a measure of size of 40 or more will have one or more selections or hits; those with a smaller measure of size may have one hit or no hits. If there is more than one hit in a larger GQ, the hits are allocated to different months for data collection (if there are more than 12 hits, then more than one hit is assigned to one or more months). All GQs in this stratum may be selected in any year regardless of whether or not they were previously selected.

The second-stage and ultimate sampling unit for larger GQs is the person. Field representatives implement the selection of people to be interviewed when they visit the GQs assigned to them each month with at least one first-stage hit. They determine the total number of residents at the GQ and use an automated listing instrument to select 10 residents to be interviewed for that month. The field representatives will return in a subsequent month (as assigned) to large GQs with more than one hit to select another group of 10 to be interviewed.

The assignment to each month of the year of sampled small GQs and one or more sampled groups of 10 people in larger GQs is similar to the procedure for housing units, in that the sampled small GQs and sampled groups of 10 people in large GQs for a state are combined, sorted, and systematically assigned to months January–December. The exceptions to the assignment procedure, due to budgetary and operational constraints, occur for correctional facilities and military barracks. While sampled state and local correctional facilities and military barracks are assigned evenly to all months in the year, all groups of 10 people in a state or local correctional facility or barracks with more than one sampled group are assigned to the same month, instead of being spread across months as is the case for other GQ types. In the case of all sampled federal prisons, all sampled groups of 10 people are assigned to September, with a period of up to 4.5 months allowed for data collection. The U.S. Bureau of Prisons generates the person samples for the federal prisons that are selected by the Census Bureau for the year; the Bureau of Prisons must also conduct security clearances for all field representatives who will conduct interviews in the sampled federal prisons.

4-C.3 Data Collection for Group Quarters

For the 2006 ACS, about 18,000 GQ facilities with one or more hits are in the sample (about 20,000 hits). Of these facilities, 850 are military facilities and 148 are federal prisons. The GQ data collection for 2006 was accomplished primarily by field representative personal visits, using an automated Group Quarters Facility questionnaire and a bilingual paper ACS questionnaire for each sampled resident. The facility questionnaire is used at the only or first visit to a GQ to collect address, contact information, and type of GQ for the sampled GQ, record up to two other GQ types for a GQ, ascertain the maximum and current population at the facility, and then generate the person-level sample. The individual GQ resident questionnaire contains the same person items as the household questionnaire but none of the housing unit questions, except for the question on receipt of food stamps.

It is clear that field representatives cannot do all of the interviewing of GQ sample persons face to face, although that is the preferred procedure. Other methods are permitted: the field representative may fill in the questionnaire by telephoning the sample person; conduct an in-person interview with a proxy, such as a relative or guardian; leave the questionnaire with the sample person to complete after ascertaining that the person is physically and mentally able to do so; or leave questionnaires with the contact person for the GQ to distribute them to sample persons and collect them after they are filled in. Any GQ contact person who is enlisted to distribute and collect questionnaires must first be sworn in as a special sworn agent of the Census Bureau, bound to protect the confidentiality of individual responses and subject to the same penalties for breach of confidentiality as regular Census Bureau employees.

4-C.4 Concerns About Group Quarters

Almost every aspect of survey operations for group quarters residents presents challenges for the Census Bureau, and successful data collection for this population requires substantial effort and resources. Feedback from ACS managers is that, after some start-up problems, the 2006 data collection for group quarters residents proceeded relatively smoothly but at considerable expense to complete a sample case. To ensure data of good quality from the GQ component of the ACS, sufficient resources must be devoted to intensive, continuing research and development to fine-tune all GQ procedures, from construction of the MAF and sampling of facilities to the collection of data from individual group quarters members, and then to rigorous control of the quality of operational procedures.

In the 2000 census, the group quarters operation was a stepchild of the household data collection operation, and poor quality of the GQ data was the result. In particular, missing data rates for most long-form-sample items on GQ questionnaires were very high (20 percent or more for four-fifths of the items and 40 percent or more for one-half of the items). The rates were much higher than missing data rates for household members and considerably higher than missing data rates for GQ residents in the 1990 census (National Research Council, 2004b:Tables 7-9, H-8). Missing data rates were particularly high for people in prisons, nursing homes, and other institutions, perhaps because of heavy reliance on administrative records for collecting the data. These and other problems in 2000 led a Committee on National Statistics panel to recommend that the Census Bureau "redesign the processes related to group quarters populations for the 2010 census, adapting the design as needed to different types of group quarters" (National Research Council, 2004b:156).

The Census Bureau has devoted considerable effort to refining its procedures for collecting data from GQ residents in the ACS, and presumably missing data rates for GQ residents, including inmates of institutions, are much reduced in the 2006 ACS compared with the 2000 long-form sample. Yet the panel is concerned about the costs of collecting high-quality GQ information relative to the benefits of the data.

The argument for collecting information on GQ residents in the ACS is so that the survey will cover the entire population similar to the long-form sample. Most national household surveys, in contrast, cover just the civilian noninstitutional population, including residents of housing units and noninstitutional GQs. The Current Population Survey Annual Social and Economic Supplement (CPS ASEC), which produces official income and poverty statistics, covers the civilian noninstitutional population plus members of the armed forces living with their families in housing units or military barracks. The CPS ASEC does not conduct interviews in college dormitories but asks parents to report college students who reside in dormitories as household members.[8]

The census will continue to obtain basic demographic information about all types of GQ residents once every 10 years for all size geographic areas. The Census Bureau's population estimates program could publish annual estimates of GQ residents—total and broken down by institutional and noninstitutional—by age, sex, race, and ethnicity for counties, cities, and townships, although the quality of these estimates is not known. National surveys have targeted some GQ populations, although they do not provide small-area estimates (for example, the periodic National Nursing Home Surveys, sponsored by the National Center for Health Statistics, and

[8]See http://www.bls.census.gov/cps/asec/smethdoc.htm.

the periodic Surveys of Inmates in Federal and State Correctional Facilities and Local Jails, sponsored by the Bureau of Justice Statistics).

The question is whether users require continuous collection of detailed long-form-type information for GQ residents for counties, cities, and smaller areas and whether their requirements are sufficiently pressing to justify the high cost of obtaining high-quality responses in the ACS. Indeed, for the institutional population, one can question the relevance of much long-form-type information. For example, what does it mean to ask a prisoner about his or her income, and how useful are the responses? Most residents of nursing homes and long-term-care hospitals likely have income from such sources as Social Security or retirement or disability benefits, but it is not clear how they or their proxies may report other income sources, such as support from family members. In fact, in 2000, fully 78 and 77 percent of prisoners and nursing home residents, respectively, had all of their income imputed because they did not answer any of the income questions. In comparison, 25 percent of household residents had all of their income imputed (National Research Council, 2004b:Tables 7-5, H-8).

The panel thinks that the Census Bureau should give serious consideration to whether long-form-sample-type data from the continuous ACS for the institutional population—and perhaps other types of GQs—is needed to an extent that justifies the costs. Dialogue with the user community could identify items that are important to collect every year on a comparable basis and items that are not needed or for which data are not likely to be of sufficient quality to be useful. Discussion with users could also determine whether it is necessary to collect any data at all for residents of some or all types of GQ. A decision to alter the universe for the ACS by excluding some or all GQ residents would require the use of an appropriate set of population estimates to use as controls for the ACS estimates. For example, household population estimates are used in the 2005 ACS estimates, and noninstitutionalized population estimates are used in other household surveys. The quality of these estimates for estimation areas (counties and groups of small counties) would need to be carefully evaluated (see Section 5-D). A decision to alter the universe for the ACS would also have implications for ACS tabulations and other data products (see Section 4-D.4 below).

4-C.5 Recommendation for Group Quarters

Recommendation 4-7: The Census Bureau should discuss with data users their requirements for detailed information from the ACS for residents of institutions and other types of GQs, particularly at the local level. The discussions should assess benefits against costs, and the results should be used to determine any changes to the GQ component of the ACS—for example, the possible deletion of institutions from the ACS universe—that would be cost-beneficial for users and stakeholders.

4-D DATA PREPARATION

This section briefly describes key procedures to prepare the ACS data products, including confidentiality protection measures (4-D.1), the collapsing of tables because of large sampling errors (4-D.2), inflation adjustments of income and housing value and costs (4-D.3), tabulation specifications with respect to the population universe and geographic areas for which various estimates are provided (4-D.4), and data quality review (4-D.5). Recommendations for research and development on these topics are contained within the applicable subsection.

4-D.1 Confidentiality Protection

4-D.1.a Confidentiality Protection Procedures

The Census Bureau uses three primary methods of disclosure avoidance to minimize the risk that someone could identify an individual respondent in the ACS data products: data swapping, categorizing variables, and top-coding. The first two methods are used for tabulations; all three methods are used for the ACS public use microdata sample (PUMS) files. The PUMS files also protect confidentiality by deleting names and addresses from the individual records, limiting geographic identification to large areas containing about 100,000 people called public use microdata areas (PUMAs), and perturbing the ages of people in households with 10 or more members. In addition, the subsampling for generating the PUMS files affords protection even if one knows a person who was in the full ACS sample because one does not know whether the person is in the PUMS subsample.

Data swapping occurs when a household has rare characteristics (such as being the only minority household in a block group). In such instances, the entire household may be swapped with a demographically similar household in a different geographic region. Only a small percentage of households are ever swapped, and they are never identified. The purpose of swapping is to ensure that users will not be able to identify a household with certainty. All data products are created from the ACS records after swapping.

Categorizing variables refers to collapsing categories of a variable within a table, or on the PUMS records, to avoid small cell sizes. For example, a table may combine some race categories, such as races other than white and black, into a single category, or a table may combine income amounts into intervals of $10,000 or more, with a wide top category, such as $100,000 or more.

Top-coding refers to assigning a value to an individual record that is the same as that assigned to other individuals, all of whom have actual values above a specified limit. For example, all individuals with wages and salaries of $100,000 or more might be assigned the value of $100,000. Top codes for the ACS PUMS files are developed separately according to the distribution of responses by state.[9]

4-D.1.b Confidentiality Protection Concerns

The panel strongly supports the protection of respondents' individual information, because a breach of confidentiality would not only undercut the Census Bureau's ability to collect information, but also break trust with respondents. At the same time, the panel is concerned that confidentiality protection not be ratcheted up without a careful consideration of the need not only to minimize disclosure risk, but also need to provide useful information for public- and private-sector decision making, research, and analysis. It is not possible to reduce the risk of disclosure to zero; the goal instead must be to minimize risks while not unduly suppressing valuable information.

Microdata Products A recent report of a panel of the Committee on National Statistics, *Expanding Access to Research Data* (National Research Council, 2005), addresses issues in balancing confidentiality and privacy protection with obtaining an adequate return on taxpayers' investment through providing users with access to rich microdata sets from government surveys. The report recommends research on techniques for providing useful, innovative public-use microdata sets that increase informational utility without increasing disclosure risk.

In the context of ACS microdata, the panel encourages the Census Bureau to revisit its decision not to include month of data collection on the PUMS as a confidentiality protection measure. Given that individual PUMS records are not identified geographically for areas with fewer than 100,000 people, it could be argued that omitting month of data collection is not necessary to protect confidentiality. Including this variable on the PUMS files would be immensely valuable for analytical purposes in light of the moving ACS reference period. For instance, knowing the month of data collection would permit data users to make their own adjustments for inflation for income amounts (see Section 4-D.3 below). It would also fa-

[9]See, for example, http://www.census.gov/acs/www/Products/PUMS/C2SS/minmaxval4.htm.

cilitate research on seasonal variations in population. If, upon investigation, it appears too risky to include the exact month of interview, then perhaps the value could be perturbed within a range of plus or minus a month (for example, a month of interview labeled as "March" might actually have occurred in February or April).

Multiyear Estimates The panel thinks that the continuous design of the ACS affords a measure of protection for respondents that the Census Bureau should take into account when considering appropriate confidentiality protections for multiyear estimates for small areas. The U.S. population is highly mobile with respect to geographic location, employment, family composition, commuting patterns, and other characteristics within and across years. Thus, the fact that 60 months of data are averaged to provide 5-year period estimates for block groups, census tracts, and small governmental jurisdictions should go a long way toward protecting individual respondents, even without additional steps to protect confidentiality. The Census Bureau, of course, will not, and should not, rely solely on averaging as a protection, but it should carefully consider the costs and benefits of each additional protection procedure and conduct research to identify the most useful protection techniques.

In this regard, the Census Bureau should consider developing selected tables with reasonably precise estimates for seasonal populations (for example, winter and summer residents) for geographic areas that experience seasonal population changes. Thought would need to be given to whether appropriate population controls can be developed for such tables or whether to use controls at all.

In addition, the Census Bureau should conduct research to determine an appropriate number of cases that need to be in the sample for a table or table cell to be released. To date, the Census Bureau appears to be using rule-based procedures for determining which tables must be deleted from publication in order to protect confidentiality. For example, the Census Bureau has developed rules for publication of worker and journey to work tabulations for traffic analysis zones and other geographic areas (Zayatz, 2005). Some of these rules appear to be reasonable, but others appear to lack a rationale.

One of these rules is that an area must have at least 10 unweighted or 60 weighted cases of workers in the sample over the year for 1-year workplace tables to be published. For 3-year and 5-year workplace tables, the corresponding minimums are, respectively, 30 unweighted or 180 weighted cases of workers in sample over the last 3 years and 50 unweighted or 300 weighted cases of workers in sample over the last 5 years. In other words, the average minimums, year by year, are the same—namely, 10 unweighted or 60 weighted cases. Assuming the 1-year period estimate minimums are reasonable, then having the same average yearly minimums for the 3-year and 5-year period estimates makes sense: even though the 3-year and 5-year period estimates are published for smaller geographic areas than the 1-year period estimates, they represent averages over longer periods of time.

A second rule is that 5-year period estimates of mode of transportation to work cross-tabulated by another variable will not be published for an area for a particular mode unless it has at least 3 unweighted workers in the sample. If it does not, then the mode must be collapsed with other modes to reach the minimum sample size requirement. Such a restriction is not imposed on the 1-year or 3-year period estimates. Given the skewed distribution of mode of transportation in the United States, whereby three-fourths of the population drives alone to work, another 10 percent carpools, and very small percentages take public transit, bicycle, walk, or work at home, this restriction may curtail the publication of needed information on transportation to work in many areas. In turn, such curtailment will handicap users who want to aggregate data for traffic analysis zones into larger areas of their own definition.

The reason for the restriction for 5-year period estimates is not clear. Mode of transportation to work is highly variable: the same individual may decide to walk to work in the summer and drive in the winter or may walk to work for 4 years and then decide to switch to a new bus line or vice versa. Collectively, the workers in a traffic analysis zone are unlikely to include the same individuals over the 5-year period because of changes in residence and employment.

The Census Bureau has time before 5-year period estimates become available in which to develop appropriate confidentiality protection strategies and techniques for transportation tables and other data products. Such strategies should seek to minimize disclosure risk in ways that recognize the protection afforded by averaging over 60 months of data. When developing confidentiality protection procedures for cross-tabulations, the Census Bureau should also, whenever possible, prefer procedures that make it possible to aggregate the data for smaller units into larger units. Thus, instead of suppressing cells of a cross-tabulation, it might be possible to use techniques that perturb the data for individual cells while preserving the marginal totals for each variable.

4-D.1.c Confidentiality Protection Recommendations

Recommendation 4-8: Because of the potential value of month of data collection for analysis of the ACS PUMS, the Census Bureau should revisit its decision to omit this variable as a confidentiality protection measure. If further research determines that including exact month of data collection would significantly increase disclosure risk, the Census

Bureau might consider perturbing the month of data collection or taking other steps to protect confidentiality. Similarly, the Census Bureau should consider developing selected summary tables that identify the season of collection (such as winter and summer) for geographic areas for which such information would be useful.

Recommendation 4-9: The Census Bureau should undertake research to develop confidentiality protection rules and procedures for tabulations from the ACS that recognize the protection afforded to respondents by pooling the data over many months. Whenever possible, the Census Bureau should prefer confidentiality protection procedures that preserve the ability to aggregate smaller geographic areas into larger, user-defined areas.

4-D.2 Collapsing Tables for Large Sampling Errors

In addition to procedures to protect confidentiality, the Census Bureau applies collapsing (or suppression) rules to the ACS 1-year and 3-year period standard tabulations that are designed to reduce the dimensions of tables, or to eliminate whole tables, that do not meet minimum standards for precision of the estimates. These collapsing rules are not applied to the 5-year period tabulations, even though the estimates will be very imprecise for small areas, because the small areas are intended to be building blocks for larger, user-defined areas.

The rules for determining which tables, or categories of tables, need to be suppressed involve examining the standard errors of every cell of a tabulation for individual tabulation areas (U.S. Census Bureau, 2006:13-10 to 13-11). For a specified table and area, the coefficient of variation (CV, the standard error of an estimate as a percentage of the estimate—see Box 2-5) is calculated for each cell of the table. If the cell entry is zero, the CV is set to 100 percent. The CV values are arrayed from high to low, and if the median CV value—the value that divides the distribution into equal halves—is greater than 61 percent, then the full table cannot be released. The categories of the table are then combined into fewer categories, and the median CV for the new table is calculated anew and the test is reapplied. If the median CV is still greater than 61 percent, then even the simpler table cannot be released (see Box 4-4 for an example).

It is difficult to evaluate this rule, but it could to lead to anomalous situations that make the data harder to use. For example, a table could be completely or partially suppressed one year and not the next year for the same geographic area, or a table could be suppressed for some, but not all, of the component areas of a large city or county. The suppression will affect small areas and minority population groups disproportionately.

BOX 4-4
Illustrative Calculation for Suppressing Table Cells with Large Sampling Error, 1-Year ACS Period Estimates

Assume a city of population 100,000, with 2,000 school-age children in a particular population group (e.g., Hispanic).

First Pass of Table

Ratio of Family Income to Poverty Threshold	Percent Chldren in Category	Coefficient of Variation (CV)
Below poverty threshold	15.0	60.4
100–149 percent of poverty	10.0	76.1
150–199 percent of poverty	10.0	76.1
200–249 percent of poverty	10.0	76.1
250–299 percent of poverty	10.0	76.1
300–349 percent of poverty	20.0	50.7
350 percent or more of poverty	25.0	43.9

What is the result?

- Median CV is 76.1.
- The table may *not* be released because the median CV is greater than 61.0.

Second Pass of Table after Combining Categories

Ratio of Family Income to Poverty Threshold	Percent Chldren in Category	Coefficient of Variation (CV)
Below poverty threshold	15.0	60.4
100–199 percent of poverty	20.0	50.7
200–299 percent of poverty	20.0	50.7
300–349 percent of poverty	20.0	50.7
350 percent or more of poverty	25.0	43.9

What is the result?

- Median CV is 50.7.
- The table may be released because the median CV is less than 61.0.

Recommendation 4-10: The Census Bureau should monitor the extent of collapsing of cells that is performed in different tables to meet minimum precision standards of 1-year and 3-year period tabulations from the ACS and assess the implications for comparisons among geographic areas and over time. After sufficient information has been gleaned about the extent of data collapsing, the Census Bureau, in consultation with data users, should assess whether its collapsing rules are sound or should be modified for one or more subject areas.

4-D.3 Inflation Adjustments

Chapter 3 discussed the procedures used by the Census Bureau to adjust income amounts for the 1-year, 3-year, and 5-year period estimates and housing value and cost amounts for the 3-year and 5-year period estimates to reflect changes in the national all-item consumer price index (CPI) over the period (see Section 3-A.2.c and Table 3-1). The discussion underlined the importance of users understanding the resulting estimates—for example, a 5-year period estimate of income or housing value is the average of all of the reported amounts over the 5 years expressed in dollars for the latest year using a national CPI adjustment. Moreover, as with any period estimate, the same inflation-adjusted average dollar amount for two areas may reflect different underlying patterns—for example, average income for 2005–2009 expressed in 2009 dollars of, say, $40,000 could result from income growth, stability, or decline over the 5-year period.

For many applications, users may prefer the Census Bureau's adjustment to latest-year dollars by using the national CPI to some other inflation adjustment or to no inflation adjustment at all. One advantage is that 1-year, 3-year, and 5-year period estimates for a large city or county will all be expressed in dollars for the same (latest) year—for example, 2009 dollars in the case of estimates for 2009, 2007–2009, and 2005–2009.

For some applications, however, users might prefer an inflation adjustment that is specific by geographic area. The problem is that area price data are limited. Currently, the Bureau of Labor Statistics (BLS) collects price data for over 100 specific areas, but it publishes price indexes for only the four regions (Northeast, Midwest, South, and West), population size classes of metropolitan statistical areas (MSAs), and the 20 largest MSAs. No price data are collected for rural areas.[10] Moreover, variation in price changes may be as great within areas for which price indexes are available as among them—for example, prices for housing and other goods may increase at a very different rate in the central city and suburbs, let alone individual neighborhoods, of an MSA. Finally, area-specific price indexes are less precise than the national all-item CPI.

For still other applications, users may require latest-year estimates for income, housing costs, or housing value. Averages of reported amounts over 3 or 5 years adjusted for inflation to the latest year are not likely to be the same as latest-year amounts. For income, this is true even for the 1-year period estimates: inflation-adjusted averages of reported income over the 23 months covered in 1-year period estimates are not likely to be the same as latest-year income estimates.

For estimating latest-year housing amounts from multiyear averages,

[10] See www.bls.gov/cpi/cpi/faq.htm.

the problem is a lack of subnational price indexes for specific items, such as housing value, rent, and different utilities or other housing costs. For income amounts, the problem is that incomes are not prices: income (in total and by component, such as wages or pension income) may increase faster (or slower) than inflation. A possibility to investigate in this context is to use estimation methods that are appropriate by type of income. For Social Security and Supplemental Security Income, it would be appropriate to use the applicable national CPI to which these payments are indexed by law. For property and self-employment income, it might be more appropriate to use an average interest rate, whereas, arguably, some types of income—specifically, public assistance and other retirement income—should not be inflated at all unless it is known that a jurisdiction has increased such payments. For wages, it could be possible to use changes at the county level in average quarterly wages for employees covered under state and federal unemployment insurance programs. These data, which are part of the BLS Quarterly Census of Employment and Wages, are released each quarter about 6–7 months after data collection.

It might be possible to develop simpler models to estimate latest-year amounts by using the published multiyear estimates. For example, by examining how well the trends in BLS county wage data estimate 1-year period income from the 5-year period estimates for large counties, a user might be able to develop a procedure for estimating latest-year incomes from the 5-year period estimates for small counties.

To determine how to produce the most helpful data on income, housing costs, and housing value, the Census Bureau should initiate a two-part discussion with users. The Census Bureau should first clearly illustrate to users the nature of the current inflation adjustment procedures. Then it should ascertain users' needs for income and housing amount information, the resultant implications for what adjustment procedures can best serve most users, and what steps to take to assist users whose needs are not satisfied by the standard procedures. Finally, the Census Bureau should consider providing tables that reflect unadjusted dollar amounts whenever it provides adjusted amounts. So doing will make clearer to users the effects of inflation and enable them to determine if another kind of adjustment would better suit their purposes.

> **Recommendation 4-11:** The Census Bureau should provide users with a full explanation of its inflation adjustment procedures and their effects on multiyear ACS estimates of income, housing costs, and housing value. It should consult with users about other kinds of income and housing amount adjustments they may need and conduct research on appropriate estimation methods (for example, methods to produce latest-year amounts from multiyear averages). It should consider pub-

lishing selected multiyear averages in nominal dollars as well as inflation-adjusted dollars.

4-D.4 Tabulation Specifications

The long-standing release plan for tabulations from the ACS includes two major elements: (1) the universe or population covered and (2) the geographic areas for which tabulations are produced. The full universe for ACS data products, beginning in 2006, will include the housing unit and GQ populations, although some tables may be published for subuniverses, such as households or the noninstitutional population. (Prior to 2006, tabulations included just the housing unit population.) For geographic areas, the available products (1-year, 3-year, and 5-year period estimates) will depend on the population size of the geographic area (refer back to Tables 2-4 and 2-5).

The Census Bureau will need to follow its plan for a number of years, not only to allow time for collection of sufficient data to begin release of 3-year period estimates in 2008 and 5-year period estimates in 2010, but also to allow both the Census Bureau and the data user community sufficient opportunity to gain experience with the various sets of tabulations. Yet the Census Bureau should not neglect to consult with users to determine if the population universe and the geographic area specifications are optimal or might be modified to produce more useful information.

With regard to population coverage, the key question is the role of GQ residents, particularly those in institutions. The Census Bureau will need to consult with users regarding appropriate universe definitions for ACS tabulations—for example, employment and income tabulations may be most useful if they are restricted to the noninstitutional population. In 2000, confidentiality concerns sometimes precluded the publication of the same tabulations separately for households and GQ residents in very small areas. Because the ACS estimates for small areas are averages over multiyear periods, confidentiality concerns could be less of a problem in this regard. Ideally, consultation with users on the most useful tabulation universes would precede and feed into the production of tables for 2006 (for release in summer 2007), which will be the first year to include GQ residents.

For the geographic area release schedule, one issue is the population size cutoff for publication of 1-year period estimates, for which the Census Bureau might consider the usefulness of lowering the current threshold of 65,000 residents to one of, say, 50,000 residents. The discussions in Chapters 2 and 3 emphasize the large sampling errors of 1-year period estimates for a small population group (such as school-age children in poverty) for geographic areas with fewer than 250,000 people, so lowering the threshold might appear to be deleterious. However, estimates for major population groups will often meet common standards of precision for areas of 50,000 population (see Table 2-8). Moreover, 50,000 is a common threshold for allocation of various types of federal assistance. Yet another advantage of lowering the threshold to provide 1-year period estimates for additional areas is that users would have more flexibility in combining the data and, consequently, would less often have to request special tabulations from the Census Bureau. For example, users could average two 1-year period estimates for a small city or county to obtain a 2-year period estimate that was more precise than the individual 1-year period estimates.

A second issue for release of geographic area tabulations concerns the feasibility of producing 3-year period estimates for user-defined statistical subareas of large cities and counties. Such subareas could be a set of aggregations of census tracts or block groups in cities and of places and towns in counties, where the city or county has at least 40,000 people (so that, at a minimum, there are two subareas, each with at least 20,000 people). If the city or county is large enough to have more than one PUMA, then the subareas could usefully nest within a PUMA to maximize the ability to relate the data for the PUMA and its subareas. (PUMA boundaries may need to be redrawn in some areas to achieve the most useful designation of subareas within PUMAs.) Finally, it may be possible to produce 1-year period estimates for large statistical subareas of PUMAs, particularly if the threshold for 1-year period estimates is lowered to 50,000 people. The Census Bureau will need to explore with users the desirability of providing additional estimates for statistical subareas of large cities and counties and weigh user needs against the feasibility of increasing the production workload for the ACS.

Recommendation 4-12: If some or all GQ residents continue to be included in the ACS, the Census Bureau should consult with users regarding the most useful population universe for tabulations, which, depending on the table, could be the entire population, the household and GQ populations separately, or the noninstitutional and institutional populations separately.

Recommendation 4-13: The Census Bureau should consider expanding the geographic areas for ACS tabulations in order to afford users greater flexibility for aggregating small areas into larger user-defined areas. Two possibilities to investigate are to lower the population threshold for 1-year period estimates to, say, 50,000, and to produce 3-year (and possibly 1-year) period estimates for user-defined statistical subareas of large cities (aggregations of census tracts or block groups) and counties (aggregations of places and towns).

4-D.5 Data Quality Review

The final step in the production and release of tabulations and other ACS data products is review by subject matter analysts to be sure there are no obvious errors or anomalies in the data. Each year the ACS processing staff and subject matter analysts must complete the entire process of preparing and reviewing data products within the span of a few months. In contrast, the preparation and review of data products from the long-form sample typically required well over a year to complete.

The volume of estimates to be reviewed each year led the Census Bureau to develop automated tools to facilitate the work of the staff. One tool, ART II, was developed in 2005 as an improved version of a similar tool (ART) used in 2003–2004. This tool automates the process of identifying statistically significant differences in estimates from one year to the next and facilitates other aspects of the review process. Other tools enable analysts and managers to track the process of review for tabulations and PUMS (U.S. Census Bureau, 2006:13-11).

We support continued efforts by the Census Bureau to automate and standardize the review process for ACS products, including not only the final review, but also review at earlier stages, such as when imputations for missing data and weighting adjustments are applied to the data records. As the time approaches when 1-year, 3-year, and 5-year period estimates must be provided for thousands of geographic areas every year (including 5-year estimates for over 200,000 individual block groups), the immensity of the review task threatens to overwhelm the analyst staff. They will run the risk of inadvertently releasing poor-quality data unless they receive a high level of technical assistance.

The Census Bureau recently identified prerelease review of demographic data, including from the ACS and other household surveys, as an important problem that merits research (Bell, 2006:10). The panel urges the Census Bureau to not only continue, but also to step up its investment of resources for automated tools, standardized protocols, and other means to facilitate an appropriate level of review of ACS data products that will ensure a high standard of quality before they are released each year. Consulting with computer software development firms and with computer scientists in academia may generate useful ideas and identify existing automated tools that are relevant to the Census Bureau's needs (see National Research Council, 2003b).

> **Recommendation 4-14:** The Census Bureau should increase its research and development on automated tools and standardized procedures to facilitate timely review and quality control of the large volume of ACS data products.

5

The Weighting of ACS 1-Year Period Estimates

As described in earlier chapters, the American Community Survey (ACS) comprises a time series of monthly samples of housing units selected each year from the Master Address File (MAF). The Census Bureau accumulates sets of monthly samples to produce 1-year, 3-year, and 5-year estimates, based on calendar years; 1-year estimates are produced only for areas with populations of 65,000 or more, 3-year estimates are produced for areas with populations of 20,000 or more, and 5-year estimates are produced for all areas (refer back to Table 2-5). This chapter presents a description and evaluation of the Census Bureau's weighting methods for producing 1-year estimates. Chapter 6 examines the weighting methods used for producing 3-year and 5-year estimates. These chapters provide a more detailed examination of the ACS weighting procedures than earlier chapters and are intended primarily for survey methodologists.

5-A OVERVIEW

As described in Chapter 2, the data collection for the ACS sample selected for a given month is spread over 3 months: mail responses are collected in the first month; computer-assisted telephone interviewing (CATI) responses are collected in the second month; and in the third month, computer-assisted personal interviewing (CAPI) responses are collected from a subsample of the housing units that have not yet responded. For purposes of analysis, the Census Bureau classifies a monthly sample as the sample units resolved in that month (the tabulation month), not the sample selected for that month, in order to make all the data for each monthly sample relate to the same time period. The units resolved in a given tabulation month comprise the mail, CATI, and CAPI responses received in that month and also the units determined in that month to be final nonresponding households, vacant housing units, and ineligible units. This procedure can be viewed as a form of nonresponse "replacement procedure" (Kish and Hess, 1959), in which sampled units resolved in the given month that were selected for prior months are treated as replacements for units selected for the given month that were resolved in later months.

Given this definition of the monthly samples, all the data used for analysis for a 1-year or multiyear period are collected during the specified calendar year or years. (An attraction of using tabulation months is that data collection is completed at the end of the year; if the monthly samples were defined in terms of sample months, it would be necessary to wait until the following February before all the data were collected for a given year.) Survey sampling weighting methods are applied to the respondents for the given period in order that valid estimates can be produced. These methods include weights to compensate for unequal selection probabilities, weighting adjustments for nonresponse, and calibration adjustments that compensate for noncoverage and can improve the precision of some survey estimates. Separate sets of weights are developed for person-level and housing unit-level analyses.

The Census Bureau has developed a nine-step weighting process for each 1-year data file, as summarized in Box 5-1. This box and the chapter text apply only to the weighting process for the housing unit population; see Box 5-2 for a brief description of the weighting process for the group quarters population.

Step 1 in Box 5-1 is the standard inverse selection probability weighting: if, say, a housing unit is selected with a probability of 1 in 10, the unit is assigned a base weight of 10, since it represents 10 housing units in the population. Subsequent steps adjust the base weights to compensate for deficiencies in the sample and to improve the precision of some estimates. These adjustments are performed within "estimation areas," which are single larger counties or combinations of smaller counties (the nonresponse adjustments in step 3 are carried out at the tract level; see below).

Steps 1 to 5 are adjustments made to the housing unit weights. The weights resulting from steps 1 to 5 apply to the household and all persons in it. Step 6 is an adjustment that is applied at the person level, leading to different weights for persons in the same household, and a revised household weight is developed in step 7. The last two steps are final adjustments to the weights.

Section 5-B describes these nine steps in more detail, and Sections 5-C and 5-D examine steps 5 and 6 more carefully. The calibration of the

BOX 5-1
The Nine-Step Weighting Process for Housing Units and Household Members in 1-Year ACS Data Files

1. *Base weights.* The base weights are the inverses of selection probabilities, including an allowance for the CAPI subsampling, computed for all selected housing units.
2. *Variation in monthly response factor.* This factor is associated with the "replacement procedure." It compensates for variations in the number of sample cases resolved across months.
3. *Noninterview factors 1 and 2.* These factors adjust for housing unit nonresponse.
4. *Mode bias noninterview factor.* This factor aims to compensate for the fact that the noninterview factors are applied to all responding households, not the households responding by CAPI.
5. *Housing unit control factor 1.* The weights developed up to step 4 are adjusted to make the weighted total of the number of housing units in an estimation area conform to an independent housing unit estimate obtained by updating counts from the last census.
6. *Population control factor.* The person weights are adjusted to make the weighted person counts for major demographic subgroups in an estimation area conform to independent population subgroup estimates obtained by updating counts from the last census.
7. *Housing unit control factor 2.* To obtain a household weight, the weight of the principal person is assigned to the household, and the housing unit weights are recalibrated to conform to the independent housing unit estimates.
8. *Adjustments to eliminate extreme weights.* If some weighting adjustments to the base weights exceed a factor of 8 in an estimation area, the weighting adjustment process is revised to eliminate such large weights.
9. *Rounding of weights.* All weights are rounded to be integers.

SOURCE: U.S. Census Bureau (2006:Ch. 11).

weights to make the sample conform to independent housing unit estimates (step 5) and to independent population estimates (step 6) raises a number of issues that require special attention.

5-B THE 1-YEAR NINE-STEP WEIGHTING PROCESS

The Census Bureau's weighting scheme for 1-year estimation starts with the standard base weights that are inverses of selection probabilities, and then makes adjustments to those weights to compensate for sample deficiencies. The adjustments are made within estimation areas, which are individual larger counties or groups of smaller counties; county size is

BOX 5-2
The Weighting Process for Residents of Group Quarters (GQ) in the 2006 ACS

1. *Base weights.* The base weights are the inverses of selection probabilities, computed for all selected GQ residents. This weight is 40 in most instances. When a GQ facility has more people than expected, a subsample of residents is selected so that only 10 people are eligible for interview. The base weights of these people equal 40 times the inverse of the subsampling factor (see Section 4-C for a description of the GQ sampling procedures).
2. *Noninterview factor.* A single factor is used in which the noninterview adjustment cells are defined by combinations of GQ types, as determined by research. Each cell must contain at least 10 people to be retained as a separate cell for the adjustment. The noninterview adjustment is carried out by cell within each state.
3. *Population control factor.* The GQ person weights are adjusted to make the weighted GQ person counts in a state conform to independent population estimates developed in the Census Bureau's postcensal population estimates program by major GQ types. These estimates start with the 2000 census state counts of GQ residents by GQ type and are updated from information provided by state partners in the Federal State Cooperative Program for Population Estimates, the Defense Department, and other agencies (http://www.census.gov/popest/topics/methodology/2005_co_char_meth.html).

SOURCE: U.S. Census Bureau (2006:11-9).

defined as the number of people living in housing units in the 2000 census. For weighting purposes the Census Bureau collapsed the 3,141 U.S. counties into 1,951 estimation areas with a minimum population size of about 16,000 persons (there are about 50 estimation areas in Puerto Rico).

An important consideration in assessing the individual adjustments is the extent to which they change the weights. In particular, substantial variation in the weighting adjustments can appreciably affect some ACS estimates, likely reducing bias, but probably also lowering precision. Information on the distribution of the weighting adjustments is provided below for some of the adjustments that were used with the 2004 ACS test survey.

5-B.1 Base Weights

Each housing unit is selected from the MAF for the ACS with a probability specified for the block in which it is located. For the 2005 ACS these probabilities range from 1.6 percent (1 in about 63) to 10 percent (1 in 10), depending on the estimate of occupied housing units for the smallest governmental unit or the census tract in which a block is located (see Table 2-3, Part A).

The base weights (step 1) for sampled housing units not subjected to CAPI subsampling are simply the inverses of the MAF selection probabilities. For housing units subject to subsampling, the overall selection probability is the product of the original selection probability from the MAF and the subsampling rate. The subsampling rate is 66.7 percent (2 in 3) for unmailable addresses; it varies between 33.3 percent (1 in 3) and 50 percent (1 in 2) for mailable addresses, with higher subsampling rates for census tracts expected to have lower mail and CATI response rates (see Table 2-3, Part B). The base weights thus vary from about 189 (a housing unit in the CAPI subsample with an initial sampling rate of about 1 in 63 and a subsampling rate of 1 in 3) to 10 (a housing unit not subject to CAPI subsampling that is selected with an initial probability of 1 in 10).

The base weights are determined by design decisions and are changed only by changing the design. The variation in the initial selection probabilities resulted from the need to satisfy precision requirements for estimates for governmental units of different sizes. The subsampling rates for CAPI interviews were determined by cost factors and the need to retain adequate sample sizes for census tracts with lower expected mail and CATI response rates. However, as illustrated in Box 4-2, the variation in base weights has the effect of lowering precision for analyses that include households or persons with differing base weights, as compared with an analysis in which the sample size is the same and the weights are constant.

5-B.2 Variation in Monthly Response Factor

The first adjustment to the base weights (step 2) arises because of the Census Bureau's decision to process the ACS monthly samples by the tabulation month in which sampled units are resolved rather than by sample month. The variation in monthly response factor (VMS) is used to correct for the imbalance in the rate of resolving sampled units across months with the aim of producing a sample that is balanced across months of the year.

To carry out the adjustment, the sum of the base weights for the units resolved in a given month (including nonresponding, vacant, and ineligible units and including the CAPI subsampling factor) is adjusted to conform to the sum of the base weights (but excluding the CAPI subsampling factor) of all units initially sampled for that month. The adjustment is made within estimation areas by applying the following simple ratio adjustment to the base weight for each resolved unit in the month in question:

$$\text{VMS} = \frac{\text{Total weight of all units sampled for the month (excluding the CAPI factor)}}{\text{Total weight of all units resolved in the month (including the CAPI factor)}}$$

To see the effect of the VMS factor, it is necessary first to recall that the initial ACS sample (before CAPI subsampling) is selected each year in two parts (see Section 4-A.1). The main sample for a given year is drawn from the appropriate segment of the MAF existing in August–September of the previous year (main sample MAF), and the supplemental sample is drawn in January–February of the given year from a segment of the new addresses subsequently added to the MAF. (There is no attempt to sample MAF addresses added during the year.) The main sample is allocated evenly across the 12 months of the given year to produce the monthly samples. However, for timing reasons, the data collection for the supplemental sample is restricted to the April–December period. The small supplemental sample is spread evenly across only these 9 months.

Consider first just the main sample. Using the base weights before CAPI subsampling, the weighted count of all units originally sampled in a given month in an estimation area is approximately equal to the number of units on the main sample MAF divided by 12 (since the sample comprises one-twelfth of the annual sample). This equivalence also holds approximately when the base weights that include the CAPI subsampling component are applied to all units originally sampled in a given month that are resolved at some time in the 3-month fieldwork period for that month's sample. However, the weighted count of all units *resolved in a month* does not necessarily equate to the number of units on the main sample MAF divided by 12 for two reasons: (1) there may be variations in the numbers of mail, CATI, and CAPI cases resolved by month; and (2) the cases resolved by CATI and CAPI in January and by CAPI in February are carryovers from the November and December monthly samples selected from the MAF for the previous year. The VMS factor is introduced to compensate for this lack of equivalence between the resolved cases and the MAF count.

The restriction of the supplemental sample to the last 9 months of the year makes the situation more complicated. In the first 3 months, when only the main sample is fielded, the VMS factor aligns the resolved cases with the main sample MAF only. The complications here are that the adjustment does not cover the units on the supplemental frame and that a number of the resolved cases are carryovers from the previous year based on a sample from that year's MAF. In the last 9 months of the year, the numerator of the VMS represents approximately one-twelfth of the main MAF plus one-ninth of the units on the supplemental frame. It thus exceeds one-twelfth of the full MAF population. In addition, a number of the cases resolved in April and May were selected only from the main sample MAF. As a consequence of these factors, the VMS adjustment does not provide a fully balanced representation over the months of the year.

Although the monthly balance is not fully achieved, the VMS does give the required representation both to units on the main MAF and to units on the supplementary frame. The failure to represent units on the supplementary frame in the first 3 months of the year is compensated by their overrepresentation in the last 9 months. In essence, over the 12 months, the VMS adjustment can be viewed as one in which 1/36 of the supplemental sample units in later months are substituted for the supplemental sample units that were not surveyed in the first 3 months. This "replacement" scheme makes the assumption that the characteristics of the replacement units are the same at the time of data collection as they were earlier on. For some characteristics—for example, occupancy status—that assumption may be questionable. While in most areas the small numbers of units on the supplemental frame make these issues unimportant, the ACS estimates could be noticeably affected in growth areas that have large numbers of new units.

In practice, the variation in the VMS factor is not great. Over all months of the 2004 ACS test survey, the value of the VMS factor for the 5th percentile is 0.87 and that for the 95th percentile is 1.24. The effect of this additional variation in the resulting weights on sampling errors is likely to be small.

A limitation of the VMS factor is that it can distort the distribution of different types of sampled cases. For example, suppose that a larger-than-average number of CAPI households is resolved in a given month. The global VMS factor compensates for this outcome by downweighting all the resolved housing units in that month, not just the CAPI housing units. To the extent that CAPI housing units have different characteristics from the rest, the monthly estimates will be biased. This limitation could be avoided by more complex adjustment factors that weight each of the types of resolved unit in the tabulation month (mail, CAPI, CATI, nonresponding, vacant, ineligible) separately to conform to the outcomes for the sample for the sample month. However, these factors would have greater variability than the VMS factors, and hence they would inflate sampling errors more.

As noted above, the VMS factor is introduced because of the decision to use the tabulation month as the basis for ACS analyses. The concern about the alternative of using the sampling month as the basis of the analyses is that responses provided about characteristics in the following 2 months will not accurately reflect those characteristics in the sample month. Moreover, the use of the sampling month would delay completion of all data collection for a given year.

Accepting the use of the tabulation month for the ACS, the utility of the VMS factor should still be investigated to assess how the current imperfect adjustment performs and whether its use warrants the (admittedly slight) increase in sampling error that it causes. In addition, the utility of this factor should be assessed in relation to the housing unit and population adjustments carried out in steps 5, 6, and 7, which all relate to a single point in time (July 1). When the population size of the area changes over the year, these adjustments are inconsistent with the VMS objective of representing the population across the year.

5-B.3 Noninterview Factors 1 and 2

The next step in the development of the weights (step 3) is to compensate for the fact that some sampled housing units do not respond to the ACS or the data collected for them are too scant to process. These housing units are dropped from the analytic data file, and the weights for responding housing units are inflated to provide representation for them. Since it is assumed that all the nonresponding units have been determined to be occupied, the adjustments are made only to the responding housing units that are occupied or temporarily occupied. Three variables often related to response rates are used in the adjustments: census tract, single-unit versus multiunit structure (building type), and month of data collection. Since the cross-classification of these three variables would create cells with very small sample sizes, the noninterview weighting adjustments are carried out in two stages. Each stage is applied to all of the occupied housing units in the file for a given calendar year.

The first stage of the noninterview adjustment (NIF1) is carried out within cells created by the cross-classification of building type and census tract, with census tracts combined if the cells contain too few responding housing units. Within each cell, the weights of responding occupied and temporarily occupied housing units are multiplied by an adjustment factor that makes the sum of the weights for responding housing units equal to the sum of the weights for responding and nonresponding housing units. The adjustments are not large: no adjustment was made for more than half the responding housing units in the 2004 ACS, and 95 percent of the adjustments were less than 1.20.

The second stage of the noninterview adjustment (NIF2) starts from the NIF1 adjusted weights and then adjusts them in the same manner within cells created by the cross-classification of building type and tabulation month, combining adjacent months if the responding sample size is too small. The 5th percentile of the NIF2 adjustments in 2004 was 0.99, and the 95th percentile was 1.10.

The noninterview adjustment process is the first step of a raking algorithm in which the process is iterated until convergence is reached. In view of the small sizes of the adjustments, no further iterations are performed. With the high weighted response rates achieved in the ACS, the noninterview adjustments should not lead to a substantial loss of precision.

5-B.4 Mode Bias Noninterview Factor

A significant drawback to the step 3 noninterview adjustments is that they increase the weights of all mail, CATI, and CAPI responding housing units to represent the nonresponding housing units, whereas the nonresponding housing units are virtually all a subset of the subsampled CAPI housing units. The Census Bureau has chosen to spread the adjustments over all responding occupied and temporarily occupied housing units because of the much smaller responding sample size that would have been available had the adjustment been confined to CAPI housing units. That smaller sample size would have severely limited the extent of control that could have been achieved on the census tract, building type, and month of data collection variables used in the step 3 adjustments. Also, restricting the adjustments to CAPI housing units would have concentrated the adjustments on responding housing units that already had larger weights because of the subsampling. Since CAPI responding housing units likely differ in some characteristics from other responding housing units, however, it seems likely that the step 3 noninterview adjustments create some bias in some estimates.

The Census Bureau introduces another adjustment (step 4), the mode bias noninterview factor (MBF), to address the bias concern with the step 3 weighting. The MBF adjustments are generally small, with only 5 percent being 0.96 or less and 5 percent being 1.03 or more, but the combined effects of their use, together with the noninterview adjustment factors in step 3, on the biases and sampling errors of ACS estimates are not transparent.

The MBF procedure, in essence, has three steps. The first step is to develop an adjustment factor for the step 2 weights—mode noninterview factor (NIFM)—similar to the adjustment factors NIF1 and NIF2 under step 3 but applied only to CAPI occupied and temporarily occupied housing units. The second step is to produce some survey estimates based on the step 2 weights with these adjustments and also the corresponding estimates with the step 3 adjustments and to calculate the MBF as a ratio of the two quantities. The third step is to multiply the step 3 weights by the MBF so that the estimates with the resultant weights conform to those produced with the NIFM-adjusted weights.

In view of the smaller sample size when adjusting only the weights of CAPI respondents, the NIFM adjustments take account of only building type and tabulation month, not census tract, within an estimation area.

When the respondent sample size in a cell of the cross-classification of these two variables is too small, adjacent months are combined. Within each cell, the NIFM is calculated as the ratio of the sum of the weights after step 2 for responding and nonresponding CAPI housing units to the equivalent sum for only the responding CAPI housing units. No adjustments are made to the step 2 weights for non-CAPI, vacant, or ineligible units. No adjustment is made for over 75 percent of housing units, and only 5 percent of the NIFM adjustments are 1.10 or larger.

The next step is to calibrate the estimates based on the weights after step 3 to those produced using the NIFM weighting adjustments. The calibration is performed within each estimation area for the cell totals of the cross-classification of tenure (owned, rented, or temporarily occupied), tabulation month, and marital status of the householder (married and widowed or single). When the sample size in a cell is deemed too small, the two marital status cells are combined. Estimates of the cell totals are produced for each cell, and then the mode bias noninterview factor for a cell is computed as the ratio of the estimated cell total using the step 2 NIFM-adjusted weights to the corresponding estimated total using the step 3 adjusted weights. In the final step, the MBF factors are applied to the step 3 weights for all occupied housing units. As noted above, the MBF adjustments are generally small.

The effects of the compensation for nonresponding housing units using the combination of steps 3 and 4 are not obvious and need to be carefully assessed. For example, it is not clear that the NIFM weighting adjustments, which are confined to CAPI cases but drop the tract-level control, lead to less biased estimates for the cells in the cross-classification of the control variables, let alone estimates for other variables. Also, since the estimates of the cell totals for the control variables under the weights developed up to this point are equated to those using the NIFM-adjusted weights, their sampling errors are those of the latter estimates. These sampling errors are likely larger than those based on the NIF1 and NIF2 (step 3) adjustments alone, because the NIFM adjustments are applied only to CAPI cases and also because CAPI cases start with higher base weights because of the subsampling. Thus, the effect of the MBF step 4 adjustments, which derive from the calibration of the NIFM weights to the step 3 weights, on estimates of the control variables and on other ACS estimates needs examination.

Given the high response rates achieved in the ACS, the nonresponse adjustments have mostly a minor impact. However, for areas with lower response rates, the adjustments may be significant. Research to compare the current adjustments with other, more standard, adjustments is warranted. For example, in some estimation areas, a raking adjustment procedure applied to the marginal totals by census tract, building type, and month of data collection and confined to CAPI cases might be more effective.

5-B.5 Housing Unit Control Factor 1

After step 4, the sum of the weights for the combination of responding housing units, vacant housing units, and ineligible units in an estimation area over the year is approximately equal to the number of units on the MAF from which the sample was selected. The next step adjusts the weights within each estimation area so that the estimated number of housing units from the ACS conforms to the independent estimate of total housing units for July 1 of the year in question produced by the Census Bureau's postcensal population estimates (PE) program. This adjustment (step 5) serves to compensate for under- or overcoverage in the MAF. The Census Bureau also argues that this step serves to make the housing unit counts consistent with the PE population controls employed in step 6.

There are several issues concerning the use of the housing unit control factor, including the quality of the PE housing unit estimates. These issues are taken up in Section 5-C below.

5-B.6 Population Control Factor

After all the preceding adjustments have been applied, each housing unit has a weight, and that weight also applies to all the persons in the housing unit. The weighting adjustment in step 6, however, leads to different weights for persons in the same household. In step 6 the person weights are adjusted so that in each estimation area the weighted sums of persons in certain sex-age-race/ethnicity subgroups in occupied housing units conform to the subgroup estimates produced by the PE program for July 1 in the year in question. The aims of the population control factor adjustment are to compensate for person noncoverage (particularly for person noncoverage within some housing units) and to improve the precision of ACS person-based estimates.

Section 5-D below discusses the use of the PE population estimates as controls in the ACS, including issues relating to the quality of these estimates.

5-B.7 Housing Unit Control Factor 2

Step 6 results in variable weights for persons within a household, raising an issue of what weight to assign to a household. The solution adopted is to start by assigning a household the weight of one of its members. The person chosen for this purpose—termed the principal person—is identified as the wife in a household in which both husband and wife are present and otherwise one of the persons who rents or owns the housing unit. The choice of principal person is based on the assumption that, if the housing unit is covered, that person is least likely to be missed in the household listing.

A housing unit is then initially allocated the weight after step 6 for its principal person (or its step 5 weight for a vacant unit). However, the sum of these weights in an estimation area will no longer agree with the PE housing unit count for the area. The second housing unit factor (step 7) is then a ratio adjustment to make the weighted sum of the weights for housing units conform to the PE estimate. An undesirable consequence of this adjustment is that the household and the principal person in it have different weights (there can also be different weights for the principal person and a spouse or unmarried partner). Research is under way to develop a weighting scheme that removes these inconsistencies (see U.S. Census Bureau, 2006:11-8; Asiala, 2007).

5-B.8 Adjustments to Eliminate Extreme Weights

The penultimate step in the process (step 8) is to identify any extremely large weight adjustments to the base weights over all the subsequent steps in the weighting process. To avoid large increases to sampling errors, the weighting adjustments are revised by collapsing adjustment cells if any overall weight adjustment factor exceeds 8.

5-B.9 Rounding of Weights

At this stage the weights developed are generally noninteger. At the final stage (step 9), for cosmetic reasons, the weights are rounded to integer values using a controlled rounding procedure. These rounding adjustments slightly increase sampling error. The impact of the rounding is greater the smaller the weight. For example, in a small governmental unit sampled at a rate of 1 in 10 each year, the rounding of a weight of 11.4 to 11 for a 1-year weight is of lesser significance than the rounding of a weight of 2.4 to 2 for a 5-year weight. An attractive feature of the rounded weights is that all results given as counts of persons or households are integers and hence all the counts and any totals of them are completely consistent. With noninteger weights, counts rounded to integer values would not be entirely consistent because of rounding errors. However, this property applies only to counts and not, for example, to percentages. The panel is not convinced that the cosmetic value of complete consistency in counts warrants the loss of precision incurred with the rounding of the weights.

5-B.10 Recommendation

The weights developed over this 9-step process play an important role in analyses of ACS 1-year data. They are designed to reduce the effects of sample bias on the ACS estimates, and they affect the precision of those estimates. The Census Bureau has conducted a substantial amount of useful research on many of these steps. However, with the data for the first year of the full ACS available, it is now time to carry out a thorough review of all steps in the process, individually and in combination, to determine if improvements can be made. The review will need to consider the impact of the current and alternative weighting schemes on the quality of a wide range of ACS estimates, covering a variety of characteristics across the full range of governmental units.

The review should start with the design decisions that lead to the sample sizes and initial base weights for governmental units of all sizes and the CAPI subsampling rates that depend on expected mail and CATI response rates. For example, the choice of subsampling rate depends on the relative costs of mail, CATI, and CAPI responses and can be informed by theoretical results on the optimum subsampling rates for initial nonrespondents developed by Hansen and Hurwitz (1946).

Although many of the adjustment factors at subsequent steps are not large, they nevertheless deserve a careful assessment to see if improvements can be made, and their combined effects need to be examined. The inconsistency in the logic for the VMS factor and the point-in-time PE housing unit and population control factors needs to be reviewed. The effects of the nonstandard nonresponse adjustments using the mode bias noninterview factor on biases and variances need examination. The case for the integer rounding of the weights needs to be critically assessed. Most importantly, the PE housing unit and particularly the population controls deserve special scrutiny, as discussed in the next two sections.

> **Recommendation 5-1:** The Census Bureau should conduct an in-depth review of the weighting scheme used for producing ACS 1-year period estimates and assess a range of alternative schemes that might improve the quality of the estimates.

5-C HOUSING UNIT CONTROLS

After step 4 in the weighting process, the weighted total number of all units sampled from the MAF for the year is equated to the number of units in the MAF from which the sample was selected. The main sample for a given year is selected in August of the previous year and a supplemental sample is then added in January to give representation to units that have been added to the MAF after the main sample was drawn. Thus, the MAF aims to cover all units included in the MAF at the time of the January sampling. Ideally, the sample would be selected monthly from all units extant at that month, but monthly sample selection is not practicable. The sample for a given year thus fails to cover households and persons in households in units that are not included in the January MAF.

Step 5 of the weighting process adjusts the weights of the eligible sampled housing units in the ACS within each estimation area so that the weighted total number of such units conforms to an independent postcensal estimate of housing units for the area produced by the Population Division of the Census Bureau in its PE program. This adjustment (housing unit control factor 1) is introduced to attempt to compensate for the failure of the ACS to cover units occurring since the January sample selection and for other deficiencies in the MAF as a sampling frame. Another reason given by the Census Bureau for the adjustment is so that any undercoverage or overcoverage of housing units in the ACS is addressed similarly to how undercoverage or overcoverage of population is addressed by the use of the PE population estimates in step 6. However, while the PE housing unit and population estimates both start from the last population census, they are developed independently thereafter, thus reducing the force of the argument for consistency.

The PE program produces a postcensal estimate of the number of housing units in each county for July 1 of each year. These postcensal estimates are produced using a component method that starts with housing unit counts from the most recent census and then adjusts them for changes in geographical boundaries and updates them with estimates of new residential construction and new residential mobile home placements and estimates of residential housing loss (http://www.census.gov/popest/housing/). The estimates are developed (but not published) at a subcounty level and then aggregated to counties and states. In jurisdictions in which building permits are required, new residential construction is taken to be residential permits issued from 2000 through the previous calendar year, allowing for a 6-month lag between permit issuance and completed construction, and an estimated 2 percent of permits that do not result in construction. The annual Survey of Construction is used to estimate the regional numbers of new housing units constructed in non-permit-issuing jurisdictions, and these numbers are then allocated across subcounty areas in proportion to their census shares of residential housing units. New mobile home placements are estimated at the county level by allocating state mobile home shipment data to counties based on their census shares. Housing loss is estimated from information provided by the American Housing Survey, with the rate of loss depending on the unit's age or whether it is a mobile home.

The housing unit controls used in the ACS weighting are thus only estimates. The utility of these controls depends on the quality of the postcensal housing unit estimates compared with the quality of the MAF. The Census Bureau has carried out an evaluation of the postcensal housing unit estimates for 2000 based on updating the 1990 census and comparing these estimates to the 2000 census housing unit counts (Devine and Coleman, 2003). Table 5-1 presents the mean absolute percentage errors (MAPEs) for the housing unit county estimates for 2000 obtained in this evaluation by 1990 size of the county. Overall the MAPE was 4.6 percent, but it varied from 1.9 percent for the largest counties to 7.3 percent for the smallest counties. The MAPE also varied by the amount of change in the number of housing units over the decade, with larger values for counties that had grown or declined considerably. The MAPEs were particularly large for small counties that had changed considerably in size (data not shown).

In assessing these results in relation to the use of the postcensal housing unit estimates in weighting adjustments in the ACS, two factors need to be borne in mind. First, the evaluation applies to estimates for 2000, 10 years after the 1990 census; as such it may be viewed as a worst-case evaluation. Second, the ACS weighting adjustments are performed in estimation areas that combine small counties, so, again, the above results may overestimate the amount of error. Nevertheless, the postcensal housing unit estimates are likely subject to appreciable error in some types of counties. Some of these errors may be random in nature, but some may be systematically biased upward or downward for certain types of counties.

An indication of magnitude of the housing unit control factor for the 2004 ACS is presented in Table 5-2. Most (84 percent) of the control fac-

TABLE 5-1 Mean Absolute Percentage Error (MAPE) of April 1, 2000, County Housing Unit Estimates Compared to April 1, 2000, Census Counts, by 1990 Size of County

1990 Housing Unit Count	Number of Counties	MAPE
All Counties	3,141	4.6
0–2,499 units	336	7.3
2,500–4,999	518	5.7
5,000–9,999	749	5.0
10,000–19,999	653	4.4
20,000–49,999	507	3.1
50,000–99,999	178	2.2
100,000 and over	200	1.9

NOTE: The percentage error is calculated for each county as: (the 2000 census housing unit count – the 2000 postcensal housing unit estimate)/the 2000 census housing unit count. The signs of these errors are dropped, and then the absolute errors are averaged across counties.

SOURCE: Devine and Coleman (2003).

TABLE 5-2 Distribution of the Housing Unit Control Factor 1 Across Counties in the 2004 ACS

Housing Unit Control Factor 1	Number of Counties	Percentage of Counties
Under 0.90	64	2.0
0.90–0.94	145	4.6
0.95–0.99	1,173	37.3
1.00–1.04	1,463	46.6
1.05–1.09	219	7.0
1.10 and over	77	2.5
Total	3,141	100.0

SOURCE: Based on data provided by the Census Bureau.

tors fall between 0.95 and 1.05 and represent minor adjustments. There are, however, a number of counties for which the factors are substantial, with extremes as low as 0.7 and as high as 1.42. More than half of the factors are 1.0 or greater, consistent with a net undercoverage in the MAF in those counties, but the factors are less than 1.0 for 44 percent of counties, suggesting a net overcoverage in the MAF for those counties, under the assumption that the postcensal estimates are accurate.

Undercoverage in the MAF arises because new housing units added after the January MAF update are not covered and also because of other housing units that are missed. Overcoverage arises if some housing units are listed more than once in the MAF, or if group quarters addresses are misclassified as housing units.[1] Both missed housing units and duplicate or misclassified listings can occur in a county. Under the assumption that the postcensal estimates are accurate, the housing unit control factors in Table 5-2 represent the net effects of overcoverage and undercoverage.

As with all weighting adjustments, the housing unit factors are based on certain assumptions that merit review by the Census Bureau. In the case of undercoverage, the factors increase the weights of MAF housing units to represent those not included in the MAF frame under an assumption that the missed housing units are missing at random. This assumption would be false if, for instance, the proportion of vacant units in the postcensal estimates is higher than in the MAF, as it might well be. In the case of overcoverage, the factors decrease the weights of MAF housing units to ad-

[1]Another potential source of coverage problems with the MAF is that some housing units are listed in the wrong county, leading to undercoverage in the county in which they should be listed and overcoverage in the county in which they are listed. However, county misclassification seems likely to be rare.

dress the problem of duplication. This procedure brings the MAF number of housing units in line with the postcensal estimates. However, in other regards it depends on an assumption that the listings of some housing units are randomly duplicated.

If the MAF was complete and not subject to duplicate listings and if the intercensal estimates were accurate, then the housing unit control factors would all be close to 1, simply adjusting the housing unit counts from the January MAF to the midyear intercensal estimates to allow for new and demolished units in the interim. (In this case the ACS housing units could be weighted to a midyear MAF count.) That many of the housing unit control factors are appreciably larger or smaller than 1 raises concerns about the quality of the MAF or the quality of the postcensal estimates or both.

Recent research by the Census Bureau (Reese, 2007) examined differences between the independent housing unit estimates for 2002–2005 and the housing unit addresses on the MAF used for the ACS in these years. The results show an increasing divergence between the two series, with the national MAF count exceeding the housing unit estimate by 2.6 percent in 2002 and rising to 4.0 percent in 2005. These results suggest a failure to completely identify and weed out duplicate, demolished, and nonresidential addresses from the MAF. The differences between the increase in the MAF and the increase in the housing unit estimates varied among counties as a function of county size, with larger differences occurring for counties with larger numbers of housing units.

The Census Bureau plans to conduct more research to gain an understanding of large discrepancies between the MAF counts and the postcensal housing unit estimates. It is, for example, possible, that the quality of the MAF differs between urban and rural areas associated with a differential updating of the frame, using the Delivery Sequence File in urban areas and the Community Address Updating System in rural areas (see Sections 4-A.4a and 4-A.4.b). As discussed in Section 4-A, a prime concern for the ACS is the continuous maintenance of a high-quality MAF for use each year throughout the decade and beyond. Weighting adjustments that attempt to compensate for deficiencies are necessarily an imperfect remedy. If this research identifies deficiencies in the MAF sampling frame, then steps should be taken to correct the frame.

At present the MAF and the postcensal estimates are developed independently. However, in the panel's view, they should be integrated to the benefit of both. For example, building permit data could be used to improve the MAF, either collected on an individual permit basis within location or simply using the current aggregates that would indicate areas in which special MAF updating is needed. Similarly, the MAF—and the ACS—could provide valuable information in updating the postcensal estimates.

Another issue that needs examination is the level at which the housing unit controls are applied. At present the controls are applied at the estimation area, but they could alternatively be applied at higher or lower levels. If the postcensal estimates are of high enough quality, they could be applied at the level of the county or a subcounty area, such as a census tract, thereby targeting the adjustments more directly to the areas where they are needed. The Census Bureau has carried out some initial research in this area (Starsinic, 2005), and the panel encourages further research along these lines.

> **Recommendation 5-2:** The Census Bureau should evaluate the quality of the postcensal housing unit estimates and the MAF sampling frame in relation to one another. In the light of this evaluation, the Census Bureau should assess the suitability of the current housing unit control factor adjustment and modify it as necessary.
>
> The Census Bureau should attempt to identify areas in which improvements can be made to the postcensal housing unit estimates and to the MAF sampling frame. In particular, it should investigate an integrated approach for developing the postcensal housing unit estimates and for continuously updating the MAF that would benefit both and reduce the variability in the housing unit control factor.

5-D POPULATION CONTROLS

After the application of the housing unit controls in step 5, each household has a weight that can be used for analysis, and that same weight could be used for each person in the household. Step 6 in the process is an adjustment to the person weights. This adjustment is used to compensate for person noncoverage in sampled households and to reduce the sampling errors for person-level estimates. The adjustments are based on the Census Bureau's PE subnational resident population estimates by age, sex, race, and Hispanic origin for July 1 each year. For these estimates the resident population in an area is defined using the decennial census "usually resident" rule as distinct from the ACS "2-month resident" rule.

As with the PE housing unit estimates, the population estimates start from the 2000 census counts and adjust for estimated changes between April 1, 2000, and July 1 of the year in question. At the outset, the 2000 census population is divided into the household population and the group quarters population. For the 2005 ACS, the population controls are based on only the household population estimates, and only the methodology for developing those estimates will be reviewed here.

The Population Estimates and Projections Area of the Population Division at the Census Bureau produces county household population estimates using a cohort-component technique that adjusts the census counts to allow for births, deaths, net international migration, net domestic migration, and net military movement during the intervening period.[2] Reliable estimates of births and deaths are obtained from vital records data. National estimates of net international migration are generated from the ACS for earlier years, and the numbers are then distributed across counties based on the distribution of noncitizen foreign-born persons in the 2000 census. Net domestic migration is estimated from Internal Revenue Service 1040 tax return records, which are matched to the Social Security Administration's Numident file in order to obtain age, sex, race, and Hispanic origin data for the tax filers and their dependents. Data on the net movement of military personnel and their dependents are provided by the Department of Defense.

A complication in developing the population estimates by race is that race in the 2000 census and the ACS is classified into six race groups (white; black; American Indian and Alaska Native; Asian; Native Hawaiian and Pacific Islander; and Some Other Race), and individuals can report multiple races, whereas most administrative data employ only four race groups (white; black; American Indian, Eskimo, or Aleut; and Asian and Pacific Islander). To address this complication, the census race categories are reduced to the four categories by eliminating the Some Other Race category and proportionately allocating all individuals into one of the four categories. The estimates are produced for the four categories and are then reallocated to the 2000 census categories for publication. For the purposes of the ACS weighting adjustments, race and Hispanic origin are combined into six categories: (1) non-Hispanic white; (2) non-Hispanic black; (3) non-Hispanic American Indian or Alaska Native; (4) non-Hispanic Asian; (5) non-Hispanic Native Hawaiian or Pacific Islander; and (6) Hispanic.

The Census Bureau has conducted an evaluation of the 2000 county total population estimates by comparing them with the 2000 census counts, in a similar way to the evaluation of the housing unit estimates described in the previous section (Blumerman and Simon, 2006). The mean absolute percentage errors displayed in Table 5-3 indicate that the level of error for the smallest counties is appreciably larger than the overall average of 3.4 percent. However, since the ACS population control adjustments are applied at the level of estimation areas—which are combinations of counties in the case of smaller counties—the average MAPE for the estimation areas should be less than that for counties. The MAPEs are also larger for counties that experienced a growth of 19.5 percent or more. Note that this evaluation applies to estimates updated to 2000 from the 1990 census and that estimates updated over shorter intervals are likely to have smaller MAPEs.

The preceding results apply to the total population estimates at the

[2]For further details, see http://www.census.gov/popest/topics/methodology/2004_co_char_meth.html.

TABLE 5-3 Mean Absolute Percentage Error (MAPE) of April 1, 2000, County Population Estimates (Official Series) Compared with April 1, 2000, Census Counts by County Population in 2000

County Population in 2000	Number of Counties	MAPE
All Counties	3,141	3.4
Under 2,500	115	6.9
2,500–4,999	177	4.3
5,000–9,999	405	3.7
10,000–19,999	651	3.3
20,000–49,999	879	3.2
50,000–99,999	390	2.6
100,000 and over	524	2.9

NOTE: The percentage error is calculated for each county as: (the 2000 census population count – the April 1, 2000, postcensal population estimate)/the 2000 census population count. The signs of these errors are dropped, and then the absolute errors are averaged across counties.

SOURCE: Blumerman and Simon (2006:Table 4).

county level. However, the ACS population controls are applied within subgroups defined by sex, age in 13 groups (0–4, 5–14, 15–17, 18–19, 20–24, 25–29, 30–34, 35–44, 45–49, 50–54, 55–64, 65–74, 75 and over), and race/ethnicity in 6 groups at the estimation area level. In practice the numerous cells in the sex by age by race/ethnicity cross-classification often have to be reduced by collapsing cells. A complex set of collapsing rules is employed, starting with collapsing categories of race/ethnicity to create collapsed cells with a minimum sample size of 10 persons and for which the weighting adjustment is less than 3.5. Subsequent collapsing of the sex by age cross-classification within collapsed race/ethnicity cells is then undertaken as necessary (U.S. Census Bureau, 2006:11-7).

The panel undertook a simple analysis in order to derive a rough indication of the level of error in the population controls being used in the ACS. For this purpose we dropped the race classification because of the problems with the differences in that classification between the population estimates and the 2000 census, retaining only Hispanic versus non-Hispanic (equivalent to collapsing the first five of the six race/ethnicity categories listed above). We also collapsed the 15–17 and 18–19 age groups into a single 15–19-year-old group. We then compared the sex by age by ethnicity cross-tabulations produced from the population estimates for 1999 with the corresponding cross-tabulations from the 2000 census for each of 1,950 estimation areas (excluding one estimation area used in the 2005 ACS that was a new county formed in 2001). The 1999 estimates were used because 2000 estimates were not published.

In a number of the cells of the cross-tabulations, the census counts were small numbers and would have led to small ACS sample sizes and the use of the cell collapsing procedure. For simplicity, rather than attempting to apply the complex collapsing procedure, cells with census counts of fewer than 500 persons are excluded in the results presented below. Table 5-4 summarizes the MAPEs resulting from this analysis, depending on which of the control variables are employed. The table also shows the MAPE for all 3,140 counties for comparison purposes. As with the MAPEs for the housing unit controls in Section 5-C above, the MAPEs for the population controls may include both random errors and systematic upward or downward biases.

As expected, the MAPE for the overall population counts is somewhat lower for estimation areas (3.1 percent) than for counties (3.6 percent). The MAPE of 3.6 percent for counties in the panel's analysis in Table 5-4 is slightly larger than the MAPE of 3.4 percent in the Census Bureau's analysis reported in Table 5-3, which compared 2000 (not 1999) population estimates to 2000 census counts. Consequently, the MAPEs in the panel's analysis for population groups are likely to be marginally larger than they would have been if 2000 population estimates could have been used.

TABLE 5-4 Mean Absolute Percentage Error (MAPE) of July 1, 1999, Estimation Area Population Estimates Compared with April 1, 2000, Census Counts, by Cells Based on Combinations of Sex, Ethnicity, and Age (Excludes Cells with Fewer Than 500 People in the 2000 Census)

Population Classification	Number of Cells Included	Number of Cells Excluded (Fewer than 500 People)	MAPE
Total population—no classification			
Estimation areas	1,950	0	3.1
(All counties)	(3,140)	(0)	(3.6)
Sex	3,900	0	3.2
Ethnicity	3,401	499	9.7
Age group	23,397	3	6.8
Sex by ethnicity	6,130	1,670	11.5
Sex by age group	46,430	370	7.2
Ethnicity by age group	28,553	18,247	9.7
Sex by ethnicity by age group	53,020	40,580	9.0

NOTE: Estimation areas are large counties and groups of smaller counties with at least 16,000 people; the median size is 55,000 people; the average size is 145,000 people; the District of Columbia is included as an estimation area (information from the U.S. Census Bureau). See note to Table 5-3 for the calculation of MAPEs.

SOURCE: Computations based on data provided by the U.S. Census Bureau.

Not surprisingly, the MAPE for estimation areas increases only slightly when the population is classified by sex only. However, the MAPEs are much larger when the population is classified by age, ethnicity, or a combination of characteristics. In particular, the population estimates classified by ethnicity are subject to appreciable error. The large number of excluded cells with the cross-classifications involving ethnicity and age groups should be noted: the MAPE values given in Table 5-4 represent only a part of the total population, and the values for the excluded part may be very different.

Table 5-5 gives the results of this analysis in a different form, presenting the distributions of the ratios of the population estimates to the census counts. When classified by sex only, the population estimates are within 10 percent of the census estimates for around 97 percent of the estimation areas. However, the corresponding percentage when the population is classified by age group falls to 77 percent, and when classified by ethnicity it falls to 67 percent. Of particular note is the 20 percent of cells with ratios less than 80 percent when the population is classified by ethnicity. Ratios of less than 80 percent occurred for non-Hispanic cells for only two estimation areas. However, the population estimates underestimated the number of Hispanics by 20 percent or more in 46 percent of estimation areas with 500 or more Hispanics. In a quarter of these areas the underestimation was at least 40 percent.

The general underestimation of the Hispanic population in the 1990s is well recognized and may not be repeated in the current decade, but this analysis does bring out the problems associated with its concentration in certain geographic areas. If the race/ethnicity classification was dropped and only the sex by age group cross-classification used in the population weighting adjustments, there would still be a quarter of the cells in which the population control was in error (over and under) by at least 10 percent, and in 4 percent of the cells the error would exceed 20 percent. This finding, of course, applies to estimates that are 9 years out from the previous census; the estimates will likely be more accurate, on average, the closer they are to the census year.

The panel has identified several alternative strategies that may serve to reduce the effects of errors in the population estimates and to deal with the extent of cell collapsing that is used with the current scheme for applying population controls. One strategy is to apply the cross-classification controls at a higher level of geography than estimation areas, with hopefully less error in the control totals. A drawback of this strategy is that ACS population estimates for counties and cities would not be consistent with the PE estimates for those areas. To ameliorate this problem, application of the cross-classification controls at a higher level of geography could be combined with the use of total population controls at the estimation area

TABLE 5-5 Percentage Ratio of July 1, 1999, Estimation Area Population Estimates to April 1, 2000, Census Counts, by Cells Based on Combinations of Sex, Ethnicity, and Age (Excludes Cells with Fewer Than 500 People in the 2000 Census)

Population Classification	Number of Cells	<80%	80%–	90%–	95%–	100%–	105%–	110%–	120%+	Total
No subclassification	1,950	0.1	2.4	14.6	61.4	20.6	0.7	0.2	0.0	100.0
Sex	3,900	0.1	2.9	15.0	58.9	21.8	0.9	0.2	0.0	100.0
Ethnicity	3,401	19.7	7.5	9.1	36.8	18.3	3.1	3.0	2.4	100.0
Age group	23,397	2.0	13.9	20.0	26.4	20.6	10.2	5.8	1.1	100.0
Sex by ethnicity	6,130	16.1	7.3	9.3	39.1	19.8	3.3	2.7	2.4	100.0
Sex by age group	46,430	2.5	14.8	19.2	24.8	19.7	11.0	6.6	1.3	100.0
Ethnicity by age group	28,553	8.5	13.1	16.9	22.4	19.2	10.3	7.0	2.5	100.0
Sex by ethnicity by age group	53,020	6.6	13.9	17.1	22.3	19.1	11.0	7.7	2.3	100.0

NOTE: Estimation areas are large counties and groups of smaller counties with at least 16,000 people; the median size is 55,000 people; the average size is 145,000 people; the District of Columbia is included as an estimation area (information from the U.S. Census Bureau).

SOURCE: Computations based on data provided by the U.S. Census Bureau.

level. Indeed, as the Census Bureau is considering, the total controls might sometimes be applied at a lower level, such as individual cities with populations of over 65,000 within estimation areas.

A strategy to reduce the amount of cell collapsing needed is to develop the weights through a raking algorithm that makes the ACS sample conform to each of the marginal distributions of the control variables, not to the joint distribution. The paper written for the panel by Jay Breidt and reproduced in Appendix B examines such alternatives.

On the issue of whether to use the race/ethnicity classification, the poor quality of the estimates in the 2000 comparison raises concerns about their comparability to what would have been obtained had the ACS interviewed the entire population. The population estimates start with the last census values and update them using administrative data. The reporting or recording of race/ethnicity in the census and in administrative data differ from each other and also from the ACS. As a result, the population estimates by race/ethnicity may not serve well as controls for the ACS sample.

The use of population controls for the population census and household surveys has a long history. It is instructive to contrast these uses with the use of population controls in the ACS. Although they appear similar, they are in fact very different. With the long-form sample, the data for the full census controls are collected for the same time and by essentially the same methods. Thus, the controls represent a poststratification adjustment, which improves the precision of the long-form estimates in a standard way. The long-form-sample controls achieve this improvement for small areas; they are applied for weighting areas, which are often as small as a block group or a census tract and never larger than a county (see National Research Council, 2004b:App. H). For household surveys, the controls are the population estimates and so subject to more error than the census counts, but for most household surveys other than the ACS the controls represent the same residence concept as the surveys; and the controls are applied at a high level of aggregation, which reduces the level of error in the population estimates. Generally, the controls for household surveys are applied for the nation as a whole by sex, age, and race/ethnicity groups and sometimes for total population by state (as, for example, in the Current Population Survey; Bureau of Labor Statistics and U.S. Census Bureau, 2002:Ch. 10).

In contrast, the population controls used in the ACS are midyear population estimates based on different residence rules and different sources than the yearly accumulation of ACS monthly samples. (For an illustrative example of the effect of population controls on areas with seasonal populations, see Section 3-C.3.) The ACS population controls should therefore not be treated as if they are poststratification controls, as is the current practice. It cannot be assumed that they necessarily improve the quality of the ACS estimates, particularly since they are applied at the estimation area level and therefore are subject to appreciable error. The panel views the use of the population controls in the ACS as a critical issue that requires major research.

Concomitantly, the panel views research on methods to improve the postcensal population estimates as a priority area for the Census Bureau. An important component of that research should be to investigate using ACS data more fully than currently in producing national estimates of international migration and particularly for estimating domestic migration. This research is even more important when considering the time series of ACS estimates. That time series will be affected at each census by the differences between the postcensal controls and the actual census counts. Section 3-G discusses this problem in the context of using the ACS estimates; see also Section 6-D.

> **Recommendation 5-3:** As a high priority, the Census Bureau should undertake research to evaluate the effect of the postcensal population controls on ACS estimates and to examine alternative methods of making an adjustment that may be superior to the one currently used (including dispensing with the population controls entirely). The Census Bureau should make users aware in ACS documentation that biases in the ACS estimates caused by errors in the population controls are not reflected in the margins of error reported with the estimates and should conduct research to examine the effects of these errors on ACS estimates.
>
> The Census Bureau should also give priority to research on ways to improve the postcensal population estimates at the county level, including estimates of internal migration and international immigration and the classification of race and ethnicity.

6

Weighting and Interpreting ACS Multiyear Estimates

The series of monthly samples that constitute the American Community Survey (ACS) can be combined in a variety of ways to produce survey estimates. The previous chapter has described the weighting methods being used to produce 1-year period estimates, combining all the data collected within a calendar year. This chapter examines the weighting methods that the Census Bureau is planning to use to produce 3-year and 5-year period estimates. The same weighting methods could also be used to produce estimates for other periods, such as 2-year and 8-year estimates. Similar methods can also be used to produce subannual estimates—perhaps without the population controls—as indeed the Census Bureau has done for areas affected by Hurricanes Katrina and Rita (see Section 1-B.2.b). As users gain experience with the ACS products, the Census Bureau should consult them about the usefulness of producing estimates for other periods.

The weights to be used in analyzing survey data need to be developed in relation to the population parameters—or estimands—to be estimated. There are several alternative estimands that can be considered when analyzing multiyear data from ACS. Three such estimands are discussed in Section 6-A.

The Census Bureau has chosen multiyear period parameters as the quantities to be estimated from ACS data accumulated over multiple years. Section 6-B reviews the weighting scheme the Census Bureau plans to use to produce estimates of the multiyear period parameters and the interpretation of these estimates.

Section 6-C then discusses the estimation of changes over time based on the multiyear period estimates, assuming no change in population size and demographic composition, geographic boundaries, or question wording over the applicable period. The complications associated with population changes are addressed in Section 6-D. Other changes that potentially affect multiyear period estimates, such as changes in geographic boundaries or question wording, and the problem of discontinuities in population and housing controls around the time of the 2010 census are addressed elsewhere in the report.

6-A ALTERNATIVE ESTIMANDS FROM MULTIYEAR DATA

During the panel's early deliberations, various alternative estimands based on 3- and 5-years of ACS data were under consideration. The discussions focused on three main forms of estimand. For the majority of applications, the most attractive estimand is the population parameter for the most recent year of the multiyear period (provided that it can be estimated with adequate precision). A second estimand is the population parameter for the middle year. The third estimand is a multiyear period parameter comparable to the 1-year period parameter. The choice between these and other parameters needs to be based not only on which is preferred from a user perspective, but also on how well the parameters can be estimated. The paper by Jay Breidt in Appendix C, commissioned by the panel, discusses methods for comparing these and other estimands.

6-A.1 Single-Year Estimands from Multiyear Data

The rationale behind the use of multiyear data for producing estimates for any single year—such as the middle year or the end year—is that the estimation can "borrow strength" from the ACS data collected in other years. The process requires a statistical model that relates the estimands across time.

Some simplifying assumptions are made in order to illustrate the key issues in developing model-dependent estimates of single-year estimands from multiyear data. It is assumed that the population of the area for which the estimate is required remains constant over the multiyear period, that the sample size is the same for each year in that period, and that the standard error of each of the 1-year estimates in the period is the same, say σ.

Let the multiyear estimate be a simple weighted combination of the 1-year estimates, denoted by $\tilde{y} = \Sigma w_i y_i$, where y_i is the 1-year estimate for year i and w_i is a weight such that $\Sigma w_i = 1$. Under the above assumptions, the variance of $\tilde{y}$ is then $\sigma^2 \Sigma w_i^2$.

The optimum choice of the w_i depends on which estimand is selected and on the way in which the 1-year parameters are assumed to vary across time. In the simplest case in which the 1-year parameters are assumed constant across time, then the optimum choice for the values for the w_i is to set them all equal (i.e., $w_i = 0.33$ for 3-year estimates, or $w_i = 0.2$ for 5-year estimates) since this choice minimizes the variance of $\tilde{y}$. In this simple case, the multiyear estimate can be viewed as an estimate for any of the years or for any combination of years.

When the 1-year parameters are assumed to vary across years, the choice of the w_i needs to take account of both the variance and bias of $\tilde{y}$ in estimating the chosen estimand. In this situation the 1-year estimate is an unbiased estimate for the estimand for any specific year, but that estimate will be too imprecise for small areas. In many cases a natural assumption to make is that the 1-year estimates for years close to the given year will be less biased in estimating the parameter for the given year than the 1-year estimates for years farther away. This assumption leads to assigning weights w_i that are largest for the given year and decline as the other 1-year estimates get farther away. Thus, for example, the weights for an estimate for the latest year of a 5-year period might be 0.06, 0.08, 0.14, 0.25, and 0.47, with greatest weight to the fifth year and declining weights for earlier years, like exponential smoothing (see Appendix C). Under the assumption made, these weights lead to a less biased estimate than equal weights, but the variance of this estimate is substantially increased. With equal weights, the variance of the 5-year estimate is $0.2\sigma^2$, whereas with this alternative weighting scheme it is 56 percent larger at $0.31\sigma^2$, and the standard error is 25 percent larger. In fact, the use of a 5-year estimator with this weighting scheme produces a final-year estimate with about the same variance as a 3-year estimator with an equal weighting scheme, which has a variance of $0.33\sigma^2$. (However, the mean square errors of these two estimators are not the same since they have different biases when used as estimators for the year 5 estimand.)

For a midyear estimand from 5-year data, the corresponding weights might be 0.14, 0.19, 0.34, 0.19, and 0.14. The pattern of these weights around the middle year is symmetric, with the weight for the middle year estimate being the largest. The variance of this weighted estimate is $0.23\sigma^2$, only 13.5 percent larger than that for the equally weighted estimate. If an estimand for a single year is required, based on this approach, a midyear estimand is thus preferable to an end-year estimand in terms of the precision of the estimate.

6-A.2 Multiyear Period Estimand from Multiyear Data

Whatever weighting scheme is adopted with multiyear data, for practical reasons a single set of weights is needed for application for all analyses. If a single-year estimand and associated weighting scheme are adopted, there will likely be characteristics for which the resulting estimates for some areas will be seriously biased due to the pattern of changes in the characteristic over the period. The use of a period estimand, averaged over the time period, avoids this concern about bias. The Census Bureau has decided to adopt the approach of period estimation for the multiyear data from the ACS, which, under the assumptions made above in Section A.1, also leads to the equal weighting scheme and hence lowest variance. However, the period estimation approach achieves its benefits by placing considerable burden on users to interpret the estimates in an appropriate manner.

As discussed in Section 3-C.1.b and later in this chapter, period estimates are difficult to interpret, and users must assess them in terms of external information about changes that have occurred during the period. For example, a 5-year period estimate of the percentage of poor families of 10 percent could reflect any of the following: a constant percentage across the 5 years; a steady increase from, say, 7 percent to 13 percent; a corresponding steady decrease; a rise and decline in the percentage across the years; and so on. To obtain an indication of the likely pattern that underlies a 5-year (or 3-year) estimate, users need to apply local knowledge of the conditions in the area over the period. They can also examine the published 1-year estimates for a larger area that contains the area of interest.

6-B MULTIYEAR PERIOD ESTIMATION

Conceptually, multiyear period estimation is the same as 1-year period estimation, merely extended over a longer period. The starting point for producing multiyear period estimates is to concatenate the 1-year data files over the 3 or 5 years involved, in the same way that 1-year estimation is based on concatenating the monthly data collected within a calendar year. Then the weighting scheme the Census Bureau is proposing for multiyear estimation from the concatenated file is broadly the same as that used for the 1-year estimation, as described in Chapter 5.

A natural and very simple way to develop weights for use with a multiyear concatenated file is to take the existing weights on each of the 1-year files and divide them by the number of years involved (3 or 5). A variant of this simple approach takes advantage of revised, updated housing and population controls for earlier years in the period, which may have become available by the time when the weights for the period estimates are being developed. Under this variant, the 1-year weights would be revised by using the applicable updated housing and population controls, and the revised weights would then be divided by 3 or 5.

The Census Bureau is planning, however, to use a similar but somewhat different approach. In its method, the first two steps in the 1-year weighting process (that is, base weights and variation in monthly response factor—see

Box 5-1) are retained, but then noninterview factors 1 and 2, the mode bias noninterview factor, and the housing unit control and population control factors are applied differently. Instead of applying these factors separately for each year, they are applied to the concatenated sample for the multiple years.

The housing unit and population controls used are the averages of the 1-year controls for the multiple years, using any revisions that have been made to the controls since the 1-year period estimates were first produced. The advantage of the Census Bureau's method is that, by pooling the sample across years first, the controls are applied to a larger sample. (This is also an advantage for the noninterview adjustments.) As a result, greater control on the population counts by age, sex, and race/ethnicity within estimation areas is possible because less collapsing of control cells is needed.

There are trade-offs to be considered between the simple approach and the Census Bureau's planned approach. The Census Bureau's approach gives greater control on demographic characteristics but lacks control over the yearly representation of the sample. The simple scheme has the benefit of ensuring that each year is represented in its right proportion in the estimation process but employs control over the demographic characteristics only to the extent that this is achieved for the 1-year estimates.

While the greater control on the demographic controls afforded by the Census Bureau's scheme appears attractive, the issue of the quality of those controls must be considered. As discussed in Section 5-D, the panel has serious concerns about the quality of the population estimates at this level of detail and recommends that the Census Bureau should carry out research on this step in the 1-year weighting process, including alternative possibilities of less detailed demographic controls, applying the controls on a marginal basis by raking, or applying the controls at a higher level of aggregation. The results of this research may lead to a population control weighting step that does not require the collapsing of control cells. In this case, the simple scheme for producing multiyear weights by dividing the 1-year weights by the number of years, or rather the variant of it with updated controls, may be preferred because it also provides temporal control. Research is needed in this area.

Another area of research on multiyear weighting is to investigate the application of the housing unit and population controls at the county level rather than at the estimation area level used in the 1-year weighting. A key concern is whether the housing unit and population estimates for small counties are of adequate quality for this use. A possible approach would be to apply a housing unit control at the county level and to use a raking algorithm to apply a total population control at the county level and demographic controls at the estimation level. (Total population and housing unit controls could also be applied for cities within counties.) Note that the preferred weighting scheme for 3-year estimation may be different from that for 5-year estimation because of the difference in sample sizes.

The Census Bureau has recently begun research into the introduction of an additional step in the weighting procedure based on linking administrative data to housing units on the MAF. This step, which involves a calibration weighting adjustment applied just to the linked units, can be applied at the tract level (Fay, 2005, 2006). The initial goal of this research was to improve the precision of 5-year estimates for tracts, but, if successful, it could improve the precision of estimates for subcounty areas more generally. Elsewhere in this report the panel has pointed out the need to improve the precision of ACS estimates for small geographical areas. Thus, this line of research warrants further investigation.

> **Recommendation 6-1:** The Census Bureau should conduct research to examine the bias and variance properties of the planned multiyear weighting scheme and compare these properties with those of some alternative schemes.

6-C ESTIMATION OF CHANGE OVER TIME

The production of estimates every year, rather than every 10 years as with the decennial census long-form sample, is a major asset of the ACS. Users will be able to study changes in estimates over the years. For areas for which 1-year estimates are provided, users will have a time series of 1-year estimates from which annual, biannual, and other changes are easily obtained, and to which more sophisticated methods of time series analysis can be applied. The Census Bureau will also provide margins of error for changes from one year to the next. With areas for which only multiyear period estimates are produced, the study of change over time is more complicated.

There are two main questions to be addressed in assessing changes in multiyear period estimates over time: (1) How is the change between two multiyear estimates to be interpreted (that is, what is the estimand)? (2) What is the precision of the estimated change? Two simplifying assumptions help to convey the essential points to be made in answering these questions. One is that each multiyear estimate for an area is a simple average of its 1-year estimates. This assumption is approximately valid for areas with populations that change little over time. The second assumption is that the precision of the 1-year estimates remains the same over time. Given the first assumption, the second assumption is a reasonable approximation, even though, when estimating a proportion (for example, the proportion poor), the magnitude of an estimate's sample error depends on the value of the proportion.

6-C.1 Interpreting Estimates of Change Between Multiyear Period Estimates

Let the time series of 1-year estimates be $y_1, y_2, y_3, \ldots, y_t, \ldots$. Let $\tilde{y}_t$ and $\tilde{y}'_t$ be 3-year and 5-year period estimates, respectively, where t denotes the last 1-year estimate in the multiyear estimate. Under the first assumption, the multiyear estimates are simply 3- or 5-year moving averages of the 1-year estimates. Thus, for example, for 3-year estimates,

$$y_3 = \frac{1}{3}(y_1 + y_2 + y_3),$$

$$y_4 = \frac{1}{3}(y_2 + y_3 + y_4),$$

$$y_5 = \frac{1}{3}(y_3 + y_4 + y_5),$$

$$y_6 = \frac{1}{3}(y_4 + y_5 + y_6), \ \ldots$$

and, for 5-year estimates,

$$y'_5 = \frac{1}{5}(y_1 + y_2 + y_3 + y_4 + y_5),$$

$$y'_6 = \frac{1}{5}(y_2 + y_3 + y_4 + y_5 + y_6), \ \ldots.$$

With the 3-year period estimates, pairs of estimates that are only 1 year apart have 2 years in common, and those that are 2 years apart have 1 year in common. There is no overlap when the pair of years is more than 2 years apart. With the 5-year period estimates, estimates that are only 1 year apart have 4 years in common, those that are 2 years apart have 3 years in common, those that are 3 years apart have 2 years in common, and those that are 4 years apart have 1 year in common. It is only when two 5-year period estimates are 5 or more years apart that there is no overlap.

The extent of overlap between two multiyear estimates determines the estimand that the difference between them is estimating. Consider the difference between two 3-year estimates, with one being $\tilde{y}_3$ and the other being $\tilde{y}_t$ with $t > 3$. With $t = 4$, because of the 2-year overlap, the difference between $\tilde{y}_4$ and $\tilde{y}_3$ is

$$y_4 - y_3 = \frac{1}{3}(y_2 + y_3 + y_4) - \frac{1}{3}(y_1 + y_2 + y_3) = \frac{1}{3}(y_4 - y_1), \qquad (1)$$

that is, one-third of the change between year 1 and year 4. With $t = 5$, with a one-year overlap, the difference between $\tilde{y}_5$ and $\tilde{y}_3$ reduces to $[(y_4 - y_1)$

$+ (y_5 - y_2)]/3$. This quantity can be viewed as estimating two-thirds of the average difference between two pairs of years that are 3 years apart—that is, years 1 and 4 and years 2 and 5. With $t = 6$, with no overlap, the difference between $\tilde{y}_6$ and $\tilde{y}_3$ is $[(y_4 - y_1) + (y_5 - y_2) + (y_6 - y_3)]/3$—that is, the average difference over three pairs of years for years that are 3 years apart. In general, for all pairs of nonoverlapping estimates, the difference between $\tilde{y}_t$ and $\tilde{y}_3$ is given by $[(y_{t-2} - y_1) + (y_{t-1} - y_2) + (y_t - y_3)]/3$—that is, the average difference for years that are $(t - 3)$ years apart. Similar results apply for differences between 5-year estimates.

These results indicate the importance of distinguishing between nonoverlapping and overlapping multiyear estimates. On one hand, with nonoverlapping estimates, the estimand can be simply viewed as an average difference across a set of 3 or 5 individual years that are apart by the number of years that the multiyear estimates are apart. On the other hand, with overlapping multiyear estimates, the estimand is only a fraction of the average 3- or 5-year difference.

To elaborate on the interpretation of differences in multiyear estimates, consider two alternative scenarios for changes in the estimands over time: one is a steady linear trend in the 1-year values, with the values increasing by, say, δ from one year to the next; the second is one in which the 1-year values stay at a constant level until year 6, at which point they increase by γ and then remain constant at the increased level after that. For simplicity of the presentation, for the latter scenario, we consider changes in 3-year estimates.

Under the first scenario, the estimand corresponding to the difference between multiyear estimates—both 3- and 5-year estimates—that are k years apart is $k\delta$, irrespective of whether the estimates are overlapping or not. Thus, the difference between adjacent multiyear estimates is an estimate of the constant annual change δ, the difference between multiyear estimates that are two years apart is an estimate of twice the annual change, and so on. These points are illustrated in Table 6-1, in which an increase in the percentage of poor families is assumed to be 1 percent per year. The 3-year estimates also increase by 1 percent per year, so that the difference between adjacent 3-year estimates is 1 percent, between 3-year estimates 2 years apart it is 2 percent, and so on. While expressing the difference

TABLE 6-1 Estimates of Poor Families in an Area Assuming a 1 Percent Annual Increase, in Percent

Year	1	2	3	4	5	6
1-Year Estimate	10	11	12	13	14	15
3-Year Estimate			11	12	13	14

TABLE 6-2 Estimates of Poor Families in an Area Assuming a 3 Percent Increase in Year 6, in Percent

Year	1	2	3	4	5	6	7	8
1-Year Estimate	10	10	10	10	10	13	13	13
3-Year Estimate			10	10	10	11	12	13

in terms of annual change is attractive, it can be misleading if the annual change is not constant, as the second scenario demonstrates (see Table 6-2). Under the second scenario, the annual population values can be represented as Y for the first 3 years and $Y + \gamma$ thereafter. The 3-year average for the first 3 years is then $Y_3 = Y$, and subsequent 3-year averages are

$$Y_4 = \frac{1}{3}[Y + Y + Y] = Y,$$

$$Y_5 = \frac{1}{3}[Y + Y + Y] = Y,$$

$$Y_6 = \frac{1}{3}[Y + Y + (Y + \gamma)] = Y + \frac{1}{3}\gamma,$$

$$Y_7 = \frac{1}{3}[Y + (Y + \gamma) + (Y + \gamma)] = Y + \frac{2}{3}\gamma,$$

$$Y_8 = \frac{1}{3}[(Y + \gamma) + (Y + \gamma) + (Y + \gamma)] = Y + \gamma, \text{ and}$$

$$Y_9 = Y_{10} = \ldots = Y + \gamma.$$

The estimand for the difference between the 3-year estimates for years 3 and 4, $\tilde{y}_4 - \tilde{y}_3$, and for years 3 and 5, $\tilde{y}_5 - \tilde{y}_3$, is thus 0 since no change has occurred over the first 5 years. The estimand for the difference between the 3-year estimates for years 3 and 6—which do not overlap—is $\gamma/3$. This difference is an average of no difference between years 1 and 4 and between years 2 and 5 and a difference of γ between years 3 and 6. Similarly, the estimand for the difference between the 3-year estimates for years 3 and 7 is $2\gamma/3$, which represents an average of no difference between years 1 and 5 and a difference of γ between years 2 and 6 and years 3 and 7. The full estimate of γ is obtained only when the 3-year estimate for year 8 is obtained, after the increase has been in effect for 3 years. This point is illustrated in Table 6-2, in which a permanent increase of 3 percent in the estimated percentage of poor families occurs in year 6. It is not until year 8 that this increase is fully reflected in the 3-year estimate, and it is only in the differences between the 3-year estimates for year 8 and for years prior to year 6 that the full 3 percent change is observed.

6-C.2 Precision of Estimates of Change Between Multiyear Period Estimates

To assess the precision of the estimates of the differences between multiyear estimates, it helps to invoke the second simplifying assumption, namely, that the variance of each of the 1-year estimates is the same, say, σ^2. Since the ACS annual samples are approximately independent of each other, the variance of a sum or difference between any two 1-year estimates is $2\sigma^2$, and, indeed, the variance of any equally weighted combination of sums and differences of k 1-year estimates is $kw^2\sigma^2$, where w is the weight attached to each estimate ($w = 1/3$ or $1/5$).

Application of these results to obtain the variances of the difference between nonoverlapping multiyear estimates is straightforward. Since there are six 1-year estimates involved in the difference between two nonoverlapping 3-year estimates ($k = 6$), and each is weighted by $1/3$, the variance and the standard error of the difference are $6\sigma^2/9 = 0.67\sigma^2$ and 0.82σ, respectively. The corresponding variance and standard error for differences between nonoverlapping 5-year estimates are $10\sigma^2/25 = 0.4\sigma^2$ and 0.63σ. By comparison, the variance and standard error of the difference between any two 1-year estimates are $2\sigma^2$ and 1.41σ.

In the case of the difference between two overlapping multiyear estimates, some of the 1-year estimates cancel out as indicated, for example, in equation (1) above. After removing the overlapping 1-year estimates in the difference, the above formula still applies: the only consequence is that the number of 1-year estimates, k, is reduced. If the multiyear estimates are 1 year apart, then $k = 2$; if they are 2 years apart, then $k = 4$; if they are 3 years apart, then $k = 6$; and if they are 4 years apart with 5-year estimates, then $k = 8$. As a result of the reduction in k with overlapping multiyear estimates, the variance and standard error of the difference are smaller than for nonoverlapping estimates. However, as noted in Section 6-C.1 above, the estimand is also different.

To aid understanding of the relationship between the different estimands and their standard errors, Table 6-3 summarizes results on standard

TABLE 6-3 Standard Errors of Estimates of Change for Various Values of the Gap Between Two Period Estimators as Multiples of the Standard Errors of a 1-Year Estimator

	Gap Between Period Estimates (Years)				
Period Estimator	1	2	3	4	5 or more
1-year	1.41	1.41	1.41	1.41	1.41
3-year	0.47	0.67	0.82	0.82	0.82
5-year	0.28	0.40	0.49	0.57	0.63

NOTE: See text for explanation of multiples.

errors for estimates of change between two period estimates that are a specific number of years apart. The table expresses the results in terms of multiples of the standard error of a 1-year estimate, σ, using the formulas given above. The multiples for the 3-year and 5-year change estimates indicate the lower standard errors that occur when using nonoverlapping multiyear estimates as compared with 1-year estimates (0.82 and 0.63, compared with 1.41). However, in a situation in which a multiyear estimate is required, the standard error of the 1-year estimate, σ, is large. Since the standard errors of the differences between both 3-year and 5-year nonoverlapping estimates are sizable proportions of σ, the standard errors of the differences will also be large. Thus, only large changes are likely to be detected as significantly different.

When the two multiyear estimates being compared overlap, there is a further reduction in the multiples, as seen on the left side of the table, but this occurs because the estimands are not comparable. Consider, for example, changes in 3-year estimates under the linear trend scenario, with the 1-year parameter increasing (or decreasing) by a constant amount, δ, each year. The estimand corresponding to the difference between two 3-year estimates with a gap of 3 years between them is 3δ, while the estimands for the differences between 3-year estimates that overlap by 1 year or 2 years are only 2δ and δ, respectively. To convert these overlapping estimates of change to estimates of the 3-year change, they need to be multiplied by 3 and 1.5, respectively. When the overlapping estimates of change are increased in this way, the standard errors of 3-year change estimates are increased by the corresponding factors given in Table 6-3. Thus, for example, the standard error of the estimate of the full 3-year change between two 3-year estimates that are 1 year apart is $3 \times 0.47\sigma = 1.41\sigma$, which is the same as that for comparing two 1-year estimates. For two 3-year estimates that are 2 years apart, the corresponding standard error is $1.5 \times 0.67\sigma = \sigma$. The fact that 3-year estimates are being compared in a situation in which a 1-year estimate with a standard error of σ is deemed too imprecise to be published implies that change estimates between overlapping 3-year estimates are very imprecise—they have standard errors as large as or larger than a 1-year estimate.

The same situation applies with estimates of differences between overlapping 5-year estimates. The most favorable case is when two 5-year estimates overlap by only 1 year. In this case, the standard error of the full estimate of the 5-year change is $(5/4) \times 0.57\sigma = 0.71\sigma$. In situations in which 5-year estimates are needed, the large size of this standard error makes comparisons of overlapping 5-year estimates rarely likely to be of interest.

The results in Table 6-3 are presented in terms of multiples of the standard errors of 1-year estimates. However, users will generally have

TABLE 6-4 Standard Errors of Estimates of Change for Various Values of the Gap Between Two Period Estimators as Multiples of the Standard Errors of the Corresponding Period Estimator

	Gap Between Period Estimates (Years)				
Period Estimator	1	2	3	4	5 or more
1-year	1.41	1.41	1.41	1.41	1.41
3-year	0.82	1.15	1.41	1.41	1.41
5-year	0.65	0.89	1.10	1.26	1.41

NOTE: See text for explanation of multiples.

available only the standard errors, or rather measures of error (based on 90 percent confidence intervals), of the 3-year or 5-year estimates. Under the assumptions made, the results in Table 6-3 are readily converted to multiples of standard errors of the multiyear estimates in Table 6-4 by noting that the standard errors of 3-year and 5-year estimates are simply $\sigma\sqrt{3}$ and $\sigma\sqrt{5}$, respectively. The multiples for the standard errors of 3-year and 5-year estimates are thus simply $\sqrt{3} = 1.73$ and $\sqrt{5} = 2.24$ times the corresponding multiples in Table 6-3, and these multiples also apply to the published measures of error. Thus, for example, for two nonoverlapping 5-year estimates (or 3-year estimates) for which the average of their measures of error is ±5 percent, the measure of error of the difference between the estimates from Table 6-4 is around ±7.1 percent (5 × 1.41), whereas for two 5-year estimates that are 2 years apart with the same average measure of error of ±5 percent, the measure of error of the difference is around ±4.5 percent (5 × 0.89).

6-C.3 Conclusions

The overall conclusions from these analyses are that estimates of change based on differences between overlapping 3-year or 5-year period estimates are generally not useful. Furthermore, even with nonoverlapping estimates, the estimates of differences will generally be fairly imprecise. Analyses of change will be most productive only when major changes have occurred or when 1-year estimates are precise enough to be published. Of course, the multiyear period estimates remain an improvement over the once-a-decade long-form sample because they are updated every year and therefore provide a more timely picture of the characteristics of an area than is possible from the long-form sample.

A final point regarding differences between multiyear period estimates is that, just like the estimates themselves, they can reflect a variety of patterns in the underlying 1-year estimates. For example, a 2 percent change between two nonoverlapping 3-year estimates could occur because all the increase—a 6 percent increase—occurred in the latest year, because a 2 percent increase occurred in the interval between the two estimates, or because of other patterns of change. Users need to be aware of the possible underlying patterns and find ways to distinguish between them based on other sources or on ACS data at other levels of aggregation.

6-D EFFECTS OF CHANGES IN POPULATION SIZE AND CHARACTERISTICS

The treatment thus far in this chapter has assumed that the population of an area has been static, or at least has changed in only minor ways, during the time period of a 3- or 5-year period estimate. While this may be a reasonable assumption for many areas, there will be some areas that experience major changes in population size or composition or both. Moreover, major changes are most likely to occur in small governmental units and census tracts, areas for which 3-year and 5-year estimates are needed.

Population changes that recur within a year, such as the seasonal patterns discussed in Section 3-C.3, affect each year of a multiyear period estimate in the same way that they affect a 1-year period estimate. The additional population changes that affect multiyear period estimates are year-to-year changes, such as population growth over time, which may be concentrated in certain demographic subgroups. The Census Bureau's planned weighting procedures for multiyear estimates reflect such changes by using the 3- or 5-year averages of the independent housing unit estimates and the independent population estimates by demographic subgroup as controls.

Users need to consider the potential effects of the planned weighting scheme on ACS estimates. For this discussion, it is useful to distinguish between two types of ACS estimates: estimates of proportions, such as the proportion of poor people, and estimates of totals, such as the number of poor people. In the case of proportions, if an area had major growth, say, an influx of young persons, the ACS multiyear period estimates of the characteristics of young people will be weighted toward their characteristics in the later years of the period, and the influx will similarly affect other estimates in which young people are included. Even if it seems likely that a characteristic of interest has remained stable over the period, the changes in the population composition will lead to differences in the proportions of the population with that characteristic over the time period. Consider, for instance, the comparison of the multiyear estimates of the unemployment rates between this area and a stable or declining area. In this case, users will need to be cognizant that the influx of young people may affect the comparison.

The situation is even more complex when the multiyear estimates are totals and the area's population changes appreciably during the period. A multiyear period estimate of a total then reflects not only changes in the prevalence of the characteristic (for example, the unemployment rate) due to temporal changes and to changes in the composition of the population, but also changes in the size of the population. As a result of these complexities, users may find multiyear estimates of totals for areas that change markedly in size to be problematic. They may prefer to develop model-based estimates of totals for the current population or for the population of the latest year of the estimation period.

In the case of counties, one possible approach for developing estimates of totals for the latest year of the estimation period (or a subsequent year) takes advantage of the county population estimates reported annually by the Census Bureau's Population Division. The simplest model assumes for the county that the prevalence rates for the characteristic of interest have remained approximately constant over the estimation period within demographic subgroups and that the population estimates by demographic subgroup for the latest year are accurate. The ACS multiyear period estimates are then used as estimates of the latest-year prevalence rates within specified demographic subgroups, and these rates are applied to the latest-year county population estimates for those subgroups. The sum of the resulting subgroup estimates then serves as an estimate of the number of persons with the given characteristic in the county in the latest year.

Although the model assumptions involved in producing such estimates are problematic, the resulting estimates may suit user needs better than the standard multiyear estimates for counties that have experienced major population changes. The extension of this simple model to produce subcounty estimates of totals introduces the further complexity that no estimates of population sizes are readily available at levels below the county. Thus an additional step is needed to produce population estimates for such areas as census tracts and small cities. Simply applying the proportion of the county's population in the area from the ACS period estimates to the current population county estimates is questionable but is a possibility if there is no better local information available.

In summary, multiyear period estimates are highly complex when the population of an area changes substantially in size or composition or both during the estimation period. Users need to pay careful attention to their interpretation in such cases and, particularly when estimating totals, they need to assess whether the estimates meet their needs.

Recommendation 6-2: The Census Bureau should consult users about the utility of the currently proposed multiyear period estimates—particularly for estimates of totals—for areas that change markedly in population size. It should investigate whether there are other forms of estimates that could be produced and would better serve user needs.

PART III

Education, Outreach, and Future Development

7

Important Next Steps

The full implementation of data collection for the American Community Survey (ACS) in 2005–2006 is a historic event for the nation's statistical system. Based on over 10 years of research and development by the Census Bureau, the ACS is intended to replace the decennial census long-form sample as a source of regularly updated demographic and socioeconomic information on the population and housing of states, counties, cities, and other governmental and statistical areas.

The panel's assessment is that the ACS will deliver on its promise to provide more timely, frequent, and complete information than the long-form sample. Given the survey's continuous design, however, ACS estimates are not the same as the long-form-sample estimates for a point in time (Census Day, April 1); instead, they represent annually updated period estimates based on 12 months of data and (once sufficient years of data are accumulated) 36 and 60 months of data. Only 60-month estimates (5-year period estimates) will be available for the smallest areas. ACS estimates also have significantly higher sampling errors than the corresponding long-form-sample estimates, a feature of particular concern for the smallest areas (small counties, cities, towns, villages, American Indian and Alaska Native areas, and school districts, as well as census tracts and block groups).

While the ACS continuous design presents challenges to users, it also affords opportunities to develop applications that go far beyond what was possible with the long-form sample. Some innovative uses of the ACS will be easier to implement than others. In the same vein, some uses of the ACS to replace long-form-sample data will be easier to implement than others.

Overall, there is no doubt that the ACS can be of great benefit to many users, not only in the short term, but also over time as the survey is improved and new measures and applications are developed.

To achieve these goals will require sustained and even expanded resources and effort on a continuing basis, not only for collection and production of the ACS data, but also for user education and outreach and methodological research and evaluation. The Census Bureau should seek adequate funding for the ACS as a top priority. The panel hopes that the user community will express its support and that Congress will provide the needed annual funding as the ACS comes fully online.

> **Recommendation 7-1:** The Census Bureau should continue to make sufficient funding of the ACS one of its top priorities. It should seek adequate funding on a continuing basis, not only for data collection and production, but also for ongoing programs of methodological research and evaluation and user outreach and education.

The Census Bureau has already devoted considerable resources to methodological research and data product design as part of the developmental work for the ACS over the past 10 years. Yet this work cannot stop with full implementation. On the contrary, the sheer volume of estimates means that the full ACS is in many ways brand new to the Census Bureau and the user community, even though the ACS concept and test data have been around for a period of years. Now is therefore the time to expand the resources for evaluation of the full production ACS and for methodological research and experimentation to improve the survey to reflect the evaluation results. Now is also the time to significantly expand the resources to educate and receive feedback from users, as over the next few years they experience for the first time the full panoply of 1-year, 3-year, and 5-year period data products from the ACS.

The purpose of this chapter is to outline this needed effort so that the ACS can evolve to meet its full potential. The chapter starts by describing an education program that is needed to inform users about what the ACS is, how to use its data products, and how interactions between the Census Bureau and the ACS user community can mutually benefit the ACS. The next section reviews the requirements for continued monitoring of basic indicators of data quality. The third section outlines areas for research and evaluation so that the ACS design, data collection, and estimation procedures can be continually improved and users can be more fully informed regarding sampling and nonsampling errors in the data. This section indicates the panel's priorities for where limited resources can be most usefully directed in the next few years. The final section briefly describes a vision of what the ACS could become as it not only supports applications that previously used the census long-form sample, but also provides the basis for improving and expanding the information that is available to understand and plan for the nation's growing, diverse communities.

7-A EDUCATING DATA USERS ABOUT THE ACS

The overriding priority for small-area data users is to adjust their perspective from having long-form-sample point-in-time estimates available once every 10 years to having 1-year, 3-year, and 5-year ACS period estimates available annually for geographic and statistical areas depending on their population size. With some exceptions, notably in the housing and transportation communities, the panel found relatively little preparation for this change on the part of data users. No doubt a key reason has been limited resources. In addition, it is often hard to imagine how to use very different kinds of data that are not yet available for most areas. Chapters 2 and 3 are intended to help users understand the key features of the ACS and to provide guidance for using the data for a range of applications, but much more work remains to be done.

While the Census Bureau has tried to facilitate the transition from the long-form sample to the ACS, the fact is that the full implementation of the ACS will be a sea change for data users. Appropriate reorientation on the part of users will not occur as a result of issuing new documentation or a new web site, essential as those elements of a data dissemination plan are. Appropriate reorientation will occur only as a result of a comprehensive education effort that is based on a plan to provide a set of best practices for data use that are well illustrated, using examples that are meaningful and that clearly explain period estimates and their differences from the point-in-time estimates that are commonly provided by other data sources. The plan must also provide for systematic feedback from users that can help the Census Bureau refine and tailor the education program to user needs. Such feedback should also benefit the Census Bureau by identifying potential problems with the data to follow up and improvements to data products that would facilitate data use.

The education and outreach plan, of which key elements are outlined below, is designed primarily for users who expect to make repeated, multiple uses of the ACS data and who will therefore need to learn about the survey in some detail. However, there are also first-time users, infrequent users, and users who lack resources for participating in educational programs (for example, users in many small governmental jurisdictions). These users need to find key estimates easily and not have to master the complexities of the data. The Census Bureau, in cooperation with the network of organizations that it enlists as partners for education and outreach (see Section 7-A.3 below), should identify ways to help these users. Being proactive in this regard will increase the value of the ACS and reduce the likelihood that users will fail to take advantage of ACS data because they find them too complex.

Approaches to help occasional users will likely also help experienced users. One approach is for the Census Bureau to work with organizations that have a mission to assist data users, such as State Data Centers, to help them develop simple data products and explanatory materials that are specifically designed for occasional or novice users. In addition, the Census Bureau itself could develop additional data products for the first-time, occasional, or resource-constrained user. These products could consist of simple tabulations that meet commonly accepted standards of precision. Similarly, simple tabulations of year-to-year change might be provided whenever there has been a significant increase or decrease in a key estimate, such as the poverty rate. The goal should be to make these products as transparent and accessible as possible, including giving them a special and prominent location on the Census Bureau's ACS web site that contains a link to the rest of the site for users who want more information.

7-A.1 Key Elements of the Education Strategy

The program of ACS education should have two major components. The first component should aim to *provide a foundation of the basics about the ACS and methods to use the data appropriately*. Users should be helped to grasp the key elements that make the ACS different from the long-form sample, the most important of which are the change from point-in-time to period estimates of characteristics, the increase in the size of sampling errors, and the opportunities and challenges that will arise with annually updated data for different time periods. After introducing data users to ACS concepts, the goal should be to educate them about the new perspectives they need to have and the new techniques they need to employ in order to make effective use of the data.

The second component should aim to *create paths for outreach to and feedback from users* that enable the Census Bureau to engage in a continuous dialogue regarding questions and issues that need to be addressed. At this stage of the program, no one, including the Census Bureau, can anticipate all of the questions and issues that will arise from the data user community. The Census Bureau will have an opportunity to accumulate a critical mass of user reactions to the 1-year and 3-year period estimates that should permit the staff to become more responsive to data users before the first 5-year period estimates are released in fall 2010. This will occur only if an adequate mechanism is in place to deliver feedback to the Census Bureau.

7-A.2 Providing a Foundation for the Basics

The Census Bureau's ACS web site (http://www.census.gov/acs/www) provides a great deal of information on all aspects of the ACS, including access to *Using Data from the 2005 American Community Survey*, a 31-page guidebook for users; the *Guide to the ACS Data Products*, an online tool for learning more about the various kinds of tables and other data products; the *ACS Data User Training Guide*, a set of PowerPoint presentations; and a voluminous *ACS Design and Methodology* document, which explains ACS operations from sampling to data release and includes facsimiles of the ACS questionnaires (U.S. Census Bureau, 2006). While helpful and necessary, these materials are not sufficient by themselves for educating data users about the ACS.

To build a foundation of knowledge that is meaningful for those who apply the data in their work, the Census Bureau needs to develop user-friendly *application-oriented* documentation and metadata, including sample applications that can be presented in paper form and on the web in the form of online tutorials. This type of documentation differs markedly from the provision of technical information. Both types of documentation are needed.

Two core features of the Census Bureau's application-oriented documentation should be, first, to provide key information to assist in the transfer to the ACS from the census long-form sample and, second, to describe methods and best practices to apply the ACS small-area data on the socioeconomic characteristics of the population for a variety of applications. Consultation with major user groups should yield instructive applications for large cities, smaller governments, rural places, transportation interest groups, and other groups that serve the data user community. One benefit of developing these kinds of examples is to enable data users, including key intermediaries, to assist the Census Bureau to establish standards and best practices for using ACS data. A recent publication, developed with input from Census Bureau staff and data users, takes this approach (Taueber, 2006). It is aimed at helping community planners access, interpret, and report on the ACS data for their areas.

The Census Bureau's consultations should include a wide range of users, including state governments, local governments (including regional and local councils of governments), not-for-profit agencies, academic researchers, the private sector, and the media. Within those sectors, applications should be developed for users with different focal interests: transportation, education, health, social services, criminal justice, economic development, and the environment. Applications should represent an assortment of typical uses: policy development, program planning, budgeting, site selection, fund allocation, and outcomes monitoring.

Basic kinds of information that users need to understand in order to make the transition from the long-form sample to the ACS, including the implications of differences for data quality and utility, include:

- Differences between the long-form-sample and ACS questionnaires: the format of the questionnaires, the application of residence concepts, the reference periods for questions, and the wording of questions.
- Differences between the long-form-sample and ACS data collection processes: an understanding of how the ACS data are collected, using different modes, from a series of monthly samples supporting annually updated 1-year, 3-year, and 5-year period estimates for different levels of geography.
- Differences between the long-form sample and the ACS in the accuracy and geographic specificity of population and housing unit controls that are applied to the estimates.
- How to compare ACS estimates to the 2000 census long-form sample (and other surveys) in light of differences between them—in particular, how to make comparisons for the 2005 ACS estimates, which pertain to households only and do not include group quarters.

Information relevant to methods and practices for using the ACS that users need to understand include:

- The provision of data from the ACS, including: the various formats for obtaining data, the geographic levels of data availability, and the trade-offs between different data products. Data access needs to be emphasized, via the American FactFinder web portal, data on CDs and DVDs, and data available from the Census Bureau's FTP sites.
- The sampling error of estimates for 12-month, 36-month, and 60-month intervals and how to interpret variability.
- How to interpret multiyear period estimates.
- How to gauge change over time using multiyear estimates and how to conduct comparisons across areas.
- Special issues for small-area data, focusing on strategies to increase precision at a small-area level, such as combining information across time and geography.

7-A.3 Building a Network for Education, Outreach, and Feedback

In order to encourage widespread and informed use of the ACS data, the Census Bureau needs to expand its infrastructure in two ways. First, it needs to establish a headquarters ACS users' staff devoted to education and outreach who would cultivate a network of trained intermediaries to assist with providing a basic ACS education to users. Second, it should form a small informal advisory group of experienced data users that meets with the ACS user staff on a regular basis in person and by conference call. The group would be a key point of contact for considering ideas to improve data products, educational materials, user outreach, and related topics.

Once a network of trained intermediaries is established, it will enable the development of a full-fledged system of regular feedback that can make the ACS education and training program—and appropriate uses of the data—grow and prosper. Feedback in the early years of implementation will assist the Census Bureau to adapt the training program to better meet user needs. In the longer term, user feedback should be a valuable source of ideas for modifying and improving the ACS to serve a wider range of applications and provide an increased return on investment in the data collection.

The Census Bureau is already reaching out to federal agencies to train and assist them in using the ACS. It has established a Federal Agency Information Program (see http://www.census.gov/acs/www/SBasics/fed.htm) through which Census Bureau staff members are available to make presentations to agency staff, provide assistance on specific applications of the ACS, such as in funding formulas, and prepare special data tabulations on a cost-reimbursable basis. The Census Bureau's work with federal agencies should be helpful to other users, such as state and local governments that interact with those agencies.

To develop an adequate network of intermediaries, however, the Census Bureau (including its regional office staff) should reach out to many organizations outside the federal government. An adequate return on the investment in the ACS can only be achieved if the small-area data are used to the widest extent possible. There are many organizations that the Census Bureau should strive to include in its network:

- the State Data Centers (SDCs, http://www.census.gov/sdc/www/);
- the Federal State Cooperative Program for Population Estimates (FSCPE, http://www.census.gov/population/www/coop/fscpe.html);
- the Census Information Centers (CICs, http://www.census.gov/clo/www/cic.html);
- college and university research institutes and data centers;
- professional associations (for example, the American Statistical Association, the Population Association of America, the American

Planning Association, the Association of Public Data Users, the American Association of Public Opinion Research, the Transportation Research Board, the American Library Association);

- major state and local government organizations (for example, the U.S. Conference of Mayors, the National Governors Association, the National Association of State Legislatures, the National Association of Counties, the National Association of Towns and Townships, the Association of Metropolitan Planning Organizations);
- local and regional councils of governments and planning agencies (for example, groups concerned with regional transportation plans and environmental issues);
- not-for-profit groups (local chapters of the United Way, Red Cross, United Hospital Fund);
- the media (for example, Investigative Reporters and Editors, Inc., and that organization's computer-assisted reporting program); and
- other for-profit and not-for-profit groups, such as market research professionals, associations of health professionals, social service agencies, and the variety of groups that serve special populations, such as the disabled, farm workers, veterans, and immigrants.

The involvement of the SDC network is of critical importance since the SDCs will be on the front line of answering information requests for 2010 census and ACS data. The SDCs must be able to effectively present ACS data and assist data users, including making the data understandable to users with a very wide range of experience and expertise. The SDC steering committee is already focusing its efforts on "training the trainers," so that SDCs have sufficient knowledge to then train the entire network of 1,800 organizations and general data users that they serve. In addition, individual SDCs have already developed helpful explanatory materials for the 2005 ACS data products (see, for example, "Ten Things to Know About the American Community Survey, 2005 Edition," prepared by the Missouri Census Data Center).[1] To move this initiative forward, the Census Bureau should support and encourage local hands-on workshops on the applications of ACS data. At least some of these workshops should be done in coordination with SDC affiliates so that best practices are provided for local users. These workshops can be the basis of an education network that, once established, can serve as an efficient information-sharing mechanism between the Census Bureau and the data user community.

In general, training courses can be developed at many different levels and for many different groups. Some can be in the form of tutorials, to be

[1]Available at http://mcdc2.missouri.edu/pub/data/acs2005/Ten_things_to_know.shtml.

presented at special workshops or sessions at the annual meetings of many key groups that are already part of the census network as listed above.

7-A.4 Working with the Media

ACS training and education should be adapted to the needs of media organizations. The media need to become a partner in explaining why the ACS is important and how the data can best be used. The media should welcome this partnership, since more frequently updated information will give them opportunities for many new stories over the decade on such topics as immigration, domestic migration, education patterns, and other topics of public concern.

The need for an active media education and partnership program is evident from the press coverage of the August 15, 2006, release of 2005 ACS data for political and statistical areas with 65,000 or more total population. This initial release provided information on age, sex, race, ethnicity, ancestry, place of birth, citizenship, year of immigration, residence last year, language spoken at home, education, disability, marital status, fertility, veteran status, and whether grandparents are caring for grandchildren in their home. A review by panel staff of 57 articles in 44 newspapers around the country published August 15-16, 2006, that used the new ACS data found that interest in the data was high but understanding of them and how to use them was often poor. The ACS was sometimes confused with other programs, such as the census and the population estimates, and understanding of how to compare the 2005 data with the 2000 long-form sample and other sources was limited (see Box 7-1).

The Census Bureau should conduct extensive analyses of news coverage of the 2005 ACS and revise and enhance its user education program and documentation accordingly, not only for the media, but also for other data users. As a top priority, guidance about comparisons of estimates from the 2005 data with the 2000 long-form sample and other sources (including the population estimates program) is clearly needed. Indeed, guidance on comparisons with the 2000 long-form sample will continue to be critical because the ACS cannot itself serve as a comparison source for estimates of change, particularly for small areas, until more years of data are released and analyzed.

7-A.5 Recommendations on User Education, Outreach, and Feedback

As with any major initiative, creating an education and outreach system to accompany the ACS will involve a significant commitment of resources from the Census Bureau and its affiliates, and from the user community as well. The ACS represents a substantial increase in the volume of informa-

BOX 7-1
Print Media Treatment of the 2005 American Community Survey

Panel staff reviewed 57 articles in 44 newspapers around the United States, published August 15–16, 2006, that featured the initial release of data from the 2005 ACS on social and demographic characteristics of areas with at least 65,000 people. The newspapers covered included major national papers, such as the *New York Times*, the *Washington Post*, and *USA Today*, other major metropolitan newspapers (for example, the *Seattle Post-Intelligencer*, the *Houston Chronicle*), and smaller newspapers (for example, the *Anchorage Daily News*, the *Lexington Herald-Leader*, the *Toledo Blade*).

Six conclusions are drawn from this review:

1. Interest is high in these data, principally because of their currency and the light they shed on such salient features of American life as increasing racial and ethnic diversity and immigration.
2. Change over time is of key interest. Three-fourths of the articles featured estimates of change from 2000 to 2005 in total population or characteristics. Most of the articles appeared to use the 2000 long-form sample as the comparison point; two articles used the Census 2000 Supplementary Survey or the 2002 ACS test survey. (Comparisons with the test surveys can be done only for areas with at least 250,000 people or, in the ACS test sites, for areas with at least 65,000 people.) *Only two articles expressed caution about ACS comparisons with the long-form sample.*
3. Population numbers are of key interest. Even though the Census Bureau emphasizes the use of the ACS for characteristics, not population counts, one-fifth of the articles explicitly focused on growth or decline in total population from 2000 to 2005. *None of the articles discussed that, for many areas, population figures*

tion available to decision makers and will become an important national asset if it is used appropriately and to its full potential. Cultivating a data user network that helps users navigate their way through this new maze of methods and issues will help the Census Bureau ensure that all will rise to the challenge of using this valuable tool of the nation's statistical system.

> **Recommendation 7-2:** The Census Bureau should develop a comprehensive program of user education, outreach, and feedback for the ACS. Two goals of the program should be (1) to educate users in the basics of the ACS, how it differs from the census long-form sample and other data sources, and appropriate methods to use the data; and (2) to develop paths for systematic feedback from users to improve the training materials, identify potential problems with the data, and sug-

(total and by age, sex, and race/ethnicity) are from the postcensal population estimates program, but for other areas (for example, cities within counties), the figures are from the ACS and have standard errors associated with them.

4. Information to help the reader understand the source of the data is sparse. One-fifth of the articles made only a glancing mention of the ACS as the data source, and one-fifth did not mention the ACS at all. Another one-fifth provided incorrect information about the nature of the ACS, confusing it with the census most often (4 articles) or with the population estimates program (1 article), or calling the ACS a mid-decade census (3 articles), a 5-year database (1 article), or a telephone survey (1 article). The remaining two-fifths of the articles provided a brief description of the ACS as a continuing monthly survey that is intended to replace the census long-form sample. *The sample size mentioned was the initial size of 3 million households per year, not the 2 million remaining after follow-up.*
5. Acknowledgment of sampling error is spotty. In all, 23 percent of the articles clearly referenced the margin of error in the ACS estimates *(two articles stated that the margin of error was so large as to render the data useless)*, 10 percent made a glancing reference to the margin of error, and the remaining 67 percent made no mention of sampling error.
6. Particularly in communities that are losing population, press articles questioned the population figures from the 2005 ACS by comparison with 2000. Three-fifths of the articles did not acknowledge a key difference between the 2005 ACS and the 2000 long-form sample that could affect such comparisons—*namely, that the ACS is limited to the household population and excluded people living in college dormitories, prisons, and other group quarters.* None of the articles mentioned another difference that could affect comparisons—*namely, the differences in the accuracy and geographic specificity of population and housing unit controls between the ACS and the long-form sample.*

gest ways to improve data products and documentation to maximize the utility of the data and facilitate data use.

Recommendation 7-3: As an integral part of its education, outreach, and feedback program for the ACS, the Census Bureau should establish a dedicated ACS user staff. That staff should partner with organizations that will assist end users, including the State Data Center network as a key partner and many other organizations and groups. The staff should work with the media to help them understand ACS data so that they can explain and showcase the value of the data to communities in an effective and accurate way.

Recommendation 7-4: The Census Bureau should establish an ongoing advisory group of experienced data users with whom to interact about user education materials, web site design, table content, and other aspects of the data products and education and outreach program for the ACS.

7-B DATA QUALITY MONITORING

A major continuing survey, such as the ACS, requires continued monitoring to ensure that data collection and production processes are performing well, to identify problem areas for investigation and development of improved processes, and to provide information to users about sampling and nonsampling errors of which they should be cognizant. For these purposes, it is essential to develop and track an appropriate set of performance measures.

Some performance measures are for use by survey managers to ensure that survey data collection and processing operations are being carried out as specified and within quality control tolerances and to flag problems for investigation. Such measures may track timely completion of check-in and data capture of mailed-back questionnaires, interviewer productivity, and the like. The panel did not review what measures the Census Bureau uses for quality control of the ACS; we trust that the Census Bureau has developed a set of appropriate measures and periodically reviews them for relevance and usefulness in identifying problems on a timely basis.

Other performance measures are useful not only to survey managers, but also to inform users of the quality of the data across areas and population groups. The Census Bureau has long experience with monitoring and maintaining the quality of its survey operations. For the ACS, it has taken a further step to put up on the ACS web site basic indicators of sampling and nonsampling errors in the data.

7-B.1 Nonsampling Error Measures

The Census Bureau currently provides four indicators of nonsampling errors, which can be accessed from the main ACS web site under "Using the Data" (http://www.census.gov/acs/www, "Quality Measures"). The four measures are sample size, survey coverage rates relative to the 2000 census-based population estimates, survey response rates (unit response), and item nonresponse rates (refer back to Box 2-4). At this time, all four indicators are available for the Census 2000 Supplementary Survey, the ACS 2001–2004 test surveys, and the 2005 ACS for the nation and states.

The panel commends the Census Bureau for providing quality measures for ACS estimates on its web site. For these measures to be useful, it is important that users of the data access them and interpret them correctly. The Census Bureau's data user advisory group and network of user education partners (see Section 7-A above) could be a valuable resource to help educate users about the meaning and value of the various indicators.

This network could also help the Census Bureau determine what additional indicators to consider adding to the web site. For example, for the 2005 ACS, it could be very useful to provide all four quality measures for individual public use microdata areas (PUMAs) to help users track basic data quality for substate areas.

Looking ahead, it would be very useful for the Census Bureau to periodically issue reports that highlight patterns of basic quality measures over time for geographic areas and population groups of interest—for example, whether (and which) item nonresponse rates are increasing or decreasing and for which areas and groups. Similarly, it would be useful for the Census Bureau to analyze unit and item nonresponse rates separately by data collection mode (mail, computer-assisted telephone interviewing, CATI, computer-assisted personal interviewing, CAPI) to see if there are patterns by geographic location or such characteristics as education level, family structure, and others. It will also be important for the Census Bureau and its network of user education partners to determine the most useful set of quality measures for the 3-year and 5-year period estimates for small areas in addition to those provided for 1-year period estimates for larger areas.

> **Recommendation 7-5:** The Census Bureau, in collaboration with user education partners, should carry out research on ways to facilitate understanding of the quality measures provided on the ACS web site. The Census Bureau and its partners should also consider what additional quality indicators—for example, some of the indicators presented at a finer level of geographic detail—would be useful to provide for the 2005 ACS and subsequent 1-year period estimates and what indicators to provide for the 3-year and 5-year period ACS estimates when those become available.

7-B.2 Sampling Errors

7-B.2.a Published Margins of Error

The Census Bureau provides a measure of sampling error for each sample-based estimate that is released in tabular form from the ACS. This measure is developed using a repeated replication method (see U.S. Census Bureau, 2006:Ch. 12). The published measure of sampling error is the margin of error around the estimate (plus or minus) at the 90 percent confidence level (1.65 times the standard error), not the commonly accepted 95 percent level, which is 1.96 times the standard error (see Box 2-5 for explanations of these terms).

The estimates of sampling error account for the variability from all the stages of weighting, including the initial sampling, the CAPI subsampling, and the population controls. Some weighting steps (see Section 5-A) are intended to reduce variability. The estimates do not account for other sources of variation, such as that introduced by imputation procedures for item nonresponse, nor for errors arising from inaccuracies in the population estimates used as controls.

Modifications are needed in the computation of the margins of error in some cases. For example, for small estimates, the margins of error shown produce a 90 percent confidence interval that includes zero, when it is not possible to obtain negative estimates (for example, ±113 for an estimate of 97 poor children in single-parent male-headed families for Ann Arbor, Michigan, from Detailed Table B17006 on the American FactFinder web site for 2005 ACS data). At the least, an explanatory note should be provided that the lower bound of the confidence interval is zero.

Sampling error measures are provided not only for population and housing characteristics, but also for estimates of total population, total housing, and basic demographic characteristics for counties, cities, and other areas that are not controlled to the census-based population or housing unit estimates. As described in Section 5-A, the ACS controls are applied for estimation areas, which are large counties and groups of small counties.

In the case of multiyear profiles of 1-year estimates, an indicator of statistical significance at the 90 percent confidence level is provided for the difference between an estimate for a specified year and the corresponding estimate for a current year. Multiyear profiles are available that compare 1-year period estimates from the C2SS and the 2001–2004 ACS test surveys for areas with 250,000 people or more; they are not being issued for the 2005 ACS, even though the 2005 ACS estimates could be compared with the estimates for 2000–2004 for large areas. They will presumably become available again beginning with the 2006 ACS.

7-B.2.b Guidance on Computing Sampling Errors

The Census Bureau provides general guidance, with just a few examples, for computing approximate estimates of standard errors for sums and differences of estimates for geographic areas and population groups that are shown in the ACS tables. (Taueber [2006[provides additional examples for local area data users on computing standard errors and other aspects of working with the ACS.) The Census Bureau's guidance is available in its publication, *Using Data from the 2005 American Community Survey* (pp. 10-11).[2] Generalized variance estimation procedures for estimating sampling errors for user-generated estimates from the public use microdata samples (PUMS) are provided in the "Accuracy Statement" for the PUMS in question.[3] In addition, for the first time, the 2005 PUMS provides replicate weights for users to calculate direct estimates of sampling error that are more precise than those from the generalized variance estimation procedures.

For standard errors of differences, the Census Bureau's guidance applies not only to comparing differences between two areas or population groups, but also to comparing differences between estimates for two points in time for which the individual standard errors are available. However, the guidance is not applicable to every calculation a user might wish to perform from the ACS tables. For example, to save space, many tables do not provide all of the aggregate categories that users may want—such as total children under age 5 with family income below the poverty level (see Detailed Table B17006). While this estimate can be obtained for an area by adding up poor children under age 5 in married-couple and single-parent families from rows in the table, there is no ready way to compute a precise standard error of the combined estimate. The reason is that the individual estimates come from the same sample and so are correlated. If an aggregate table were available that provided the desired sum, then the standard error would be available, but there is not such a table for this example. The guidance alludes to this problem but does not explain it.

7-B.2.c Recommendation for Sampling Error Documentation

Given the lack of technical sophistication of many users of the ACS data, the Census Bureau needs to evaluate its presentation of sampling errors to be most helpful to the widest range of users. A helpful first step would be to provide 95 percent margins of error for consistency with commonly accepted survey practice. It would also be helpful to provide margins of error that do not include zero, although this would require a different technique to estimate the standard error and a different format for presenting the information.

An even more ambitious step would be to rethink the presentation of tables on the ACS web site. As suggested in Section A above, the Census Bureau could identify key estimates that meet common standards of precision, such as having a standard error that is 10 percent or less of the estimate.

---

[2]Available by clicking on "Survey Methodology" or "Accuracy of the Data" from any ACS table accessed through the American FactFinder web site (http://www.factfinder.census.gov/).

[3]See, for example, http://factfinder.census.gov/home/en/acs_pums_2005.html.

These quite precise estimates would be highlighted for users who want to know what they can confidently learn from the data and are daunted by the array of table cells with large margins of error.

In addition, the Census Bureau should review its guidance for calculating standard errors for user-constructed estimates of sums and differences. The documentation should provide many more examples for a range of applications to make clear how the guidance can be used. It should also emphasize more strongly when the guidance is not readily applicable.

> **Recommendation 7-6:** The Census Bureau, in consultation with data users and statistical methodologists, should evaluate its presentation of sampling errors of estimates that are published on the ACS web site and also its descriptions of methods for computing approximate estimates of sampling errors for estimates for which sampling errors are not published. Steps should be identified to improve the usability and ease of comprehension of information on sampling errors.

7-C PRIORITIES FOR ASSESSMENT AND IMPROVEMENT OF SURVEY QUALITY

In addition to monitoring basic quality measures, a major continuing survey such as the ACS requires periodic, in-depth assessments of data quality on a wide range of dimensions across time and among population groups and geographic areas. The benefits of such assessments accrue not only to data users, who can gain deeper understanding of the value and challenges of the data, but also to survey managers who require information to help them identify areas for methodological research and subsequent survey improvement.

7-C.1 Quality Profile

A comprehensive survey evaluation is referred to as a quality profile. Such a document brings together and analyzes the magnitudes of and contributions to sampling and nonsampling errors from various survey component processes for estimates from a survey, generally, and, when possible, for specific questionnaire items. A quality profile also typically includes comparisons of selected survey estimates with estimates from other surveys or administrative records. Examples of quality profiles include those developed for the American Housing Survey (Chakrabarty, 1996); the Residential Energy Consumption Survey (Energy Information Administration, 1996); the Schools and Staffing Survey (Kalton et al., 2000); and the Survey of Income and Program Participation (U.S. Census Bureau, 1998).

A quality profile for the ACS would be complex to prepare and require significant time and effort to pull together, analyze, and present information on all of the topics that should be included. Nonetheless, work on an ACS quality profile needs to begin now, building on the evaluation studies that were conducted of the Census 2000 Supplementary Survey and the 2000 long-form sample. The first step is to develop the framework, or outline, for the quality profile. The outline could then be used to plan a research program for assessing specific aspects of the ACS and using the results as the basis of a program of survey improvement. As research findings are accumulated, they can form the basis for chapters of the quality profile. To be most useful not only for users, but also for survey managers, the various chapters should be issued and updated on an ongoing basis. If staff resources are insufficient to manage the profile materials, the Census Bureau could seek outside assistance for the work.

The topics to include in an outline of an ACS quality profile fall under two main categories. The first category includes reports of what is known about sampling and nonsampling errors for estimates of interest, including differences in the magnitude of errors for geographic areas and population groups. The second category includes the results of analyses to determine the sources of various types of error in the estimates, particularly the effects of the various components of the survey design and operations, such as sample design, data collection mode, questionnaire design, weighting, imputation, and others. Results of experiments with alternative methods should also be included.

More specifically, the outline might cover such headings as:

- Sources of nonsampling and sampling errors and their extent and effects:
  - o Sampling frame: completeness, currency, and accuracy of the Master Address File (MAF) for housing units and group quarters in geographic areas; assessments of the quality and usefulness of various MAF updating operations.
  - o Sample design: effects on standard errors of estimates of different initial sampling rates, particularly among states with different numbers and types of small jurisdictions and among similar-sized small jurisdictions; benefits and drawbacks of alternative designs.
  - o Sample design: effects on standard errors of CAPI subsampling rates.
  - o Data collection mode: patterns of response by mode for population groups and geographic areas; effects of mode differences on precision and bias for questionnaire items; correlates of mode differences; results of experiments to reduce mode differences.

- o Questionnaire design and wording: effects on response rates, response variance, and response bias for content items; results of experiments with alternative wording.
- o Residence rules: how respondents and interviewers interpret the 2-month residence rule compared with the decennial census usual residence rule and the effects on population coverage in the ACS.
- o Weighting: effects of each weighting stage on the precision of 1-year period estimates.
- o Population and housing unit controls: accuracy of controls at different levels of geography and for population groups and geographic areas; how their use affects 1-year, 3-year, and 5-year period estimates.
- o Imputation: patterns of item imputation for geographic areas and population groups; the effects of imputation on precision and bias of estimates.
- o Inflation adjustments (if retained): accuracy of methods for inflating income and housing dollar amounts for 1-year, 3-year, and 5-year period estimates for geographic areas and population groups; pros and cons of alternative methods.
- o Confidentiality protection: extent of data suppression to protect confidentiality for geographic areas and population groups; risks and benefits of alternative protection methods.
- o Table collapsing for precision: extent of collapsing for geographic areas and population groups; pros and cons of alternative collapsing schemes.
- o Variance estimation: estimates of the variance not accounted for due to item imputation and other sources.

- Comparability of ACS estimates with other data sources:
  - o Comparability of aggregate estimates for as many content items as possible, taking account of differences between the ACS and the comparison source(s).
  - o Consistency of microlevel data from matching studies of ACS records with records from an administrative system (for example, Food Stamp Program records or Social Security records).
  - o Regression analyses of correlates of differences between the ACS and other sources.
- Regularly repeated, summary assessments of precision (variance) for geographic areas and population groups.
- Regularly repeated, summary assessments of measurement error (bias) for key content items, drawing on all available information.

To measure the magnitudes of various kinds of errors and to analyze sources of error, various methods are available. They include

- aggregate comparisons of ACS estimates with estimates from other surveys or administrative records;
- exploratory, graphical, and regression analyses to identify geographical and other patterns in the data that suggest hypotheses for further analysis (for an example, see National Research Council [2004b:186-193], which reports on graphical and regression analyses of 1990 and 2000 census tract mail return rates by geographic area and population characteristics);
- microlevel matches of individual ACS records with records from other sources (for examples, see Coder, 1991, 1992, which report on exact matches of Internal Revenue Service (IRS) earnings records with the Current Population Survey (CPS) and Survey of Income and Program Participation, respectively);
- reinterviews of samples of ACS respondents (reinterviews are included in the ACS Methods Panels, see Section 7-C.2 below);
- designed experiments using cognitive testing and other structured interview techniques with small samples;
- designed experiments with large samples of households (the ACS methods panels provide examples—see Section 7-C.2 below); and
- sensitivity and other simulation analyses with existing data.

The different methods have advantages and disadvantages in terms of the time and resources required to carry them out, the questionnaire items for which they are feasible, the robustness of their results in terms of sampling and nonsampling errors, and whether they contribute to understanding sources of error and not just the magnitudes of error in the ACS estimates.

In designing an ongoing assessment program for the ACS and selecting priority topics for research in the short term and longer term, the Census Bureau must balance important uses of the data against feasibility and resource constraints. Input from the Census Bureau's network of user education partners should be helpful in this regard. In turn, it will be important for educating users to provide the results of data quality assessments not only in technical reports but also in user-friendly formats. Because users of the ACS will undoubtedly want to know the distributions of data quality assessments across time and among geographic areas and population groups and not simply U.S. or state totals, Census Bureau analysts will need to become facile with modern graphical analysis tools and exploratory data analysis techniques. The Census Bureau historically has not made much use of these methods, but they are essential for identifying and displaying temporal, spatial, and demographic patterns of interest from a data set as large as the ACS (see, e.g., National Research Council, 2001:App. B). In turn, the ability to more readily identify data quality patterns should facilitate planning for in-depth research and evaluation to identify ways to improve the ACS.

Recommendation 7-7: The Census Bureau should develop and publish an ongoing quality profile for the ACS to inform users of the survey's data quality, to guide the development of a continuing program of data quality assessments, and to identify areas for survey improvement. The Census Bureau should seek input from users on priority topics for assessment and design reports that they would find to be useful additions to the technical reports.

BOX 7-2
2006 and 2007 American Community Survey Methods Panels

"Methods panel" is a term used by the Census Bureau to refer to samples of households that are used for testing and experimentation for a continuing household survey. For the ACS, the Census Bureau fielded a 2006 methods panel (see *Federal Register*, vol. 70, no. 45, March 9, 2005: 11609-11610). It is planning to field a 2007 methods panel later in the year (see *Federal Register*, vol. 71, no. 94, Tuesday, May 16, 2006:28302-28305).

2006 ACS Methods Panel

The 2006 ACS Methods Panel (also known as the 2006 ACS Content Test) was designed to test new questionnaire content to be considered for inclusion in the ACS in 2008 and modification of existing content to improve response. The test included four stages:

1. Determination, with input from federal agency stakeholders, of eligible content for the test.
2. Cognitive laboratory pretesting, expert reviews, and other methods to develop alternative versions of the eligible questions. Eleven of 25 existing housing questions, 15 of existing population questions, and 3 new population questions were identified for inclusion in stage 3.
3. National sample field test of about 50,000 housing unit addresses. About half the sample served as the control panel, receiving the existing ACS questionnaire; the other half served as the test panel, receiving alternative versions of the questionnaire. Mailed out to all sample addresses were advance letters, questionnaires, and reminder postcards, followed by second questionnaires to

7-C.2 Methods Panels

The Census Bureau recently began a program to field large samples of households, called methods panels, as the vehicle for large-scale experimentation with features of the ACS. The 2006 Methods Panel included 50,000 households and was used to test alternative wording for existing and new questions. A 2007 Methods Panel, which is to include almost 70,000 households, is planned to test not only question wording and questionnaire format, but also strategies to improve mail response (see Box 7-2).

The Census Bureau is to be commended for initiating the ACS methods panels. The program should be continued because of the continuing need for large-scale experimentation on questionnaire format, question wording, instructions for reporting residence, the effects of data collection mode, and other aspects of the ACS data collection. The need for continuing large-scale experimentation exists because federal data requirements from the ACS can be expected to evolve over time, as socioeconomic conditions and concerns change. Also, respondent behavior may change in ways that affect

---

nonrespondents. After 4 weeks, nonrespondents were followed up by CATI; 4 weeks later, remaining nonrespondents were followed up by CAPI. There was no telephone questionnaire assistance or telephone edit follow-up, which could have influenced respondents' answers. After data collection, a subsample of mail, CATI, and CAPI respondents who furnished a telephone number were followed up by CATI to measure simple response variability and response bias by comparing answers from the first interview (by mail, CATI, or CAPI) and the second CATI interview.

4. Analysis of results and recommendations for new and revised content for the ACS beginning in 2008—expected in early 2007.

2007 ACS Methods Panel

The 2007 ACS Methods Panel is designed with two tracks:

1. The first track will address new and improved content, including a new question on field of bachelor's degree and a modified format for the basic demographic questions (age, sex, race, ethnicity, household relationship). Four different questionnaires will be mailed to a total of 30,000 housing units, with CATI and CAPI follow-up and a CATI content reinterview.
2. The second track will address ways to increase mail response and thereby contain costs. One strategy for testing is to make another mailing to nonrespondents for which a telephone number is lacking (three different mailing pieces will be sent to 6,000 housing units each). Another strategy for Puerto Rico and targeted areas of the United States with the lowest levels of mail response is to include a motivational piece in the questionnaire package. Two different mailing pieces will be sent to 10,000 housing units each in the targeted areas in the United States.

data quality and costs (for example, mail response could decline), which would require testing of new ways to improve response.

The Census Bureau should carefully evaluate its experience with the 2006 ACS Methods Panel with regard to costs and statistical power for the intended analyses. It may be that some testing can be done with fewer sample cases.

> **Recommendation 7-8:** The Census Bureau should continue to seek funding with which to implement methods panels (large samples of households) for experimentation with questionnaire design, question wording, residence rules, data collection mode, and other features of the ACS. The methods panels should be conducted annually so that the survey can be kept current in meeting data needs and collecting responses in the most efficient and effective ways.

7-C.3 The Panel's Priorities for Assessment

Even with the significant resources that the panel believes should be provided for ACS research and evaluation (see Recommendations 7-1 and 7-8), the program cannot investigate every aspect of this detailed, complex survey and certainly not on the same time schedule. It is important to establish priorities in consultation with methodologists and data users.

In Chapters 4, 5, and 6, the panel identified areas for research and evaluation. The panel's complete list covers many aspects of ACS data collection, processing, estimation, and data. Acknowledging the need for prioritization, the needed research and evaluation topics are grouped into two categories below: high priority and other. Note that the priority categorization does not necessarily imply a time frame in which the research should be completed. Some high-priority analyses will require extended work, while others can be more quickly completed. Some analyses may be one-time efforts; other will need to be repeated on a continuing basis.

Many high-priority analyses are not costly in that they do not involve field data collection, or the costs can be shared with other programs in the Census Bureau. The panel recognizes, however, that Census Bureau analysts have many responsibilities, and the panel encourages the Bureau to augment its staff resources to the extent possible through fellowships, internships, and other collaborative arrangements with outside researchers.

7-C.3.a High-Priority Areas for ACS Research and Evaluation

The panel has identified seven areas as high priority for evaluation, followed by research and development to improve the ACS on the basis of the results: sample size and allocation; the MAF; population controls;

residence rules; estimates of change; comparisons with other surveys and administrative records; and the development of automated tools for data quality review. Each of these areas is important not only for the usefulness of the ACS, but also for its credibility with users as a satisfactory replacement for the rich (but outdated) small-area information that was previously provided by the census long-form sample. Failure to address these seven topics could harm the quality of the ACS data and make it difficult for users to adapt their long-form-sample applications to this new survey with its continuous design.

Sample Size and Allocation A critically important issue for assessment, which requires a combination of research, consultation with users, and consideration of budget resources, is the ACS sample size and its allocation across the various governmental units (see Recommendation 4-4). The panel is concerned about the much larger sampling errors of ACS estimates compared with long-form-sample estimates, particularly for estimates for small governmental units, which, unlike census tracts and block groups, do not lend themselves to combination into larger areas. It seems imperative to develop strategies for improving the precision of the ACS estimates. The costs and benefits of alternative approaches can be evaluated using low-cost simulation methods; no new data collection will be required. Whether a solution can be found that is acceptable to users and to Congress (for funding) is not clear, but the effort to explore alternatives, including trade-offs (for example, perhaps giving up school district oversampling to increase the sample for other small jurisdictions) should be made. At a minimum, users should be fully informed of the trade-offs and the implications of alternative approaches for a range of applications. They should also be given specific guidance on strategies for increasing the precision of estimates by collapsing categories and combining estimates over time and across geographic areas.

Master Address File Research to evaluate and improve the MAF is critical for the completeness and accuracy not only of the 2010 census, but also of the ACS. Errors in the MAF can lead to omission of households, duplication of households, and assigning households to incorrect geographic locations. MAF research and development can be costly in that it often involves field work to identify problems and evaluate alternative approaches for improvement. Consequently, it may not be feasible to carry out much MAF research in the immediate future that is not part of the 2010 census planning.

Major work on the 2010 census MAF will not begin until late in the decade, when a complete block canvass and local review are conducted. However, beginning with the 2005 ACS, systematic examination of the differences between the housing unit controls that are used for the ACS and the MAF could contribute importantly to MAF evaluation and improvement in its coverage (see Reese, 2007). In particular, identification of large differences, positive and negative, could provide the basis for targeted field evaluations to determine reasons for discrepancies and suggest methods to improve the MAF in areas with particular kinds of address problems, such as small multiunit structures (see Recommendations 4-1, 4-2, 4-3, and 5-2). Ideally, research and development on the MAF would proceed on a continuous basis after 2010 so that the ACS MAF is as kept as up to date and accurate as possible.

Population Controls Another critically important area for assessment is the accuracy and application of the population control adjustments to the survey weights. The adjustments may adversely affect the accuracy of estimates for some kinds of areas, such as those experiencing seasonal population fluctuations or rapid population growth or decline. They also will not capture differential rates of population growth in small areas within estimation areas (large counties and groups of small counties). ACS estimates produced with population controls for a census year will likely differ—sometimes substantially—from the census counts for many areas, producing discontinuities in time series of ACS estimates.

Full evaluation of the current procedures for producing the controls, as well as of alternative procedures that are under development (see Section 7-D.4 below and Recommendation 5-3), requires 2010 census counts for comparative assessment. However, work can proceed now to design the evaluation program. Moreover, it could be helpful to conduct more extensive analyses that compare the 1999 population estimates with the 2000 census counts (see Section 5-D). In addition, analysis should be conducted, beginning with the 2005 ACS data, of how much difference the controls make to the ACS survey weights and to identify systematic patterns of large upward and downward adjustments that merit investigation. Also, research should be conducted to assess the effects of errors in the population controls on ACS estimates of characteristics, and users should be made aware of the results.

Evaluation of the population controls requires research that should be low cost, although given the many responsibilities of Census Bureau staff, the Bureau may want to arrange for outside researchers to work collaboratively with Bureau analysts. Additional resources will be required for work to improve the methods for producing the population (and housing) controls on the basis of evaluation results and to implement new methods on a production basis. However, the costs can be spread over several Census Bureau programs, not just the ACS, given the many uses of the population estimates.

Residence Rules For many purposes, including comparisons with the 2010 census and with the annual population estimates, it is critical to conduct research to understand the implementation of the 2-month residence concept in the ACS and its effects on estimates for geographic areas and population groups. Experiments should be included in the ACS methods panels to determine how respondents interpret the 2-month residence rule in deciding whom to include and not include on the questionnaire and how their responses differ when they are asked to apply the census usual residence rule (see Recommendation 4-6). Such research could identify needed changes to question wording and instructions for reporting residence that would make reporting more consistent with the rules. The Census Bureau plans—and the panel supports—a program of annual methods panels, so that there should be little additional cost of the recommended research.

Estimates of Change A major focus for many data users in using the ACS is to examine estimates of change—from the preceding year, from the last census—for geographic areas and population groups of interest. The ACS provides successive 1-year and (once the necessary data are accumulated) 3-year and 5-year period estimates, but not direct estimates of change. As discussed in Chapters 3 and 6, using period estimates to track trends over time, particularly the 3-year and 5-year estimates, is not straightforward and the interpretation may often be unclear. Users will need specific, detailed guidance on how to work with the period estimates for time-trend analyses if they are not to be frustrated in their use of the ACS.

Comparisons with Other Data Sources It is important that the Census Bureau periodically compare selected ACS estimates with the corresponding estimates from other surveys and administrative records—for example, comparing ACS estimates of income and employment with those from the CPS and the IRS Statistics of Income, or comparing ACS estimates of housing characteristics with those from the American Housing Survey and administrative records. The Census Bureau established a precedent for this kind of work when it performed a large number of aggregate comparisons between estimates from the Census 2000 Supplementary Survey and the 2000 long-form sample; these comparisons helped establish the validity of the ACS (see Section 2-B).

It is often difficult to develop valid comparisons given that data sources differ in details of definitions, data collection operations, and other features. Moreover, analysts cannot assume that a particular comparison source is a gold standard of truth, as all data sets contain errors. Nonetheless, when well executed, aggregate comparisons can document differences in estimates and suggest reasons for differences. In turn, these findings can stimulate further research on which data source—the ACS or another—appears to be more accurate and ways to improve the ACS, the other source, or both. In addition, such comparisons are important to establish face validity of the ACS for users who have long relied on other data sources.

Automated Tools and Standardized, Well-documented Procedures for Data Quality Review While the Census Bureau has made strides in this area, it should conduct further testing and implementation of tools and procedures that can facilitate careful and timely review of the quality of ACS estimates by Bureau analysts (see Recommendation 4-14). When multiple estimates—1-year, 3-year, and 5-year period estimates for geographic areas and population groups—begin to pour out of the data collection and processing system (beginning in 2008 for 3-year period estimates and 2010 for 5-year period estimates), the Census Bureau must be in a position to cope with them. Users will expect the Census Bureau to keep to its announced schedule of releasing all estimates within 8–10 months of the end of data collection and, at the same time, to minimize obvious errors in the estimates (for example, assigning a group quarters to an incorrect geographic location or misaligning the decimal place in coding income). Having the best automated tools and documented procedures possible will be essential to enable the Census Bureau's analysts to do a good job of data quality review under tight time schedules and constrained staff resources.

7-C.3.b Other Areas for ACS Research and Evaluation

In addition to the seven high-priority topics discussed in Section 7-C.3.a above, the panel believes that six other areas are important to include in the ACS research program. Work in these areas should move forward to the extent that resources permit.

Four of the six areas involve research and consultation with users that, if possible, would be useful to complete in time to make decisions on whether to change certain features of the ACS 3-year and 5-year period data products before these products are first released. The required research in each of these four areas could be largely based on low-cost simulations of the advantages and disadvantages of alternative approaches:

- Determination of the universe for the survey—specifically, whether to drop some or all group quarters from the ACS to save resources, and, if some or all group quarters are retained, which tables to present for the total population, household population, and group quarters population to be most useful for users (see Recommendations 4-7 and 4-12).
- Refinement of confidentiality protection procedures for 3-year and 5-year period estimates to recognize the protection afforded by

averaging over 36 and 60 months of data and consideration of including month of interview in the PUMS (see Recommendations 4-8 and 4-9).

- Assessment of the inflation adjustments for 1-year, 3-year, and 5-year period income and housing value and cost estimates to determine if the current procedures best serve the needs of users and the costs and benefits of alternative procedures, including no adjustments at all (see Recommendation 4-11). In addition, guidance should be developed to help users interpret ACS dollar estimates with the current inflation adjustment procedures, and provisions should be made to provide unadjusted estimates to users who need them them.
- Determination of geographic areas for publication (see Recommendation 4-13): Does it makes sense—considering user needs, feasibility, and effects on precision of estimates—to reduce the population threshold for 1-year period estimates from 65,000 to 50,000 and to develop and publish 3-year (and possibly 1-year) period estimates for components of PUMAs?

The other two areas that would benefit from research involve data collection modes and weighting adjustments:

- Experimentation on the response effects of the different data collection modes used in the ACS—mailout-mailback, CATI, and CAPI (see Recommendation 4-5). This topic is important because of the large proportion of responses that the ACS obtains from CAPI or CATI and not the original mailout mode and the likelihood that mode of collection differentially affects responses. Mode effect experiments could be included in an ACS methods panel.
- Assessment of the effects of the various steps in the weighting process for producing 1-year period estimates (that is, the steps other than the housing unit and population controls) (see Recommendation 5-1). Although not as important as research on the population and housing controls, analysis of the other weighting steps could be useful to identify possible ways to simplify the process and modify one or more steps to improve the precision and accuracy of the ACS estimates.

The quality profile outline provided in Section 7-C.1 above lists other topics for research and evaluation in addition to those the panel specifically addressed. Although these topics were not singled out by the panel, they should not be lost sight of when the Census Bureau is allocating research resources. In particular, two related topics that warrant investigation whenever resources become available are methods to impute missing data responses and methods to include the variability from item imputation in addition to sampling error in estimating the standard errors of ACS estimates. Item nonresponse is less of a problem than it was in the 2000 long-form sample, but the effects of the imputation procedures should still be investigated. The Census Bureau should also investigate the utility of using more sophisticated imputation methods than those currently being employed. It should evaluate alternative methods for including the variability from item imputation in the estimates of the sampling errors for estimates from the ACS (see Bell [2006] for research on including imputation in variances for estimates from the 2000 census).

Finally, it will be important for the Census Bureau to have a process for periodically reviewing its research and evaluation priorities and adjusting them as appropriate. It may be that an area thought to be of pressing concern appears less so upon initial investigation, whereas an area that was not high priority to begin with becomes of increasing concern for uses of the data. Close consultation with users and monitoring of ACS data quality will help the Census Bureau keep its research and evaluation program on track.

The Census Bureau will also need to periodically reevaluate its research priorities in light of available funding and staff. The Census Bureau should plan its research and evaluation program from the beginning to involve both intramural projects by its own staff and extramural work by outside researchers. In this way, it can better ensure that there are always highly qualified researchers actively assessing the ACS even if in-house staff are pulled away on production and other priorities.

7-C.3.c Recommendation for Research Priorities

> **Recommendation 7-9:** The Census Bureau should assign priority to the following topics for research and development: sample size and allocation; the MAF; population controls; residence rules; estimates of change with multiyear averages; comparisons with other surveys and administrative records; and the development of automated tools for data quality review of ACS products.

7-D A VISION FOR THE FUTURE

At the present time, the ACS is viewed by the Census Bureau and data users primarily as a replacement for the long-form sample. While the panel agrees with that thrust in the short term, neither the Census Bureau nor the user community should lose sight of the vast potential for the ACS to contribute to new and improved measurement over the longer term. The continuous design of the ACS, which may initially challenge users to adapt their long-form-sample-based applications to the new data, provides the platform for developing important innovative applications for the future.

There are at least five ways in which the ACS could contribute to new and improved measurement. They involve (1) more timely and accurate measures of key indicators for small geographic areas by combining information from the ACS, other surveys, and administrative records; (2) measures of seasonal population fluctuations and multiple residences; (3) cost-effective, up-to-date data collection for rare populations; (4) improved population estimates; and (5) improved estimates from other household surveys (other surveys may also help improve the ACS).

7-D.1 Small-Area Estimates

The planned ACS estimates for geographic areas involve accumulating and averaging 12, 36, and 60 months of data, depending on population size. For small counties, cities, and other areas for which 3-year or 5-year period estimates are provided, many users would very likely prefer continuously updated 1-year estimates for the latest year rather than estimates that represent an average over a longer period of time.

Modern small-area estimation methods that borrow information across time, geography, and data sources could be used to develop indirect 1-year period estimates for key indicators, such as poverty, unemployment, food stamp participation, and others, for all counties and cities (not just those with fewer than 65,000 people). Statistical models could use data from the ACS and relevant administrative records to generate indirect estimates that would likely improve on the direct ACS estimates in precision, accuracy, and currency. Depending on the availability of administrative records, the indirect estimates might lag the latest release of the period estimates, although models could possibly be developed to project the indirect estimates forward 1 or 2 years to represent the latest year.

Small-area estimation models that use the ACS could also incorporate estimates from other surveys when those surveys are believed to provide estimates of higher quality than the ACS estimates. For example, the Current Population Survey (CPS) very likely provides more accurate measures of labor force, employment, and unemployment status than the ACS (see Section 2-B.2.e). The CPS includes a more detailed set of questions and has other design features, such as a fixed reference week for measurement, to reduce nonsampling error.[4] Although the CPS sample size, even when accumulated for 12 months, does not support precise estimates for subnational

---

[4]See http://www.bls.gov/cps/cps_over.htm#overview.

areas except for a few large states, it could play an important role in small-area model-based estimates from the ACS by providing controls so that the ACS estimates reflect the best available national and regional estimates.

A substantial amount of work needs to be carried out to make indirect estimation a reality for the ACS. The Census Bureau has already taken some important initial steps (see Bell, 2006; Chand and Alexander, 1997; Huang and Bell, 2005). In addition, the Census Bureau's Small Area Income and Poverty Estimates (SAIPE) models and its Small Area Health Insurance Estimates (SAHIE) models are closely related to the models that might be worthwhile to develop for the ACS, as are the models used for the Bureau of Labor Statistics' Local Area Unemployment Statistics (LAUS) program.[5]

The SAIPE models of poverty and median income for states, counties, and school districts currently use data from the CPS Annual Social and Economic Supplement (CPS ASEC), federal income tax records, food stamp records, the 2000 census long-form sample, and census-based population estimates. Census Bureau researchers have conducted work on the potential for school lunch program records, earned income tax credit records, and Medicaid records to improve the SAIPE models. The SAHIE models of health insurance coverage for states and counties currently use data from the CPS ASEC, federal income tax records, food stamp records, Medicaid records, and census-based population estimates.

The LAUS models of employment and unemployment for states and a few other large areas currently use data from the monthly CPS (current and historical estimates); the monthly Current Employment Statistics (CES) program, which surveys a large number of nonfarm business establishments; and state unemployment insurance (UI) records. The LAUS estimates for smaller areas, such as counties and cities, are constructed through a building-block approach that uses data from the CPS, the CES program, state UI systems, and the 2000 census long-form sample.

Presumably, the inclusion of the ACS in all of these models, which are designed to improve the CPS estimates, could result in small-area estimates that are more precise than the current model-based estimates. As noted above, models could also be developed to improve the ACS direct estimates by producing more precise small-area estimates that represent a current (or recent) time period instead of averages over a longer time period.

Three caveats are in order. First, it is not clear how strong a predictive model can be developed that would improve on the ACS period estimates for many of the characteristics of interest. Second, the effort required to

[5]For SAIPE, see National Research Council, 2000a, 2000b; http://www.census.gov/hhes/www/saipe/saipe.html; for SAHIE, see http://www.census.gov/hhes/www/sahie/index.html; for LAUS, see http://www.bls.gov/lau/lauov.htm.

generate a set of indirect estimates for one characteristic, say, the poverty rate, may not provide much information for the development of indirect estimates for another characteristic, say, employment. Third, if a multivariate approach is used to exploit the correlations that exist among ACS estimates, the complexity of the modeling task is greatly increased. Consequently, a program to develop a large number of indirect estimates would take substantial time and resources. Yet the payoffs could be great from selected indirect estimates that are continuously updated for such purposes as fund allocation.

7-D.2 Seasonal and Multiple Residences

The long-form sample could not provide information on seasonal fluctuations in population, which characterize many localities, because it was conducted at a point in time and asked only about the location of the respondent's usual residence. In contrast, the ACS is conducted continuously and asks respondents to employ a 2-month residence rule. The current data processing and estimation system for the ACS ignores the month-by-month information, producing instead period estimates for 1, 3, and 5 years that are controlled to census-based population and housing unit estimates as of July 1 of a specific year. However, the Census Bureau's use of monthly data to produce pre– and post–Hurricane Katrina and Rita profiles for affected areas in the Gulf Coast demonstrates that it could be not only feasible, but also very valuable to produce such profiles for other areas.

To investigate the feasibility of producing part-year data for specified areas on a regular basis, the Census Bureau should conduct research on the extent to which the ACS monthly data exhibit significant seasonal variations in total population and key characteristics for localities expected to have such variations. It would be important to inform this analysis from the results of the test recommended by this panel and the Panel on Residence Rules in the Decennial Census on how respondents record their residence using the ACS 2-month rule compared with the census usual residence rule. This test may identify responses that do not accord with the 2-month rule that can be ameliorated by changes in question wording and instructions for the ACS.

The outcome of research on seasonal residence could be special data products for areas that have significant seasonal fluctuations, which would represent a major addition to the stock of useful information for them. One problem concerns sample size, given that seasonal change may be evident only for small areas. To the extent that seasonal patterns tend to be repeated each year, it would be possible, and likely essential, to combine multiple years of data in order to produce sufficiently precise estimates of part-year populations for affected areas.

In addition to estimates of seasonal population fluctuations, the ACS could be a vehicle for information about multiple residences more generally—for example, people with weekday and weekend homes or students away at college or boarding school. Questions that may be needed to improve reporting of residence using the 2-month rule, such as whether a household member has another residence, could also provide useful information on multiple residences. Such information would be valuable not only for planning and research, but also for designing coverage improvement programs for the decennial census.

7-D.3 Surveying Rare Populations

The census long-form sample has historically provided the basis for follow-up surveys for specific, relatively small, or "rare," populations, such as scientists and engineers and low-income minorities. By using the long-form sample to identify a population of interest for follow-up after the census, targeted postcensus surveys could be more cost-effective than nontargeted stand-alone surveys, which require much larger sample sizes to capture enough cases of the rare population of interest.

The ACS can similarly provide the basis for sampling a small targeted population by serving as the initial screener to identify specific households or persons for interview. (ACS data can also be used to identify areas with a higher percentage of the target population for selecting a sample, using more current data than the long-form sample.) The ACS has the advantage that it can be used for this purpose more often than once a decade, although care will need to be taken to minimize respondent burden and provide for informed consent for any follow-on survey.

There is a procedure for identifying and testing new questions to be included in the ACS, which could potentially expand its use as a screener. For example, a question on field of bachelor's degree is planned for testing in the 2007 Methods Panel. If the question is added to the ACS, it will be used to target a sample of people in science and engineering fields to support the work of the National Science Foundation. Of course, there is a limit on how many questions can be added to the ACS without an adverse effect on response rates and public perception of the survey, unless some questions can be identified for deletion. Moreover, all ACS questions are mandatory, which makes it incumbent on the Census Bureau to consider the response burden of any new questions very carefully.

7-D.4 Improving Population Estimates

There is a pressing need for the Census Bureau to conduct research on methods to improve the estimates of population by age, sex, race, and ethnicity that are used as controls for the ACS and serve so many other important purposes, such as providing factors for fund allocation formulas, controls for other household surveys, and denominators for vital rates. Information from the ACS on place of birth, citizenship, and year of immigration is already used to generate estimates of net migration from abroad for the population estimates program, and the Census Bureau is interested in examining other components of the estimates that might benefit from ACS information. For example, ACS estimates might supplement IRS tax records to estimate internal migration at the county level and perhaps for smaller geographies.

The ACS could also possibly improve the population estimates and its own coverage of population and housing through linkages with the Census Bureau's E-StARS program (see Section 4-A.4). The E-StARS Master Address Auxiliary File could be used to improve the MAF, which would in turn improve the ACS coverage of housing units. (At present, the MAF provides input to E-StARS, but there is no feedback loop back to the MAF.) Going a next step, ACS estimates of occupancy rates and persons per household could possibly be used with an improved MAF count to generate an alternative set of population estimates to compare with the estimates that are produced from the current component method (see Section 5-C). Yet another approach is to use E-StARS to provide population controls for subcounty areas within the framework of the existing population estimates. The Census Bureau has begun work along these lines, which should be pursued.

Critical to making progress toward improved population estimates is for the Census Bureau to design and conduct an extensive evaluation program of alternative estimation methods and data sources in conjunction with the 2010 census. In planning and evaluating its research, the Census Bureau should involve knowledgeable users and producers of population estimates, such as the members of the Federal State Cooperative Program for Population Estimates.

7-D.5 Improving Survey Estimates

Most items on the ACS questionnaire are covered in other household surveys, often in much more detail. For example, as noted in Section 7-D.1 above, the monthly CPS, which provides the nation's official measure of unemployment, includes additional questions about work status beyond those used in the ACS to determine each respondent's labor force situation. Other surveys that overlap with the ACS include the American Housing Survey, the CPS Annual Social and Economic Supplement, the Survey of Income and Program Participation, the National Health Interview Survey, and the National Household Travel Survey. These other surveys not only obtain extensive information about their primary topic, but also typically include a large number of additional variables for use in analysis. However, they rarely provide state, or substate, estimates.

A critical question for the future of federal statistics is in what ways the ACS can contribute to and in what ways it can borrow strength from the other major national household surveys. At this early stage of implementation of the ACS, it would be foolish to think about dropping or curtailing another survey because its content overlaps with the ACS. Instead, what is needed is in-depth research to compare estimates, determine the strengths and weaknesses of each, and develop methods to improve both the ACS and other surveys. Each will undoubtedly continue to have an important role to play—the ACS primarily by providing small-area estimates and other household surveys primarily by supporting rich, multivariate policy analysis and basic social science research. The challenge will be to integrate the ACS and other surveys in ways that strengthen them all.

One way in which the ACS can help other household surveys involves the MAF sampling frame. Assuming that the advent of the ACS will lead to continuous updating and improvement of the MAF (see Section 4-A.4), it should be possible to update the sampling frames for other surveys more than once a decade. Indeed, the Census Bureau plans to adopt the MAF as the sampling frame for its other household surveys. In addition, responses to the ACS could be used to identify population groups of interest for oversampling in other surveys.

With regard to improved estimates for overlapping content items, the ACS could likely help other surveys—and vice versa—in several ways. For example, if research establishes that ACS estimates of a particular item are comparable with those for that item in another survey, the ACS could provide valuable controls for the other survey. But if research establishes that the ACS estimates are less accurate than those from another survey, estimates from the other survey might be used to calibrate the ACS estimates for small subgroups in some simple model-based way.

For key items of national importance, it might become possible to use the ACS, other surveys, and administrative records to develop the best estimates for the nation, states, and, possibly, substate areas. These estimates could be published as independent time series, similar to the estimates of gross domestic product, which draw on many data sources.

7-D.6 Recommendation for Future Research and Development

Recommendation 7-10: As part of its research and development program for the ACS, the Census Bureau should dedicate a portion of resources to pursue innovative, longer term projects. While short-term research and development must focus on the ACS as a replacement for the census long-form sample, research must also address how the ACS can improve the nation's information on population and housing in ways that were not possible with the long-form sample and may not even be envisioned today.

References

Alexander, C.H. (1993). *A continuous measurement alternative for the U.S. census.* Report CM-10, October 28. U.S. Census Bureau, U.S. Department of Commerce, Washington, DC.

______ (1997). Impact of multiyear averaging of data from the American Community Survey. Pp. 644-649 in *Proceedings of the Joint Statistical Meetings, ASA Section on Survey Research Methods.* Alexandria, VA: American Statistical Association.

______ (1998, August). *Recent developments in the American Community Survey.* Paper prepared for the Joint Statistical Meetings, Dallas. U.S. Census Bureau, U.S. Department of Commerce, Washington, DC.

______ (2001). Still rolling: Leslie Kish's "rolling samples" and the American Community Survey. In *Proceedings of Statistics Canada Symposium 2001—Achieving data quality in a statistical agency: A methodological perspective.* Ottawa: Statistics Canada.

Anderson, M.J. (2000). *Encyclopedia of the U.S. census.* Washington, DC: CQ Press.

Asiala, M.E. (2004). *Specifications for selecting the ACS 2005 main HU sample (ACS-S-40).* Draft memorandum to L. McGinn from R. Singh. U.S. Census Bureau, U.S. Department of Commerce, Washington, DC.

______ (2005). *American Community Survey research report: Differential sub-sampling in the computer assisted personal interview sample selection in areas of low cooperation rates.* Memorandum to R. Singh from D. Hubble. U.S. Census Bureau, U.S. Department of Commerce, Washington, DC.

______ (2007, February 21). *ACS weighting and estimation research.* PowerPoint presentation. U.S. Census Bureau, U.S. Department of Commerce, Washington, DC.

Bell, W.R. (2006). *Using statistical models to combine CPS and ACS income and poverty estimates with other data (administrative records).* Draft paper, May 8. Prepared for the Workshop Exploring Possible Development of One Set of Integrated Income and Poverty Estimates, May 30-31. U.S. Census Bureau, U.S. Department of Commerce, Washington, DC.

Biemer, P.P., and Lyberg, L.E. (2003). *Introduction to survey quality.* Hoboken, NJ: John Wiley and Sons, Inc.

Blumerman, L.M., and Simon, C.R. (2006). *Subnational population estimates for the 1990s: An evaluation.* Draft paper. Population Division, U.S. Bureau of the Census, U.S. Department of Commerce, Washington, DC.

Bureau of Labor Statistics and U.S. Census Bureau (2002, March). *Current Population Survey design and methodology.* Technical Paper 63RV. Washington, DC: U.S. Department of Labor and U.S. Department of Commerce. Available: http://www.bls.census.gov/cps/tp/tp63.htm.

Bureau of Transportation Statistics (1996, April). *Implications of continuous measurement for the uses of census data in transportation planning.* Washington, DC: U.S. Department of Transportation.

Chakrabarty, R., assisted by G. Torres (1996). *The American Housing Survey—A quality profile.* Current Housing Reports, H121/95-1. Office of Policy Development and Research, U.S. Department of Housing and Urban Development and U.S. Census Bureau. Washington, DC: U.S. Department of Commerce.

Chand, N., and Alexander, C.H. (1997). *Small area estimation with administrative records and continuous measurement.* Paper prepared for the Joint Statistical Meetings, Anaheim, CA. U.S. Census Bureau, U.S. Department of Commerce, Washington, DC.

Citro, C.F. (2000a). Federal agency uses of census data. Pp. 216-217 in M.J. Anderson (Ed.), *Encyclopedia of the U.S. census.* Washington, DC: CQ Press.

_____ (2000b). Long form. Pp. 272-277 in M.J. Anderson (Ed.), *Encyclopedia of the U.S. Census.* Washington, DC: CQ Press.

_____ (2000c). Sampling for content. Pp. 323-325 in M.J. Anderson (Ed.), *Encyclopedia of the U.S. census.* Washington, DC: CQ Press.

Coder, J. (1991). *Exploring nonsampling errors in the wage and salary income data from the March Current Population Survey.* Paper prepared for the Allied Social Sciences Association/Society of Government Economists meeting, Washington, DC. U.S. Census Bureau, U.S. Department of Commerce, Washington, DC.

_____ (1992). Using administrative record information to evaluate the quality of the income data collected in the Survey of Income and Program Participation. In *Proceedings of Statistics Canada Symposium 92—Design and analysis of longitudinal surveys.* Ottawa: Statistics Canada.

Dean, J., and Peterson, A. (2005). Updating the Master Address File: Analysis of adding addresses via the Community Address Updating System. In *Proceedings of the Joint Statistical Meetings, ASA Section on Government Statistics.* Alexandria, VA: American Statistical Association.

de Leeuw, E.D. (2005). To mix or not to mix data collection modes in surveys. *Journal of Official Statistics, 21*(2), 233-255.

Desplanques, G. (2003). *The new census.* Paris, France: National Institute of Statistics and Economic Studies.

Devine, J.E., and Coleman, C.D. (2003). *People might move but housing units don't: An evaluation of the state and county housing unit estimates.* Population Division Working Paper Series No. 71. U.S. Census Bureau, U.S. Department of Commerce, Washington, DC.

Dillman, D. (2000). *Mail and Internet surveys: The tailored design method,* 2nd edition. New York: John Wiley and Sons, Inc.

Energy Information Administration (1996). *Residential energy consumption survey quality profile.* Prepared by T.B. Jabine. Washington, DC: U.S. Department of Energy.

Fay, R.E. (2005). Model-assisted estimation for the American Community Survey. Pp. 3016-3023 in *Proceedings of the Joint Statistical Meetings, ASA Section on Survey Research Methods.* Alexandria, VA: American Statistical Association.

_____ (2006). Using administrative records with model-assisted estimation for the American Community Survey. In *Proceedings of the Joint Statistical Meetings*. Alexandria, VA: American Statistical Association.

Groves, R., Fowler, F., Couper, M., Lepkowski, J., Singer, E., and Tourangeau, R. (2004). *Survey methodology*. Hoboken, NJ: John Wiley and Sons, Inc.

Hansen, M.H., and Hurwitz, W.N. (1946). The problem of nonresponse in sample surveys. *Journal of the American Statistical Association*, *41*, 517-529.

Hefter, S.P. (2005a, June 16). *Specifications for selecting the main and supplemental housing unit address samples for the American Community Survey*. 2006 American Community Survey Sampling Memorandum Series #ACS06-S-3, draft memorandum to L. McGinn from R.P. Singh. U.S. Census Bureau, U.S. Department of Commerce, Washington, DC.

_____ (2005b, May 19). *Specifications for selecting the American Community Survey group quarters sample*. 2006 American Community Survey Sampling Memorandum Series #ACS06-S-6, draft memorandum to L. McGinn from R.P. Singh. U.S. Census Bureau, U.S. Department of Commerce, Washington, DC.

Horvitz, D.G. (1986, May 1). *Statement to the Subcommittee on Census and Population, Committee on Post Office and Civil Service, U.S. House of Representatives*. Research Triangle Park, NC: Research Triangle Institute.

Huang, E.T., and Bell, W.R. (2005). *An empirical study on using ACS supplementary survey data in SAIPE state poverty models*. Working paper. Statistical Research Division, U.S. Census Bureau, U.S. Department of Commerce, Washington, DC.

Israel, S. (2006). *Disability in the San Francisco Bay area: Comparing data between Census 2000 and the American Community Survey 2000-2004*. Working paper. Planning Section, Metropolitan Transportation Commission, Oakland, CA.

Kalton, G., Levine, D.B., Waksberg, J., and Helmick, J. (1998, June 29). *The American Community Survey—The quality of rural data: Report of a conference*. Rockville, MD: Westat.

Kalton, G., Winglee, M., Krawchuk, S., and Levine, D. (2000). *Quality profile for SASS rounds 1-3: 1098-1005: Aspects of the quality of data in the Schools and Staffing Surveys (SASS)*. Washington, DC: U.S. Department of Education, National Center for Education Statistics.

Kish, L. (1992). Weighting for unequal P_i. *Journal of Official Statistics*, *8*, 183-200.

Kish, L., and Hess, I. (1959). A "replacement" procedure for reducing the bias of nonresponse. *The American Statistician*, *49*, 520-538.

Maples, J.J. (2004, August). *Investigating the use of IRS tax data in the SAIPE school district poverty estimates*. Presentation to the Joint Statistical Meetings. U.S. Census Bureau, U.S. Department of Commerce, Washington, DC.

Martin, E., and Gerber, E. (2003, August 5). *Methodological influences on comparability of race measurements: several cautionary examples*. Paper prepared for the Joint Statistical Meetings, San Francisco. U.S. Bureau of the Census, U.S. Department of Commerce, Washington, DC.

National Research Council (1995). *Modernizing the U.S. census*. Panel on Census Requirements in the Year 2000 and Beyond, B. Edmonston and C. Schultze, eds. Committee on National Statistics. Commission on Behavioral and Social Sciences and Education. Washington, DC: National Academy Press.

_____ (2000a). *Small-area estimates of school-age children in poverty—Evaluation of current methodology*. Panel on Estimates of Poverty for Small Geographic Areas, C.F. Citro and G. Kalton, eds. Committee on National Statistics. Commission on Behavioral and Social Sciences and Education. Washington, DC: National Academy Press.

______ (2000b). *Small-area income and poverty estimates—Priorities for 2000 and beyond.* Panel on Estimates of Poverty for Small Geographic Areas, C.F. Citro and G. Kalton, eds. Committee on National Statistics. Commission on Behavioral and Social Sciences and Education. Washington, DC: National Academy Press.

______ (2001). *The 2000 census: Interim assessment.* Panel to Review the 2000 Census, C.F. Citro, D.L. Cork, and J.L. Norwood, eds. Committee on National Statistics. Division of Behavioral and Social Sciences and Education. Washington, DC: National Academy Press.

______ (2003a). *Statistical issues in allocating funds by formula.* Panel on Formula Allocations, T.A. Louis, T.B. Jabine, and M.A. Gerstein, eds. Committee on National Statistics. Division of Behavioral and Social Sciences and Education. Washington, DC: The National Academies Press.

______ (2003b). *Survey automation: Report and workshop proceedings.* Oversight Committee for the Workshop on Survey Automation, M.L. Cohen, D.L. Cork, R. Groves, and W. Kalsbeek, eds. Committee on National Statistics. Division of Behavioral and Social Sciences and Education. Washington, DC: The National Academies Press.

______ (2004a). *Reengineering the 2010 census: Risks and challenges.* Panel on Research on Future Census Methods, D.L. Cork, M.L. Cohen, and B.F. King, eds. Committee on National Statistics. Division of Behavioral and Social Sciences and Education. Washington, DC: The National Academies Press.

______ (2004b). *The 2000 census: Counting under adversity.* Panel to Review the 2000 Census, C.F. Citro, D.L. Cork, and J.L. Norwood, eds. Committee on National Statistics. Division of Behavioral and Social Sciences and Education. Washington, DC: The National Academies Press.

______ (2005). *Expanding access to research data: Reconciling risks and opportunities.* Panel on Access to Research Data, Committee on National Statistics. Division of Behavioral and Social Sciences and Education. Washington, DC: The National Academies Press.

______ (2006). *Once, only once, and in the right place: Residence rules in the decennial census.* Panel on Residence Rules in the Decennial Census, D.L. Cork and P.R. Voss, eds. Committee on National Statistics. Division of Behavioral and Social Sciences and Education. Washington, DC: The National Academies Press.

Nelson, C. (2006, May 30). *What do we know about differences between CPS and ACS income and poverty estimates?* Paper prepared for the Workshop Exploring Possible Development of One Set of Integrated Income and Poverty Estimates. U.S. Bureau of the Census, U.S. Department of Commerce, Washington, DC.

Office for National Statistics (2003). *Alternatives to a census: Rolling census.* Census Strategic Development Review Information Paper. London, England: Office for National Statistics.

ORC Macro (2002, September). *The American Community Survey: Challenges and opportunities for HUD.* Paper prepared under contract for the U.S. Department of Housing and Urban Development, Washington, DC.

Palumbo, T. (2005, May 11). *Toward improving the quality of labor force data from the American Community Survey.* PowerPoint presentation. U.S. Bureau of the Census, U.S. Department of Commerce, Washington, DC.

Posey, K.G., and Welniak, E. (1998, March). *Income in the ACS: Comparisons to the 1990 census.* Prepared for the American Community Survey (ACS) symposium held at the Bureau of the Census, Suitland, Maryland. Last revised: June 17, 1999. U.S. Census Bureau, U.S. Department of Commerce, Washington, DC.

Reese, A.J. (2007, February 21). *A comparison of housing unit estimates to the American Community Survey Master Address File.* PowerPoint presentation. U.S. Census Bureau, U.S. Department of Commerce, Washington, DC.

Resnick, D., and Obenski, S. (2006). Methods and file acquisitions supporting the expanded use of administrative records. Paper prepared for the Joint Statistical Meetings, Seattle. U.S. Census Bureau, U.S. Department of Commerce.

Ruggles, S. (2000). IPUMS (Integrated Public Use Microdata Series). Pp. 264-267 in M.J. Anderson (Ed.), *Encyclopedia of the U.S. census.* Washington, DC: CQ Press.

Schneider, P. (2004). *Content and data quality in census 2000.* Census 2000 Testing, Experimentation, and Evaluation Program, Topic Report 12 (revised, January 22). U.S. Census Bureau. Washington, DC: U.S. Department of Commerce.

Starsinic, M. (2005). American Community Survey: Improving reliability for small area estimates. Pp. 3592-3599 in *Proceedings of the Joint Statistical Meetings, ASA Section on Survey Research Methods.* (CD-ROM). Alexandria, VA: American Statistical Association.

Stern, S., and Brault, M. (2005, January 28). *Disability data from the American Community Survey: A brief examination of the effects of a question redesign in 2003.* Working paper. Housing and Household Economic Statistics Division, U.S. Census Bureau, U.S. Department of Commerce, Washington, DC.

Taeuber, C.M. (2006). *American Community Survey data for community planning.* Victoria, British Columbia, Canada: Trafford Publishing.

Turek, J., Denmead, G., and James, B. (2005). *Poverty estimates in the ACS and other income surveys: What is the impact of methodology?* Working paper. Office of the Assistant Secretary for Planning and Evaluation, U.S. Department of Health and Human Services.

U.S. Census Bureau (1998). *Survey of income and program participation quality profile 1998*, 3rd edition. Prepared by Westat, G. Kalton, project director. Washington, DC: U.S. Department of Commerce. Available: http://www.bls.census.gov/sipp/workpaper/230/pdf.

______ (2001, July). *Meeting 21st century demographic data needs—Implementing the American Community Survey, Report 1: Demonstrating operational feasibility.* Washington, DC: U.S. Department of Commerce.

______ (2002a). *2002 census of governments, government organization.* Vol. 1, no. 1. Washington, DC: U.S. Department of Commerce.

______ (2002b, May). *Meeting 21st century demographic data needs—Implementing the American Community Survey, Report 2: Demonstrating survey quality.* Washington, DC: U.S. Department of Commerce.

______ (2003, December). *Meeting 21st century demographic data needs—Implementing the American Community Survey, Report 3: Testing the use of voluntary methods.* Washington, DC: U.S. Department of Commerce.

______ (2004a, May). *Meeting 21st century demographic data needs—Implementing the American Community Survey, Report 4: Comparing general demographic and housing characteristics with census 2000.* Washington, DC: U.S. Department of Commerce.

______ (2004b, May). *Meeting 21st century demographic data needs—Implementing the American Community Survey, Report 5: Comparing economic characteristics with census 2000.* Washington, DC: U.S. Department of Commerce.

______ (2004c, June). *Meeting 21st century demographic data needs—Implementing the American Community Survey, Report 7: Comparing quality measures: Comparing the American Community Survey's three-year averages and census 2000's long form sample estimates.* Washington, DC: U.S. Department of Commerce.

______ (2004d, June). *Meeting 21st century demographic data needs—Implementing the American Community Survey, Report 8: Comparison of the ACS 3-year average and the census 2000 sample for a sample of counties and tracts.* Washington, DC: U.S. Department of Commerce.

______ (2004e, June). *Meeting 21st century demographic data needs—Implementing the American Community Survey, Report 9: Comparing social characteristics with census 2000.* Washington, DC: U.S. Department of Commerce.

______ (2004f, July). *Meeting 21st century demographic data needs—Implementing the American Community Survey, Report 10: Comparing selected physical and financial housing characteristics with census 2000.* Washington, DC: U.S. Department of Commerce.

______ (2005, December). *Census 2000, public use microdata sample (PUMS), United States, technical documentation.* Washington, DC: U.S. Department of Commerce.

______ (2006, May). *Design and methodology—American Community Survey.* Washington, DC: U.S. Department of Commerce. Available: http://www.census.gov/acs/www.

U.S. General Accounting Office (1998, March). *2000 census: Preparations for dress rehearsal leave many unanswered questions.* GAO/GGD-98-74. Washington, DC: U.S. Government Printing Office.

U.S. General Services Administration (2006). *Catalog of domestic federal assistance.* Washington, DC. Available: http://www.gas.gov/cfda.

U.S. Government Accountability Office (2004, October). *American Community Survey—Key unresolved issues.* GAO-05-82. Washington, DC: U.S. Government Printing Office.

Zaslavsky, A.M., and Schirm, A.L. (2002). Interactions between survey estimates and federal funding formulas. *Journal of Official Statistics, 18*(3), 371-391.

Zayatz, L.V. (2005, December 13). *ACS base tables on workplace and means of transportation review.* Memorandum for L. McGinn. U.S. Census Bureau, U.S. Department of Commerce, Washington, DC.

APPENDIX A

Acronyms and Abbreviations

ACS	American Community Survey
AHS	American Housing Survey
APDU	Association of Public Data Users
BEA	Bureau of Economic Analysis
BLS	Bureau of Labor Statistics
C2SS	Census 2000 Supplementary Survey
CAPI	Computer-assisted personal interviewing
CATI	Computer-assisted telephone interviewing
CAUS	Community Address Updating System
CES	Current Employment Statistics Program (of the Bureau of Labor Statistics)
CI	Confidence interval
CIC	Census Information Center
CNSTAT	Committee on National Statistics
CPI	Consumer Price Index
CPI-U-RS	Consumer Price Index, Urban Consumers, Research Series
CPS	Current Population Survey
CPS ASEC	Current Population Survey Annual Social and Economic Supplement
CTPP	Census Transportation Planning Package
CV	Coefficient of variation
DMAF	Decennial Master Address File
DSF	Delivery Sequence File (of the U.S. Postal Service)

e-StARS	Electronic Statistical Administrative Records System (of the U.S. Census Bureau)
FSCPE	Federal State Cooperative Program for Population Estimates
FTP	File transfer protocol
GQ	Group quarters
HUD	U.S. Department of Housing and Urban Development
IRS	Internal Revenue Service
LAUS	Local Area Unemployment Statistics Program (of the Bureau of Labor Statistics)
LUCA	Local Update of Census Addresses
MAF	Master Address File (of the U.S. Census Bureau)
MBF	Mode bias factor (ACS weight adjustment)
MOE	Margin of error
NHIS	National Health Interview Survey
NIF1	Noninterview factor 1 (ACS weight adjustment)
NIF2	Noninterview factor 2 (ACS weight adjustment)
NIF3	Noninterview factor 3 (ACS weight adjustment)
NRC	National Research Council
OMB	U.S. Office of Management and Budget
PE	Population Estimates Program (of the U.S. Census Bureau)
PRCS	Puerto Rico Community Survey
PUMA	Public use microdata area
PUMS	Public use microdata sample
SAHIE	Small Area Health Insurance Estimates Program (of the U.S. Census Bureau)
SAIPE	Small Area Income and Poverty Estimates Program (of the U.S. Census Bureau)
SDC	State Data Center
SE	Standard error
SF	Summary file
SIPP	Survey of Income and Program Participation
StARS	Statistical Administrative Records System (see e-StARS)
TAZ	Traffic analysis zone
UI	Unemployment insurance records (of state governments)
USPS	U.S. Postal Service
VMS	Variation in monthly response (ACS weight adjustment)

B

Controlling the American Community Survey to Postcensal Population Estimates

F. Jay Breidt
Colorado State University

ABSTRACT

The U.S. Census Bureau has proposed the use of postcensal population estimates as population controls for the American Community Survey at a fine level of geographic and demographic stratification. These population estimates are known to be imperfect. Bias and variance of post-stratification estimators with imperfect population controls at various levels of aggregation are considered. The bias and variance are computed with respect to the "model" (including data generation and postcensal population estimation) or with respect to the "design" (including coverage, sampling, response and postcensal population estimation). Bias and variance depend in a complex way on the interactions of postcensal population estimation errors with undercoverage error and nonresponse. Numerical examples illustrate that in the presence of imperfect postcensal population estimations, control at higher levels of aggregation may be better in terms of bias than control at a fine level of post-stratification.

B–1 INTRODUCTION

B–1.1 Background

The estimation procedures used with the American Community Survey (ACS) include a poststratification step that employs postcensal population estimates by demographic strata as controls. The controls are applied within estimation areas that consist of larger counties or combinations of smaller counties. The National Research Council's Panel on the Functionality and Usability of Data from the American Community Survey (ACS) asked me to evaluate these plans, comparing them to direct estimates that forgo the use of these controls, and comparing them to estimates controlled at higher levels of demographic and/or geographic aggregation. The panel's charge for me was to evaluate these options from a theoretical, not an empirical, perspective. In particular, the charge was to look at the bias and variance of these estimators, for both population estimation and estimation of other characteristics, at different levels of aggregation.

This paper proposes a simple theoretical framework under which bias and variance can be computed. Cochran (1977), following Stephan (1941), discusses the effect of non-random, imperfect estimates of population controls on mean square error under repeated sampling, with full coverage and response. By contrast, I propose stochastic models for: data generation ("model"); coverage, sampling, and response ("design"); and population estimation. Within this stochastic framework, I consider estimation with or without controls, the level of aggregation for those controls, and whether or not those controls have errors. In Section B–2, I compute bias and variance with respect to model and population estimation or with respect to design and population estimation and discuss the implications of those computations. Numerical examples of bias for a simple, artificial population are computed in Section B–3. Finally, I discuss the results and outline a few directions for further investigation in Section B–4.

I focus on bias and variance because the computations can then be performed without knowledge of distributional properties beyond second-order moments. In practice, second-moment properties might feasibly be estimated from data. Risks with respect to losses other than squared error loss might be computed under stochastic mechanisms similar to those considered here. For example, losses that weight bias more heavily might be of interest in certain funding formulas.

To begin, note that population controls are helpful in mitigating all three errors of nonobservation: coverage error, sampling error, and nonresponse. To illustrate this, I focus on a theoretical framework that includes a probability sample of elements from a finite population, with stochastic mechanisms for frame inclusion and for nonresponse, and with weighting that reflects the probability sampling and poststratification. (Nonresponse adjustments internal to the sample are not explicitly considered here.) This framework is simplified considerably from that of the ACS, which samples households (clusters) instead of people (elements) and includes up to 11 factors for weighting. This framework should, however, provide insight into the issues surrounding the use of imperfect population controls in the ACS.

B–1.2 Notation and Estimators

Let U denote a finite population. Let y_i denote a generic study variable of interest corresponding to the ith element. Let $F_i = 1$ if element i is in the frame, 0 otherwise; $S_i = 1$ if element i is sampled, 0 otherwise; and

$R_i = 1$ if element i responds, 0 otherwise. Assume probability sampling of frame elements with inclusion probability $\pi_i F_i$ for element i, where (for notational convenience) $\pi_i > 0$ for all $i \in U$. Let $U = \cup_{g,h} U_{gh}$ denote a two-way stratification of the population of interest, with disjoint cells U_{gh}, and let C_{gh} denote a perfect census count of the elements in cell (subpopulation) g, h. In practice, the counts $\{C_{gh}\}$ are unknown and are projected from past census data using techniques of demographic analysis. Let D_{gh} denote the population estimate in cell g, h. Furthermore, define the row margin census counts and population estimates,

$$C_{g\bullet} = \sum_h C_{gh}, \quad D_{g\bullet} = \sum_h D_{gh};$$

column margin census counts and population estimates,

$$C_{\bullet h} = \sum_g C_{gh}, \quad D_{\bullet h} = \sum_g D_{gh};$$

and overall census count and population estimate,

$$C_{\bullet\bullet} = \sum_{g,h} C_{gh}, \quad D_{\bullet\bullet} = \sum_{g,h} D_{gh}.$$

Consider estimation of the population total $T = T(y_i) = \sum_{g,h} \sum_{i \in U_{gh}} y_i$. Define the cell indicator

$$z_{kli} = \begin{cases} 1, & i \in U_{kl}, \\ 0 & \text{otherwise} \end{cases}$$

and note that $T(z_{kli}) = C_{kl}$ is the count in cell k, l; $T(z_{kli} y_i)$ is the total of y in cell k, l; $T(\sum_h z_{ghi})$ is the count in row g ($U_g = \cup_h U_{gh}$); $T(\sum_h z_{ghi} y_i)$ is the total of y in row g; etc. Thus, both counts and totals at various levels of aggregation are implicitly included in the discussion that follows.

Three kinds of estimators of T are of interest. The first is the poststratification estimator (PSE) with cell controls:

$$\hat{T}_{\text{cell}} = \sum_{g,h} D_{gh} \frac{\sum_{i \in U_{gh}} F_i S_i R_i y_i / \pi_i}{\sum_{i \in U_{gh}} F_i S_i R_i / \pi_i}. \tag{B.1}$$

Plugging in z_{kli} for y_i in (B.1), we have that $\hat{T}_{\text{cell}} = D_{kl}$, so that the sample is controlled to the postcensal population estimates in every cell. The PSE with cell controls therefore is similar to the Census Bureau's plan to control at a fine level of demographic stratification within estimation areas.

The second estimator is the PSE with control on one margin. Without loss of generality, take this as the row margin:

$$\hat{T}_{\text{marg}} = \sum_g D_{g\bullet} \frac{\sum_h \sum_{i \in U_{gh}} F_i S_i R_i y_i / \pi_i}{\sum_h \sum_{i \in U_{gh}} F_i S_i R_i / \pi_i}. \tag{B.2}$$

This estimator is controlled to the postcensal population estimates for the row margins, $\{D_{g\bullet}\} = \{\sum_h D_{gh}\}$. This estimator represents the alternative of controlling to population estimates at a higher level of aggregation, such as controlling for age and sex within estimation areas but not controlling for race or Hispanic origin.

The final estimator has control only to the overall population estimate, $D_{\bullet\bullet}$:

$$\hat{T}_{\text{overall}} = D_{\bullet\bullet} \frac{\sum_{g,h} \sum_{i \in U_{gh}} F_i S_i R_i y_i / \pi_i}{\sum_{g,h} \sum_{i \in U_{gh}} F_i S_i R_i / \pi_i}. \tag{B.3}$$

This will be referred to as the overall-control estimator and represents the option of direct estimates that forgo the use of all but the overall control.

Some additional notation is useful in describing properties of the estimators. Denote the empirical mean, variance, and coefficient of variation of a variable $\{x_i\}$ on the set A by

$$\begin{aligned}
\texttt{ave}_A(x_i) &= \frac{1}{|A|} \sum_{i \in A} x_i \\
\texttt{var}_A(x_i) &= \frac{\sum_{i \in A} x_i^2 - \left(\sum_{i \in A} x_i\right)^2 / |A|}{|A| - 1} \\
\texttt{cv}_A(x_i) &= \frac{\sqrt{\texttt{var}_A(x_i)}}{\texttt{ave}_A(x_i)}.
\end{aligned}$$

Similarly, denote the empirical covariance and correlation between variables x_i and y_i on the set A by

$$\begin{aligned}
\texttt{cov}_A(x_i, y_i) &= \frac{\sum_{i \in A} x_i y_i - \left(\sum_{i \in A} x_i\right)\left(\sum_{i \in A} y_i\right) / |A|}{|A| - 1} \\
\texttt{corr}_A(x_i, y_i) &= \frac{\texttt{cov}_A(x_i, y_i)}{\sqrt{\texttt{var}_A(x_i)}\sqrt{\texttt{var}_A(y_i)}}.
\end{aligned}$$

In the computations that follow, the variable x_i is often piecewise constant over subsets of A. For example, consider a variable x_i defined as $x_i = b_{gh}$ for $i \in U_{gh}$, under the two-way stratification $U = \cup_{g,h} U_{gh}$. Then

$$\texttt{ave}_U(x_i) = \frac{1}{|U|} \sum_{i \in U} x_i = \frac{1}{C_{\bullet\bullet}} \sum_{g,h} \sum_{i \in U_{gh}} b_{gh} = \frac{1}{C_{\bullet\bullet}} \sum_{g,h} C_{gh} b_{gh}.$$

B–1.3 Assumptions

Assume that the postcensal population estimation errors $\{D_{gh} - C_{gh}\}$ are independent random variables, independent of the design variables $\{F_i, S_i, R_i\}$ and of the data $\{y_i\}$. These independence assumptions might be questionable in a real application. For example, it seems likely that a subpopulation with serious coverage issues might also suffer from large postcensal population estimation errors for many of the same reasons, so that frame variables and estimation errors would be correlated. Furthermore, estimation errors in different age categories of the same race/sex group might well be correlated. Nevertheless, computations under this simple stochastic model may provide some useful insights into the problems with imperfect controls.

Let

$$\delta_{gh} = \mathrm{E}\left[D_{gh} - C_{gh}\right], \quad \mathrm{Var}\left(D_{gh} - C_{gh}\right) = \tau^2_{gh}.$$

The special case of perfect population controls is obtained with $\delta_{gh} \equiv 0$ and $\tau^2_{gh} \equiv 0$. Factors that may affect these biases and variances in the ACS application could include time since last census, demographic grouping (age/race/sex/Hispanic origin), geographic grouping, and interactions (e.g., young people in college-dominated estimation areas).

Properties of the estimators can be derived under assumptions on the errors in the population estimates, along with assumptions on the design or the data-generating model. Expectation with respect to the data-generating (superpopulation) model will be differentiated from expectation with respect to the design by adding a subscript m to the expectation operator. The population estimation error distribution will be included in either design or model expectations.

The data-generating model assumptions are that $\{y_i\}$ are independent with common mean $\mathrm{E}_m[y_i] = \mu_{gh}$ and common variance $\mathrm{Var}_m(y_i) = \sigma^2_{gh}$ for $i \in U_{gh}$. Since y_i is a generic study variable, this assumption would ideally hold for any choice of study variable y_i, so that the poststratification is appropriate for all study variables. In this case, no design assumptions are required.

The design assumptions are the following:

- $\{R_i\}$ are conditionally independent given $\{S_i, F_i\}$, with $\mathrm{E}[R_i \mid S_i, F_i] = \rho_{gh} S_i F_i$ for $i \in U_{gh}$;
- $\mathrm{E}[S_i \mid F_i] = \pi_i F_i$ for all i;
- $\{F_i\}$ are independent, with $\mathrm{E}[F_i] = \phi_{gh}$ for $i \in U_{gh}$.

If these design assumptions hold, then the poststratification is appropriate for all study variables, regardless of their data-generating mechanisms.

Note that either of these sets of assumptions imply that the cell-level stratification is *sufficient* (fine enough) to account for differences in coverage and response, and so it is an appropriate level of aggregation for post-stratification adjustments. It is possible that the cell-level stratification is too fine, but if in fact the coverage and response vary from cell to cell, then the cell-level stratification is *necessary* in the sense that estimators controlled at higher levels of aggregation have bias due to variation in coverage and response, as demonstrated in the next section.

For simplicity in variance computations, I also make the assumption that the probability sampling design is Poisson sampling, that is, $\{S_i\}$ given $\{F_i\}$ are independent. This assumption is not used in bias computations. Expressions computed using this assumption are compact and interpretable. Computations could be done under other designs as well.

B–2 PROPERTIES OF THE ESTIMATORS

This section examines properties of the PSE with cell-level controls, margin-level controls, and overall population controls, looking at the bias and variance of each under the model assumptions and under the design assumptions. Computations under the model assumptions are exact, while those under the design require some large sample approximations. Inference under the design is widely accepted by survey practitioners, although both model-based and design-based inference play important roles in official statistics. (Many survey texts, including Särndal, Swensson, and Wretman, 1992, and Brewer, 2002, discuss inference and illustrate computations under model and under design.) In this paper, the derived expressions are perfectly parallel under model or design, so the choice of model or design is not critical.

B–2.1 Model Properties of PSE with Cell-Level Controls

With respect to both the data-generating model and the distribution of the postcensal population estimation errors, the PSE with cell-level controls has bias

$$\begin{aligned} \mathrm{E}_m\left[\hat{T}_{\text{cell}} - T\right] &= \sum_{g,h}(C_{gh}+\delta_{gh})\frac{\sum_{i\in U_{gh}} F_i S_i R_i \mu_{gh}/\pi_i}{\sum_{i\in U_{gh}} F_i S_i R_i/\pi_i} - \sum_{g,h} C_{gh}\mu_{gh} \\ &= \sum_{g,h}\delta_{gh}\mu_{gh}, \end{aligned}$$

so that the estimator is unbiased only if postcensal population estimation biases are orthogonal to cell means. In general, this will not be the case, so the PSE with cell-level controls will be biased. It is useful to express this bias as

$$\begin{aligned}\mathrm{E}_m\left[\hat{T}_{\text{cell}}-T\right] &= \left[(C_{\bullet\bullet}-1)\mathtt{corr}_U\left(\frac{\delta_{gh}}{C_{gh}},\mu_{gh}\right)\mathtt{cv}_U\left(\frac{\delta_{gh}}{C_{gh}}\right)\times\right.\\ &\qquad\left.\mathtt{cv}_U\left(\mu_{gh}\right)\mathtt{ave}_U\left(\frac{\delta_{gh}}{C_{gh}}\right)\mathtt{ave}_U\left(\mu_{gh}\right)\right]\\ &\quad+\mathtt{ave}_U\left(\frac{\delta_{gh}}{C_{gh}}\right)\sum_{g,h}C_{gh}\mu_{gh},\end{aligned}$$

using an extension of the argument in equation (15.6.3) of Särndal, Swensson, and Wretman (1992). The above expression assumes that $\mathtt{ave}_U\left(\frac{\delta_{gh}}{C_{gh}}\right)\neq 0$, $\mathtt{ave}_U\left(\mu_{gh}\right)\neq 0$, $\mathtt{var}_U\left(\frac{\delta_{gh}}{C_{gh}}\right)\neq 0$, and $\mathtt{var}_U\left(\mu_{gh}\right)\neq 0$, so that the correlation and coefficients of variation are well defined. In the rest of this paper, it is always assumed implicitly that any correlations and coefficients of variation are well defined.

Then the relative bias is

$$\begin{aligned}\mathrm{RelBias}_m\left(\hat{T}_{\text{cell}}\right) &= \left[\frac{(C_{\bullet\bullet}-1)}{C_{\bullet\bullet}}\mathtt{corr}_U\left(\frac{\delta_{gh}}{C_{gh}},\mu_{gh}\right)\mathtt{cv}_U\left(\frac{\delta_{gh}}{C_{gh}}\right)\times\right.\\ &\qquad\left.\mathtt{cv}_U\left(\mu_{gh}\right)\mathtt{ave}_U\left(\frac{\delta_{gh}}{C_{gh}}\right)\right]+\mathtt{ave}_U\left(\frac{\delta_{gh}}{C_{gh}}\right)\\ &\simeq \mathtt{ave}_U\left(\frac{\delta_{gh}}{C_{gh}}\right)\left\{1+\left[\mathtt{cv}_U\left(\frac{\delta_{gh}}{C_{gh}}\right)\mathtt{cv}_U\left(\mu_{gh}\right)\times\right.\right.\\ &\qquad\left.\left.\mathtt{corr}_U\left(\frac{\delta_{gh}}{C_{gh}},\mu_{gh}\right)\right]\right\}.\end{aligned} \tag{B.4}$$

A key feature of this bias is the correlation term between postcensal population estimation biases and cell means. If the biases are nearly constant from cell to cell, then $\mathtt{cv}_U\left(\delta_{gh}/C_{gh}\right)\simeq 0$, and the correlation term contributes little to the bias. If the biases differ from cell to cell, then the correlation term gives a signed contribution to the bias. On one hand, for example, suppose that the errors in the postcensal population estimates cross-classified by age, race/ethnicity, and sex result in an overestimate of the the number of individuals with high income but do not overestimate elsewhere. Then $\mathtt{cv}_U\left(\delta_{gh}/C_{gh}\right)\neq 0$, and the correlation is positive (overestimate, high income), so a positive term is contributed to the bias. On the other hand, suppose that the postcensal population estimates overestimate the number of individuals with low income but do not overestimate elsewhere. Then

$\mathrm{cv}_U\left(\delta_{gh}/C_{gh}\right) \neq 0$, and the correlation is negative (overestimate, low income), so a negative term is contributed to the bias.

With respect to both the data-generating model and the distribution of the postcensal population estimation errors, the PSE with cell-level controls has variance

$$\begin{aligned}\mathrm{Var}_m\left(\hat{T}_{\text{cell}}\right) &= \sum_{g,h} \mathrm{E}\left[D^2_{gh}\right] \sigma^2_{gh} \frac{\sum_{i\in U_{gh}} F_i S_i R_i/\pi_i^2}{\left(\sum_{i\in U_{gh}} F_i S_i R_i/\pi_i\right)^2} + \sum_{g,h} \mu^2_{gh}\tau^2_{gh} \\ &= \sum_{g,h}\left\{(C_{gh}+\delta_{gh})^2 + \tau^2_{gh}\right\} \sigma^2_{gh} \frac{\sum_{i\in U_{gh}} F_i S_i R_i/\pi_i^2}{\left(\sum_{i\in U_{gh}} F_i S_i R_i/\pi_i\right)^2} \\ &\quad + \sum_{g,h} \mu^2_{gh}\tau^2_{gh}. \end{aligned} \qquad \text{(B.5)}$$

This is not necessarily larger than the variance with perfect controls. For example, with $D_{gh} \equiv 0$, the estimator has zero variance ($\delta_{gh} \equiv -C_{gh}$, $\tau^2_{gh} \equiv 0$) but large bias.

Note that

$$\begin{aligned} \frac{\sum_{i\in U_{gh}} F_i S_i R_i/\pi_i^2}{\left(\sum_{i\in U_{gh}} F_i S_i R_i/\pi_i\right)^2 / C_{gh}} &= \frac{\sum_{i\in U_{gh}} F_i S_i R_i/\pi_i^2 - \frac{1}{C_{gh}}\left(\sum_{i\in U_{gh}} F_i S_i R_i/\pi_i\right)^2}{(C_{gh}-1)\frac{1}{C_{gh}(C_{gh}-1)}\left(\sum_{i\in U_{gh}} F_i S_i R_i/\pi_i\right)^2} + 1 \\ &\simeq \frac{\mathtt{var}_{U_{gh}}(F_i S_i R_i/\pi_i)}{\left(\mathtt{ave}_{U_{gh}}(F_i S_i R_i/\pi_i)\right)^2} + 1 \\ &= \left(\mathtt{cv}_{U_{gh}}(F_i S_i R_i/\pi_i)\right)^2 + 1. \end{aligned}$$

Thus, in the case of perfect controls ($\delta_{gh} \equiv 0$, $\tau^2_{gh} \equiv 0$), we have from (B.5) that

$$\begin{aligned}\mathrm{Var}_m\left(\hat{T}_{\text{cell}}\right) &\simeq \sum_{g,h} C_{gh}\sigma^2_{gh}\left\{\left(\mathtt{cv}_{U_{gh}}(F_i S_i R_i/\pi_i)\right)^2 + 1\right\} \\ &\geq \sum_{g,h} C_{gh}\sigma^2_{gh} = \mathrm{Var}_m(T).\end{aligned}$$

The lower bound is obtained if $F_i \equiv 1$, $\pi_i \equiv 1$ (hence $S_i \equiv 1$), and $R_i \equiv 1$, that is, a census from a perfect frame with full response. In all other cases, undercoverage, sampling error, and nonresponse increase the variance.

B–2.2 Design Properties of PSE with Cell-Level Controls

The PSE is nonlinear in the design variables, so its expectation and variance under the design are approximated from the usual Taylor series linearization,

$$\hat{T}_{\text{cell}} \simeq \sum_{g,h} (C_{gh} + \delta_{gh})\,\bar{y}_{gh} + \sum_{g,h} \bar{y}_{gh}\,(D_{gh} - C_{gh} - \delta_{gh}) + \sum_{g,h} \frac{C_{gh} + \delta_{gh}}{C_{gh}\phi_{gh}\rho_{gh}} \sum_{i\in U_{gh}} \frac{F_i S_i R_i}{\pi_i}\,(y_i - \bar{y}_{gh})\,. \tag{B.6}$$

The design bias is then approximately

$$\mathrm{E}\left[\hat{T}_{\text{cell}}\right] - T \simeq \sum_{g,h} (C_{gh} + \delta_{gh})\,\bar{y}_{gh} - \sum_{g,h} C_{gh}\bar{y}_{gh} = \sum_{g,h} \delta_{gh}\bar{y}_{gh}$$

so that the PSE is biased under the stated design assumptions, unless postcensal population estimation biases happen to be orthogonal to cell means. By the same argument used for the relative model bias, the relative design bias is then approximated as

$$\mathrm{RelBias}\left(\hat{T}_{\text{cell}}\right) \simeq \mathtt{ave}_U\left(\frac{\delta_{gh}}{C_{gh}}\right)\left\{\mathtt{corr}_U\left(\frac{\delta_{gh}}{C_{gh}}, \bar{y}_{gh}\right)\mathtt{cv}_U\left(\frac{\delta_{gh}}{C_{gh}}\right)\mathtt{cv}_U\left(\bar{y}_{gh}\right) + 1\right\}, \tag{B.7}$$

which has the same interpretation as the relative bias under the model.

From (B.6), the design variance of the PSE is approximately

$$\mathrm{Var}\left(\hat{T}_{\text{cell}}\right) \simeq \sum_{g,h} \frac{(C_{gh} + \delta_{gh})^2}{C_{gh}^2} \sum_{i\in U_{gh}} \frac{(1 - \phi_{gh}\pi_i\rho_{gh})}{\phi_{gh}\pi_i\rho_{gh}}\,(y_i - \bar{y}_{gh})^2 + \sum_{g,h} \bar{y}_{gh}^2\tau_{gh}^2, \tag{B.8}$$

which parallels the model variance in (B.5). A key part of this expression involves squared deviations from subpopulation means, so that if the subpopulations are more homogeneous than the population as a whole, the design variance will tend to be smaller than that of a direct estimator. In addition, the first component of this variance is zero if $\phi_{gh} \equiv 1$, $\pi_i \equiv 1$, and $\rho_{gh} \equiv 1$; that is, a census from a perfect frame with full response would have zero design variance in the case of perfect controls. In all other cases, undercoverage, sampling error, and nonresponse increase the variance because the corresponding probabilities appear in the denominator of the first term.

B–2.3 Model Properties of PSE with Margin-Level Controls

Define $A_{gh} = \sum_{i \in U_{gh}} F_i S_i R_i / (\pi_i C_{gh})$, the estimated observation probability in cell g, h. Let $U_g = \cup_h U_{gh}$. The model bias of the PSE with margin-level controls is

$$\begin{aligned}
\mathrm{E}_m\left[\hat{T}_{\mathrm{marg}} - T\right] \\
&= \sum_g \left\{ (C_{g\bullet} + \delta_{g\bullet}) \frac{\sum_h C_{gh} A_{gh} \mu_{gh}}{\sum_h C_{gh} A_{gh}} - \frac{\sum_h C_{gh} A_{gh} \sum_h C_{gh} \mu_{gh}}{\sum_h C_{gh} A_{gh}} \right\} \\
&= \sum_g \mathtt{corr}_{U_g}\left(A_{gh}, \mu_{gh}\right) \mathtt{cv}_{U_g}\left(A_{gh}\right) \mathtt{cv}_{U_g}\left(\mu_{gh}\right) \mathtt{ave}_{U_g}\left(\mu_{gh}\right) (C_{g\bullet} - 1) \\
&\quad + \sum_g \delta_{g\bullet} \left\{ \frac{C_{g\bullet} - 1}{C_{g\bullet}} \mathtt{corr}_{U_g}\left(A_{gh}, \mu_{gh}\right) \mathtt{cv}_{U_g}\left(A_{gh}\right) \mathtt{cv}_{U_g}\left(\mu_{gh}\right) \mathtt{ave}_{U_g}\left(\mu_{gh}\right) \right. \\
&\qquad \left. + \mathtt{ave}_{U_g}\left(\mu_{gh}\right) \right\}
\end{aligned}$$

where it is understood that $\{A_{gh}\}_{i \in U_g}$ is a data set of $C_{g\bullet}$ values: A_{g1} repeated C_{g1} times, A_{g2} repeated C_{g2} times, and so forth. Similarly for $\{\mu_{gh}\}$. The relative bias is then computed as

$$\begin{aligned}
\mathrm{RelBias}_m\left(\hat{T}_{\mathrm{marg}}\right) \\
&= \sum_g \mathtt{corr}_{U_g}\left(A_{gh}, \mu_{gh}\right) \mathtt{cv}_{U_g}(A_{gh}) \mathtt{cv}_{U_g}(\mu_{gh}) \frac{\mathtt{ave}_{U_g}(\mu_{gh})(C_{g\bullet} - 1)}{\sum_{g,h} C_{gh} \mu_{gh}} \\
&\quad + \sum_g \frac{\delta_{g\bullet}}{C_{\bullet\bullet}} \frac{\mathtt{ave}_{U_g}\left(\mu_{gh}\right)}{\mathtt{ave}_U\left(\mu_{gh}\right)} \left\{ \frac{C_{g\bullet} - 1}{C_{g\bullet}} \mathtt{corr}_{U_g}\left(A_{gh}, \mu_{gh}\right) \right. \\
&\qquad \left. \times\, \mathtt{cv}_{U_g}\left(A_{gh}\right) \mathtt{cv}_{U_g}\left(\mu_{gh}\right) + 1 \right\} \\
&\simeq \sum_g \frac{C_{g\bullet}}{C_{\bullet\bullet}} \frac{\mathtt{ave}_{U_g}(\mu_{gh})}{\mathtt{ave}_U(\mu_{gh})} \mathtt{ave}_{U_g}\left(\frac{\delta_{gh}}{C_{gh}}\right) \\
&\quad + \sum_g \frac{C_{g\bullet}}{C_{\bullet\bullet}} \frac{\mathtt{ave}_{U_g}(\mu_{gh})}{\mathtt{ave}_U(\mu_{gh})} \mathtt{corr}_{U_g}\left(A_{gh}, \mu_{gh}\right) \mathtt{cv}_{U_g}(A_{gh}) \mathtt{cv}_{U_g}(\mu_{gh}) \\
&\quad + \sum_g \left\{ \frac{C_{g\bullet}}{C_{\bullet\bullet}} \frac{\mathtt{ave}_{U_g}(\mu_{gh})}{\mathtt{ave}_U(\mu_{gh})} \mathtt{corr}_{U_g}\left(A_{gh}, \mu_{gh}\right) \mathtt{cv}_{U_g}(A_{gh}) \right. \\
&\qquad \left. \mathtt{cv}_{U_g}(\mu_{gh}) \mathtt{ave}_{U_g}\left(\frac{\delta_{gh}}{C_{gh}}\right) \right\}. \qquad \text{(B.9)}
\end{aligned}$$

The first term is bias attributable solely to population estimation error. It would appear even in the absence of any bias due to undercoverage or nonresponse. The second term is bias due only to undercoverage and nonresponse; this term appears even in the absence of postcensal population estimation errors. The final term represents the contribution to the bias from the interaction between coverage/nonresponse bias and postcensal population estimation bias.

The second two bias terms reflect the fact that nonresponse bias and undercoverage bias are not adequately adjusted for at the row margin level, if the response and coverage actually vary from cell to cell within the row. If there is no variation in the estimated observation probabilities $\{A_{gh}\}$ within a particular row, then $\text{cv}_{U_g}(A_{gh}) = 0$, and undercoverage/nonresponse in that particular row contributes no bias because controlling at the row margin was an appropriate adjustment. If there is variation within a particular row, then the bias is determined by the amount of correlation between estimated observation probabilities and cell means. For example, if response and coverage is high in cells that have higher average incomes than other cells in the row, then the correlation between probabilities and cell means is positive and the bias is positive. If response and coverage is high in low-income cells compared with other cells, then the correlation between probabilities and cell means is negative and the bias is negative.

The model variance of the PSE with imperfect margin controls is

$$\begin{aligned}\text{Var}_m\left(\hat{T}_{\text{marg}}\right) &= \sum_g \left\{\tau^2_{g\bullet} + (C_{g\bullet} + \delta_{g\bullet})^2\right\} \frac{\sum_h \sigma^2_{gh} \sum_{i\in U_{gh}} F_i S_i R_i / \pi_i^2}{\left(\sum_h \sum_{i\in U_{gh}} F_i S_i R_i / \pi_i\right)^2} \\ &\quad + \sum_g \tau^2_{g\bullet} \left(\frac{\sum_h C_{gh} A_{gh} \mu_{gh}}{\sum_h C_{gh} A_{gh}}\right)^2. \qquad \text{(B.10)}\end{aligned}$$

B–2.4 Design Properties of PSE with Margin-Level Controls

The PSE with margin-level controls is nonlinear in the design variables, so its expectation and variance under the design are approximated from the usual Taylor series linearization,

$$\begin{aligned}\hat{T}_{\text{marg}} &\simeq \sum_g (C_{g\bullet} + \delta_{g\bullet}) \frac{\sum_h C_{gh} \phi_{gh} \rho_{gh} \bar{y}_{gh}}{\sum_h C_{gh} \phi_{gh} \rho_{gh}} \\ &\quad + \sum_g \frac{\sum_h C_{gh} \phi_{gh} \rho_{gh} \bar{y}_{gh}}{\sum_h C_{gh} \phi_{gh} \rho_{gh}} (D_{g\bullet} - C_{g\bullet} - \delta_{g\bullet}) \\ &\quad + \sum_g \frac{C_{g\bullet} + \delta_{g\bullet}}{\sum_h C_{gh} \phi_{gh} \rho_{gh}} \sum_h \sum_{i\in U_{gh}} \frac{F_i S_i R_i}{\pi_i} \left(y_i - \frac{\sum_h C_{gh} \phi_{gh} \rho_{gh} \bar{y}_{gh}}{\sum_h C_{gh} \phi_{gh} \rho_{gh}}\right). \qquad \text{(B.11)}\end{aligned}$$

The relative design bias is then given approximately by

$$
\begin{aligned}
&\text{RelBias}\left(\hat{T}_{\text{marg}}\right) \\
&\simeq \sum_g \frac{C_{g\bullet}}{C_{\bullet\bullet}} \frac{\texttt{ave}_{U_g}(\bar{y}_{gh})}{\texttt{ave}_U(\bar{y}_{gh})} \texttt{ave}_{U_g}\left(\frac{\delta_{gh}}{C_{gh}}\right) \\
&+\sum_g \frac{C_{g\bullet}}{C_{\bullet\bullet}} \frac{\texttt{ave}_{U_g}(\bar{y}_{gh})}{\texttt{ave}_U(\bar{y}_{gh})} \texttt{corr}_{U_g}\left(\phi_{gh}\rho_{gh}, \bar{y}_{gh}\right) \texttt{cv}_{U_g}(\phi_{gh}\rho_{gh}) \texttt{cv}_{U_g}(\bar{y}_{gh}) \\
&+\sum_g \left\{ \frac{C_{g\bullet}}{C_{\bullet\bullet}} \frac{\texttt{ave}_{U_g}(\bar{y}_{gh})}{\texttt{ave}_U(\bar{y}_{gh})} \texttt{corr}_{U_g}\left(\phi_{gh}\rho_{gh}, \bar{y}_{gh}\right) \texttt{cv}_{U_g}(\phi_{gh}\rho_{gh}) \texttt{cv}_{U_g}(\bar{y}_{gh}) \right. \\
&\qquad \left. \times \texttt{ave}_{U_g}\left(\frac{\delta_{gh}}{C_{gh}}\right) \right\}. \qquad \text{(B.12)}
\end{aligned}
$$

The interpretation of this relative design bias directly parallels the relative model bias described above. The first term is bias attributable solely to population estimation error. It would appear even in the absence of any bias due to undercoverage or nonresponse. The second term is bias due only to undercoverage and nonresponse; this term appears even in the absence of postcensal population estimation errors. The final term represents the contribution to the bias from the interaction between undercoverage/nonresponse bias and postcensal population estimation bias. If $\{\phi_{gh}\rho_{gh}\}$ does not depend on h within row g, then a single control for row g is appropriate, $\texttt{cv}_{U_g}(\phi_{gh}\rho_{gh}) = 0$, and undercoverage/nonresponse within row g contributes no bias to the overall estimator.

From (B.11), the design variance of the estimator is approximately

$$
\begin{aligned}
&\text{Var}\left(\hat{T}_{\text{marg}}\right) \\
&\simeq \left\{ \sum_g \left(\frac{C_{g\bullet} + \delta_{g\bullet}}{\sum_h C_{gh}\phi_{gh}\rho_{gh}} \right)^2 \right. \\
&\qquad \left. \sum_h \sum_{i \in U_{gh}} \frac{\phi_{gh}\rho_{gh}\left(1 - \phi_{gh}\pi_i\rho_{gh}\right)}{\pi_i} \left(y_i - \frac{\sum_h C_{gh}\phi_{gh}\rho_{gh}\bar{y}_{gh}}{\sum_h C_{gh}\phi_{gh}\rho_{gh}} \right)^2 \right\} \\
&+\sum_g \tau_{g\bullet}^2 \left(\frac{\sum_h C_{gh}\phi_{gh}\rho_{gh}\bar{y}_{gh}}{\sum_h C_{gh}\phi_{gh}\rho_{gh}} \right)^2. \qquad \text{(B.13)}
\end{aligned}
$$

B–2.5 Model Properties of PSE with Overall-Level Control

The properties of the PSE with overall-level control can be derived from the previous results for the PSE with margin control by noting that controlling on the overall count is analogous to controlling on the row margin when there is a single row. Thus the overall-control estimator has relative model bias given approximately by

$$\begin{aligned}\text{RelBias}_m\left(\hat{T}_{\text{overall}}\right) & \\ \simeq \mathtt{ave}_U\left(\frac{\delta_{gh}}{C_{gh}}\right) & \\ + \mathtt{corr}_U\left(A_{gh},\mu_{gh}\right)\mathtt{cv}_U(A_{gh})\mathtt{cv}_U(\mu_{gh}) & \\ + \mathtt{corr}_U\left(A_{gh},\mu_{gh}\right)\mathtt{cv}_U(A_{gh})\mathtt{cv}_U(\mu_{gh})\mathtt{ave}_U\left(\frac{\delta_{gh}}{C_{gh}}\right). & \quad \text{(B.14)}\end{aligned}$$

The first term in the relative bias is present even in the absence of any undercoverage/nonresponse bias. The second and third terms in the relative bias reflect the fact that nonresponse bias and undercoverage bias are not adequately adjusted for at the population level, if the response and coverage actually vary from cell to cell. If there is no variation in the estimated observation probabilities $\{A_{gh}\}$, then $\mathtt{cv}_U(A_{gh}) = 0$, and there is no undercoverage/nonresponse bias because controlling at the population level was in fact the appropriate adjustment. In the more likely scenario of some variation, however, there is bias determined by the amount of correlation between estimated observation probabilities and cell means.

The model variance of the overall-control estimator is

$$\begin{aligned}\text{Var}_m\left(\hat{T}_{\text{overall}}\right) &= \left\{\tau^2_{\bullet\bullet} + (C_{\bullet\bullet} + \delta_{\bullet\bullet})^2\right\}\frac{\sum_{g,h}\sigma^2_{gh}\sum_{i\in U_{gh}}F_iS_iR_i/\pi_i^2}{\left(\sum_{g,h}\sum_{i\in U_{gh}}F_iS_iR_i/\pi_i\right)^2} \\ &\quad + \tau^2_{\bullet\bullet}\left(\frac{\sum_{g,h}C_{gh}A_{gh}\mu_{gh}}{\sum_{g,h}C_{gh}A_{gh}}\right)^2. \quad \text{(B.15)}\end{aligned}$$

B–2.6 Design Properties of the Overall-Control Estimator

Again using the fact that the overall-control estimator is exactly like an estimator with row control and a single row, the overall-control estimator has relative design bias given approximately by

$$\begin{aligned}\text{RelBias}\left(\hat{T}_{\text{overall}}\right) & \\ \simeq \mathtt{ave}_U\left(\frac{\delta_{gh}}{C_{gh}}\right) & \\ + \mathtt{corr}_U\left(\phi_{gh}\rho_{gh},\bar{y}_{gh}\right)\mathtt{cv}_U(\phi_{gh}\rho_{gh})\mathtt{cv}_U(\bar{y}_{gh}) & \\ + \mathtt{corr}_U\left(\phi_{gh}\rho_{gh},\bar{y}_{gh}\right)\mathtt{cv}_U(\phi_{gh}\rho_{gh})\mathtt{cv}_U(\bar{y}_{gh})\mathtt{ave}_U\left(\frac{\delta_{gh}}{C_{gh}}\right). & \quad \text{(B.16)}\end{aligned}$$

The interpretation of this relative design bias directly parallels the relative model bias described above. The first term is bias attributable solely to estimation error. It would appear even in the absence of any bias due to undercoverage or nonresponse. The second term is bias due only to undercoverage and nonresponse; this term appears even in the absence of postcensal population estimation errors. The final term represents the contribution to the bias from the interaction between undercoverage/nonresponse bias and postcensal population estimation bias.

The design variance of the overall-control estimator is approximately

$$\begin{aligned}
\operatorname{Var}\left(\hat{T}_{\text{overall}}\right) \\
&\simeq \left\{ \left(\frac{C_{\bullet\bullet} + \delta_{\bullet\bullet}}{\sum_{g,h} C_{gh}\phi_{gh}\rho_{gh}} \right)^2 \right. \\
&\quad \left. \sum_{g,h} \sum_{i \in U_{gh}} \frac{\phi_{gh}\rho_{gh}\left(1 - \phi_{gh}\pi_i\rho_{gh}\right)}{\pi_i} \left(y_i - \frac{\sum_{g,h} C_{gh}\phi_{gh}\rho_{gh}\bar{y}_{gh}}{\sum_{g,h} C_{gh}\phi_{gh}\rho_{gh}} \right)^2 \right\} \\
&\quad + \tau_{\bullet\bullet}^2 \left(\frac{\sum_{g,h} C_{gh}\phi_{gh}\rho_{gh}\bar{y}_{gh}}{\sum_{g,h} C_{gh}\phi_{gh}\rho_{gh}} \right)^2 . \qquad \text{(B.17)}
\end{aligned}$$

B–3 NUMERICAL EXAMPLES

Consider a simple artificial population of size $C_{\bullet\bullet} = 1{,}000$ that is poststratified into a two-way table of four equally sized cells with the following characteristics:

	Column 1	Column 2	Row Totals
Row 1	$C_{11} = 250$ $A_{11} = 0.7$ or $\phi_{11}\rho_{11} = 0.7$ $\mu_{11} = 2$ or $\bar{y}_{11} = 2$	$C_{12} = 250$ $A_{12} = 0.8$ or $\phi_{12}\rho_{12} = 0.8$ $\mu_{12} = 1$ or $\bar{y}_{12} = 1$	$C_{1\bullet} = 500$
Row 2	$C_{21} = 250$ $A_{21} = 0.9$ or $\phi_{21}\rho_{21} = 0.9$ $\mu_{21} = 10$ or $\bar{y}_{21} = 10$	$C_{22} = 250$ $A_{22} = 1.0$ or $\phi_{22}\rho_{22} = 1.0$ $\mu_{22} = 2$ or $\bar{y}_{22} = 2$	$C_{2\bullet} = 500$
Column Totals	$C_{\bullet 1} = 500$	$C_{\bullet 2} = 500$	$C_{\bullet\bullet} = 1{,}000$

I focus exclusively on bias in these numerical examples for two main reasons. First, relative bias can be computed without any assumptions on the covariance structure of the postcensal population estimation errors. The independence assumptions under which the variances are computed in Section B–2 may not be plausible in the ACS application, as discussed earlier. Second, mitigation of bias due to undercoverage errors is usually the most important reason for poststratification.

For given levels of postcensal population estimation biases $\{\delta_{gh}\}$, the exact biases and relative biases for the various estimators, under either the model or design assumptions, can be computed directly from these values for cell counts and totals, margin (row or column) counts and totals, and the overall counts and totals. That is, by choosing A_{gh}, μ_{gh} from the above table, we obtain the relative model bias, and by choosing $\phi_{gh}\rho_{gh}, \bar{y}_{gh}$, we obtain the relative design bias. The computation is identical in either case. For example, the exact bias of $\hat{T}_{\text{overall}}$ as an estimator of the overall total given 10 percent relative bias in all of the postcensal population estimates is

$$1.10 * 1{,}000 * \frac{250 * 0.7 * 2 + 250 * 0.8 * 1 + 250 * 0.9 * 10 + 250 * 1.0 * 2}{250 * 0.7 + 250 * 0.8 + 250 * 0.9 + 250 * 1.0}$$
$$-(250 * 2 + 250 * 1 + 250 * 10 + 250 * 2) = 520.59,$$

so that the exact relative bias is

$$\frac{520.59}{250 * 2 + 250 * 1 + 250 * 10 + 250 * 2} = \frac{520.59}{3{,}750} = 0.14.$$

The corresponding approximate relative biases can then be computed from equations (B.14) under the model or (B.16) under the design.

In the numerical examples that follow, I consider three different settings for the postcensal population estimation biases: unbiased ($\delta_{gh}/C_{gh} \equiv 0$), equally biased across cells ($\delta_{gh}/C_{gh} \equiv 0.1$), and unequally biased across cells ($\delta_{11}/C_{11} = -0.1$, $\delta_{12}/C_{12} = 0.0$, $\delta_{21}/C_{21} = 0.2$, $\delta_{22}/C_{22} = 0.3$). Exact relative biases are tabled; approximate relative biases are accurate to three decimal places in this particular example and are not tabled. Note also that relative bias is scale-invariant in the sense that

$$\frac{\mathrm{E}_m\left[\hat{T}(cy_i) - T(cy_i)\right]}{\mathrm{E}_m\left[T(cy_i)\right]} = \frac{\mathrm{E}_m\left[\hat{T}(y_i) - T(y_i)\right]}{\mathrm{E}_m\left[T(y_i)\right]},$$
$$\frac{\mathrm{E}\left[\hat{T}(cy_i)\right] - T(cy_i)}{T(cy_i)} = \frac{\mathrm{E}\left[\hat{T}(y_i)\right] - T(y_i)}{T(y_i)}$$

for $c \neq 0$. Since the bias for estimation of a cell total can be computed with study variable $\mu_{kl} z_{kli}$ or $\bar{y}_{kl} z_{kli}$ in place of $y_i z_{kli}$, the relative bias for estimation of cell totals is the same as the relative bias for estimation of cell counts in every cell, as shown in the following tables. This does not hold for rows, columns, or the overall total.

The unbiased case corresponds to perfect population controls, for which the cell-level PSE is unbiased under the model and approximately unbiased under the design, as shown in Table B.1. The PSEs with margin-level controls and with no controls are generally biased, however, because there is variation in the coverage/response from cell to cell. Exceptions are the unbiased estimates of row counts with $\hat{T}_{\text{row}}$, column counts with $\hat{T}_{\text{column}}$, and overall population count with any estimator.

For the equally biased case, Table B.2 shows that the cell-level PSE has the same relative bias for both counts and totals in all cells and margins. This bias arises solely from the postcensal population estimation bias since $\text{cv}_U\left(\delta_{gh}/C_{gh}\right) = 0$ in equations (B.4) and (B.7).

The margin-level PSEs and the overall-control estimator are all biased, both due to the postcensal population estimation biases and the variability of coverage and response from cell to cell. The relative biases are equal for estimation of row counts with $\hat{T}_{\text{row}}$, column counts with $\hat{T}_{\text{column}}$, and overall population count with any estimator, but generally the relative biases vary. It is interesting to note that for estimation of the overall total, the estimator with smallest relative bias is $\hat{T}_{\text{row}}$, followed by $\hat{T}_{\text{cell}}$, $\hat{T}_{\text{overall}}$, and finally $\hat{T}_{\text{column}}$. Thus, controlling at a higher level of aggregation can be better than controlling at the cell level, or it can be worse than not controlling at all, depending on which margin is chosen. This example illustrates the complexities in deciding among various levels of control.

The final example, in Table B.3, has varying postcensal population estimation biases from cell to cell. In this case, there are some large relative biases throughout the table. Each of the estimators outperforms the others for some estimand, and so no approach dominates. For estimation of the overall total, the best estimator uses no controls, and the worst is the cell-level PSE. This example indicates that controlling at a higher level of aggregation (margins) can be better than controlling at the cell level. In this particular case, neither margin-level PSE beats the overall-control estimator. This example illustrates once again the complexities in deciding among various levels of control.

B–4 DISCUSSION

The derivations and numerical examples described in earlier sections illustrate some important points that may be relevant to the use of population estimates as controls in weighting for the American Community Survey. First, the poststratification estimator that is controlled at the cell level is unbiased only if the postcensal population estimation biases (in correctly specified cells) are orthogonal to cell means, a situation that is unlikely to occur in practice. The cell-level PSE is likely to be biased for many ACS estimates.

Second, the margin-level PSE has biases arising from both nonobservation errors (undercoverage and nonresponse) and postcensal population estimation errors, as does the overall-control estimator. These different sources

TABLE B.1 Exact Relative Biases for Case of Unbiased Postcensal Population Estimates

	Column 1				Column 2				Row Sums			
	$\hat{T}_{cell}$	$\hat{T}_{row}$	$\hat{T}_{column}$	$\hat{T}_{overall}$	$\hat{T}_{cell}$	$\hat{T}_{row}$	$\hat{T}_{column}$	$\hat{T}_{overall}$	$\hat{T}_{cell}$	$\hat{T}_{row}$	$\hat{T}_{column}$	$\hat{T}_{overall}$
Row 1												
Count	0	−0.067	−0.13	−0.18	0	0.067	−0.11	−0.059	0	0	−0.12	−0.12
Total	0	−0.067	−0.13	−0.18	0	0.067	−0.11	−0.059	0	−0.022	−0.12	−0.14
Row 2												
Count	0	−0.053	0.13	0.059	0	0.053	0.11	0.18	0	0	0.12	0.12
Total	0	−0.053	0.13	0.059	0	0.053	0.11	0.18	0	−0.035	0.13	0.078
Column Sums												
Count	0	−0.060	0	−0.059	0	0.060	0	0.059	0	0	0	0
Total	0	−0.055	0.083	0.020	0	0.057	0.037	0.098	0	−0.033	0.074	0.035

NOTE: Shown are exact relative biases for estimators controlled to postcensal population estimates at cell or margin (row or column), or controlled only to overall postcensal population estimate ("overall"), in the case of unbiased postcensal population estimates ($\delta_{gh}/C_{gh} \equiv 0$). Approximate relative biases from formulae in text are equivalent to three decimal places with exact relative biases.

TABLE B.2 Exact Relative Biases for Case of Equally Biased Postcensal Population Estimates

	Column 1				Column 2				Row Sums			
	$\hat{T}_{\text{cell}}$	$\hat{T}_{\text{row}}$	$\hat{T}_{\text{column}}$	$\hat{T}_{\text{overall}}$	$\hat{T}_{\text{cell}}$	$\hat{T}_{\text{row}}$	$\hat{T}_{\text{column}}$	$\hat{T}_{\text{overall}}$	$\hat{T}_{\text{cell}}$	$\hat{T}_{\text{row}}$	$\hat{T}_{\text{column}}$	$\hat{T}_{\text{overall}}$
Row 1												
Count	0.1	0.027	−0.038	−0.094	0.1	0.17	−0.022	0.035	0.1	0.1	−0.030	−0.029
Total	0.1	0.027	−0.038	−0.094	0.1	0.17	−0.022	0.035	0.1	0.076	−0.032	−0.051
Row 2												
Count	0.1	0.042	0.24	0.16	0.1	0.16	0.22	0.29	0.1	0.1	0.23	0.23
Total	0.1	0.042	0.24	0.16	0.1	0.16	0.22	0.29	0.1	0.061	0.24	0.19
Column Sums												
Count	0.1	0.034	0.1	0.035	0.1	0.17	0.1	0.16	0.1	0.1	0.1	0.1
Total	0.1	0.040	0.19	0.12	0.1	0.16	0.14	0.21	0.1	0.064	0.18	0.14

NOTE: Shown are exact relative biases for estimators controlled to postcensal population estimates at cell or margin (row or column), or controlled only to overall postcensal population estimate ("overall"), in the case of equally biased postcensal population estimates ($\delta_{gh}/C_{gh} \equiv 0.1$). Approximate relative biases from formulae in text are equivalent to three decimal places with exact relative biases.

TABLE B.3 Exact Relative Biases for Case of Unequally Biased Postcensal Population Estimates

	Column 1				Column 2				Row Sums			
	$\hat{T}_{cell}$	$\hat{T}_{row}$	$\hat{T}_{column}$	$\hat{T}_{overall}$	$\hat{T}_{cell}$	$\hat{T}_{row}$	$\hat{T}_{column}$	$\hat{T}_{overall}$	$\hat{T}_{cell}$	$\hat{T}_{row}$	$\hat{T}_{column}$	$\hat{T}_{overall}$
Row 1												
Count	−0.1	−0.11	−0.081	−0.094	0	0.013	0.022	0.035	−0.05	−0.05	−0.03	−0.029
Total	−0.1	−0.11	−0.081	−0.094	0	0.013	0.022	0.035	−0.067	−0.071	−0.047	−0.051
Row 2												
Count	0.2	0.18	0.18	0.16	0.3	0.32	0.28	0.29	0.25	0.25	0.23	0.23
Total	0.2	0.18	0.18	0.16	0.3	0.32	0.28	0.29	0.22	0.21	0.20	0.19
Column Sums												
Count	0.05	0.035	0.05	0.035	0.15	0.16	0.15	0.16	0.1	0.1	0.1	0.1
Total	0.15	0.13	0.14	0.12	0.2	0.22	0.19	0.21	0.16	0.15	0.15	0.14

NOTE: Shown are exact relative biases for estimators controlled to postcensal population estimates at cell or margin (row or column), or controlled only to overall postcensal population estimate ("overall"), in the case of unequally biased postcensal population estimates ($\delta_{11}/C_{11} = -0.1, \delta_{12}/C_{12} = 0.0, \delta_{21}/C_{21} = 0.2, \delta_{22}/C_{22} = 0.3$). Approximate relative biases from formulae in text are equivalent to three decimal places with exact relative biases.

of bias interact in complex ways, so that it is difficult to characterize the appropriate level of aggregation for poststratification. Numerical examples show that cell control, row control, column control, or no control may be best depending on the parameter settings and population quantity of interest. Control on both margins through some kind of raking procedure was not treated here but is worthy of further consideration.

The results in this paper indicate that the Census Bureau's plan for control at a fine level of demographic stratification within estimation areas may be problematic. It may yield estimators with bias properties worse than no controls at all.

This paper is only a first step in evaluation of the possible effects of errors in postcensal population controls on ACS estimates. Research is needed in a number of directions. First, the numerical results are limited and the parameter values in that limited study were chosen to illustrate potential problems, which may or may not occur in real ACS data. For example, the artificial population has cell means that vary by a factor of 10, which may or may not be realistic in ACS applications. It is necessary to explore a range of parameter values (response probabilities, coverage probabilities, cell means, postcensal population estimation bias, etc.) that are plausible in real ACS applications to determine whether or not the potential problems identified in this paper are real problems for the ACS.

Second, the numerical experiments focus exclusively on bias, because bias is a major reason for poststratification and because the independence assumptions under which variances are derived in this paper are possibly unrealistic. Certainly bias is critical, and in many applications it dominates variance. Ultimately, the interest is mean squared error, the sum of squared bias and variance. To study variance analytically, it is necessary to make some assumptions about the covariance structure for the various types of errors (for example, assumptions about correlations among postcensal population estimation errors in different cells, or between postcensal population estimation errors and frame imperfections). These assumptions should be guided by a careful consideration of the ACS and the methods of postcensal population estimation. Analytic computations under these assumptions could be supplemented or replaced by simulations.

Finally, this paper does not explore the full complexity of the weighting factors used for the ACS, so the issue of bias would need further study, both analytical and empirical, in this more complex setting.

REFERENCES

Brewer, K. (2002). *Combined survey sampling inference: Weighing Basu's elephants.* London: Arnold.

Cochran, W.G. (1977). *Sampling techniques*, 3rd edition. New York: Wiley.

Särndal, C.-E., Swensson, B., and Wretman, J. (1992). *Model assisted survey sampling*, New York: Springer.
Stephan, F.F. (1941). Stratification in representative sampling. *Journal of Marketing*, 6, 38–46.

C

Alternatives to the Multiyear Period Estimation Strategy for the American Community Survey

F. Jay Breidt
Colorado State University

ABSTRACT

A class of estimation strategies that includes simple moving averages and direct estimates as special cases is evaluated for small area estimation in the American Community Survey. The evaluation is based on both operational and theoretical considerations. Operationally, the estimation strategies considered are feasible in a massive-scale production environment. Theoretically, the estimation strategies are compared using simple decision-theoretic tools, which suggest good compromise strategies that borrow strength across time in a robust way. Strategies outside the class considered here can be evaluated with these same decision-theoretic tools.

C–1 INTRODUCTION

C–1.1 Proposed Multiyear Estimation Strategy for the ACS

The publication plans for estimates from the American Community Survey (ACS) are to produce 1-year estimates for areas with populations of 65,000 persons or more, 3-year estimates for areas with 20,000 persons or more, and 5-year estimates for all areas (all governmental units, census tracts, and block groups). When the survey is fully established, 1-year, 3-year, and 5-year estimates will be produced every year based on the latest prior year or set of prior years. The National Research Council's Panel on the Functionality and Usability of Data from the American Community Survey (ACS) asked me to address the multiyear estimation strategy for the ACS. The weighting scheme currently proposed by the Census Bureau involves pooling the survey data across the 3 or 5 years. The weights will be developed starting with the inverse selection probabilities of sampled households.

A number of adjustments will then be made including nonresponse adjustments and poststratification adjustments to postcensal housing unit and population estimates produced by the Census Bureau's Population Division. The housing and population controls are averages of the 1-year controls for the multiple years. Details of the weighting procedures are given in Chapters 5 and 6 of the panel's report.

The multiyear estimates produced by the Census Bureau's weighting scheme can be viewed as period estimates: they represent averages that reflect both changing characteristics and changes in the area's populations across the years. The limitation of these estimates, and changes in them over time, is that they can be difficult to interpret and may not suit user needs. The panel therefore invited me to investigate other estimation strategies for the multiyear data, and in particular the use of several years of data to produce an estimate for a single year (e.g., year 3 or 5 from 5 years of ACS data) in place of the period estimate. The fact that any strategy that is adopted would have to implemented in a massive production environment imposes constraints: It needs to be simple and require no auxiliary data; and each unit (household or person) should have only one analysis weight within the given 3- or 5-year data set in order to enable a wide range of consistent analyses across variables and areas of different sizes.

There are strong arguments that one can make in support of a strategy that uses different weights for producing single-year estimates for areas of different sizes from multiyear data. For example, there are definite advantages to borrowing strength over more years for areas with small populations and for variables that are more stable in time. However, ACS data users would likely find the non-uniformity an unwelcome complication and possibly undesirable, at least during the start up phase of ACS. The paper therefore focuses exclusively on uniform strategies.

Two simplifying assumptions are made throughout the paper in order to convey the essence of a principled strategy for producing single-year estimates with desirable properties from multiyear data. The first is that the population size of an area does not change over the 3- or 5-year estimation period. This assumption may hold reasonably well for many areas, but there will be areas for which it does not hold. The second assumption is that the 1-year estimates for each of the years in the period have the same variance. With the assumption of a constant population size, the ACS sample size in an area is likely to be approximately the same each year. Thus, if the element variances are about the same across the years, the second assumption will hold approximately. While element variances may often be reasonably equal, that will not always be the case.

Under these two assumptions, and ignoring the nonresponse and calibration weighting adjustments, the Census Bureau's period estimate reduces to a simple average of the 1-year estimates for the period. In order to produce estimates for a given year within the period, this paper examines a strategy that assigns differential weights to the 1-year estimates, with largest weight given to the year in question and the next larger weights to the neighboring years. Each year the ACS will update the 3- and 5-year estimates by replacing the earliest year by the latest year. For this reason, the estimates produced are described here as moving average (MA) estimates.

Although the simple MA is operationally convenient and easily understood, a number of questions arise regarding its appropriateness. It can be applied to any sort of characteristics across any geographic and demographic domains, but does it give efficient estimates for those characteristics on those domains? Is the method defensible? Is it a principled approach for obtaining estimates, with some theoretical justification? Can it be extended in a logical way to novel estimation problems?

C–1.2 Signal-Plus-Noise Model

To address these questions, I begin with a general model for annual ACS estimates, written as a time series $\{Y_t\}$. Assume the classical signal-plus-noise formulation

$$Y_t = \theta_t + e_t$$

where the signal, θ_t, represents the true unobserved population characteristic in year t, and e_t represents both sampling and variable nonsampling error (the estimates are assumed to be unbiased). See, for example, Scott and Smith (1974) as well as Binder and Hidiroglou (1988) and the references therein. On one hand, presumably, the sampling error would have some negative correlation by design, since the ACS rolls through the population, avoiding selection of the same households month-to-month. On the other hand, the variable nonsampling error would be expected to have some positive month-to-month correlation (e.g., due to nonresponse follow-up with common computer-assisted telephone and personal interviewing staffs from month to month). Assume that at the annual level of aggregation considered here, these correlations are negligible, so that $\{e_t\}$ is uncorrelated.

It is convenient to let

$$\boldsymbol{\theta}_t = (\theta_{t-m+1}, \ldots, \theta_t)^T$$

denote the vector consisting of the m most recent annual true values of the characteristic of interest. Furthermore, consider the random vector of m ACS annual estimates,

$$\boldsymbol{Y}_t = (Y_{t-m+1}, \ldots, Y_t)^T$$

so that

$$\boldsymbol{Y}_t = \boldsymbol{\theta}_t + \boldsymbol{e}_t,$$

with $e_t \sim (0, \sigma_e{}^2 I)$, where I is the $m \times m$ identity matrix.

I consider linear estimators of θ_t given by some known $m \times m$ matrix S multiplied by Y_t: $\hat{\theta}_t = SY_t$. (Note that such estimators use only data from the m-year time window, for direct comparability with the ACS 3- and 5-year estimates. Given the various dynamics that the ACS will be subject to, it makes sense to limit the data used to a small number of years, although of course some information is lost by this restriction.)

Various population characteristics might be of interest, several of which can be written as linear functions $\mathbf{z}^T\theta_t$ for some known $m \times 1$ vector $\mathbf{z}$. Examples of linear functions for $m = 5$ include $\mathbf{z}^T = [0,0,0,0,1]$ for current level, $\mathbf{z}^T = [0,0,1,0,0]$ for midpoint level, and $\mathbf{z}^T = [1/5,1/5,1/5,1/5,1/5]$ for temporal average. Each of these linear functions can be estimated in the obvious way as

$$\widehat{\mathbf{z}^T\theta_t} = \mathbf{z}^T S Y_t.$$

Estimates of change are more complicated, since there are at least two obvious estimation strategies. The first is to define, for $k < m$, the k-year change as a linear function of θ_t, with

$$\mathbf{z}^T\theta_t = [0,\ldots,0,-1,0,\ldots,0,1]\theta_t,$$

and estimate k-year change by

$$\widehat{\mathbf{z}^T\theta_t} = \mathbf{z}^T S Y_t$$

exactly as above. Note that this "current" estimator uses only data from the current m-year time window.

The second estimation strategy for change is to compute the difference between the published level estimate for year t and year $t - k$. Note that this "final" estimator uses data from both the current m-year time window and the lagged m-year time window. For example, in estimation of 1-year change, the "current" estimate of change is computed from only the current 5-year window, while the "final" estimate of change is computed from consecutive 5-year windows. The final change estimate is presumably the one that would be published in order to maintain consistency with published level estimates.

C–1.3 Classes of Estimation Strategies

The matrix S can be chosen in a number of ways. It could be constructed from principles of filter design commonly used in time series: for example, one can construct a filter that passes a quadratic trend without distortion while eliminating certain seasonal components or attenuating noise at certain frequencies (Brockwell and Davis, 1991, Chapter 1).

More formally, strategies may be based on temporal models, either stochastic or deterministic. If the temporal model is stochastic, then optimal filters can be computed using well-known principles; for example, the Kalman filter does these computations recursively for large classes of linear state-space models.

If the state-space model is a random walk plus noise (a special case of a process that is integrated of order one, or I(1)), then the Kalman filter becomes equivalent to exponential smoothing as $m \to \infty$ (Harvey, 1989, p. 440). If the state-space model is a particular type of local linear trend (a special case of an I(2) process), then the Kalman filter yields double exponential smoothing (Harvey, 1989, p. 177).

The temporal model might, however, be deterministic, specifying only that the true values θ_t evolve as a smooth unknown function of time. In this case, methods from nonparametric regression, such as local polynomial kernels or smoothing splines, could be used to derive estimation strategies. It is interesting to note the connections between the stochastic and deterministic cases. First, exponential smoothing can be derived as a special case of nonparametric regression: the Nadaraya-Watson kernel smoother (zero-th order polynomial) with a particular form of half-kernel (Gijbels, Pope, and Wand, 1999). Second, smoothing splines can be derived as the optimal filtering solution for a local linear trend stochastic model (Durbin and Koopman, 2001, p. 61).

Other strategies might be devised based on spatial or spatiotemporal considerations, but it seems difficult to develop spatial methods applicable in a large-scale production environment. On one hand, defining spatial neighborhoods would be difficult, since they would vary substantially across space. On the other hand, defining temporal neighborhoods is straightforward. In addition, there may be political complications that arise from borrowing strength across governmental units that are not raised by temporal averaging, because of the interest in comparisons among governmental units.

C–1.4 Framework for Comparing Estimation Strategies

It is tempting to use existing theory for either Kalman filtering in the stochastic case or nonparametric regression in the deterministic case, to evaluate the performance of various estimation strategies. However, these methods typically assume that $m \to \infty$, which is not the case in the ACS application. I use other techniques to evaluate the performance of estimation strategies.

Given the large number of possible estimation strategies, one needs a principled approach to comparing them and choosing a reasonable compromise among them. In comparing these strategies theoretically, it is critical to keep in mind the operational constraint that it is not feasible or desirable to develop separate estimation strategies for each governmental unit of interest. The strategies must be quite generic.

With this in mind, I propose a simple decision-theoretic framework for comparing strategies. I focus on squared error loss

$$\left(\mathbf{z}^T\theta_t - \mathbf{z}^T S Y_t\right)\left(\mathbf{z}^T\theta_t - \mathbf{z}^T S Y_t\right)^T,$$

for which the corresponding risk is the prediction mean squared error (MSE) $\mathbf{z}^T\Omega\mathbf{z}$, where

$$\Omega = E(\theta_t - SY_t)(\theta_t - SY_t)^T.$$

C–2 METHODS

This framework for comparing estimation strategies is best illustrated by focusing on a particular class of strategies. I use the class of I(1) strategies as formulated by William Bell in a presentation at a 1998 Committee on National Statistics workshop on the ACS (National Research Council, 2001). The I(1) strategies are derived from a random walk plus noise model, but this derivation is not important in what follows. The strategy to be evaluated might be derived from a formal model or might be entirely ad hoc, but in either case it can be evaluated with the methods to be described.

Define

$$\Delta_m = \begin{bmatrix} -1 & 1 & 0 & \cdots & 0 \\ 0 & -1 & 1 & \cdots & 0 \\ \vdots & \vdots & \ddots & \ddots & \vdots \\ 0 & 0 & \cdots & -1 & 1 \end{bmatrix}_{(m-1)\times m},$$

with the subscript suppressed when the dimension is clear, and let

$$A = \alpha I_{m-1} + \Delta\Delta^T,$$

with $\alpha \geq 0$ prespecified. Define the smoother matrix

$$S = I_m - \Delta^T A^{-1} \Delta.$$

Note that the rows of S sum to 1 for any choice of α, since $\Delta\mathbf{1} = 0$.

A few numerical examples of S demonstrate the breadth of the I(1) estimation strategies. Consider the case $m = 5$. With $\alpha = 0$, the smoother matrix is

$$S = \begin{bmatrix} 0.2 & 0.2 & 0.2 & 0.2 & 0.2 \\ 0.2 & 0.2 & 0.2 & 0.2 & 0.2 \\ 0.2 & 0.2 & 0.2 & 0.2 & 0.2 \\ 0.2 & 0.2 & 0.2 & 0.2 & 0.2 \\ 0.2 & 0.2 & 0.2 & 0.2 & 0.2 \end{bmatrix}.$$

In this case, the estimate of current level becomes

$$\begin{aligned}(0,0,0,0,1)S\boldsymbol{Y}_t &= (0.2,0.2,0.2,0.2,0.2)\boldsymbol{Y}_t\\ &= \frac{Y_{t-4}+Y_{t-3}+Y_{t-2}+Y_{t-1}+Y_t}{5},\end{aligned}$$

which corresponds to the 5-year period estimate proposed by the Census Bureau.

The other extreme is obtained as $\alpha \to \infty$, in which case the smoother matrix is

$$S = \begin{bmatrix} 1 & 0 & 0 & 0 & 0 \\ 0 & 1 & 0 & 0 & 0 \\ 0 & 0 & 1 & 0 & 0 \\ 0 & 0 & 0 & 1 & 0 \\ 0 & 0 & 0 & 0 & 1 \end{bmatrix}.$$

In this case the estimate of current level is simply

$$(0,0,0,0,1)S\boldsymbol{Y}_t = (0,0,0,0,1)\boldsymbol{Y}_t = Y_t,$$

the direct estimate in the current year. This is the estimator produced by the Census Bureau for areas of more than 65,000 persons.

Between these two extremes lies a continuum of smoothing possibilities. Consider, for example, $\alpha = 0.4206232$, in which case the smoother matrix is

$$S = \begin{bmatrix} 0.4738 & 0.2525 & 0.1374 & 0.0800 & 0.0563 \\ 0.2525 & 0.3587 & 0.1951 & 0.1137 & 0.0800 \\ 0.1374 & 0.1951 & 0.3350 & 0.1951 & 0.1374 \\ 0.0800 & 0.1137 & 0.1951 & 0.3587 & 0.2525 \\ 0.0563 & 0.0800 & 0.1374 & 0.2525 & 0.4738 \end{bmatrix}.$$

The estimate of current level in this case is

$$\begin{aligned}&(0,0,0,0,1)S\boldsymbol{Y}_t\\ &\quad = (0.0563,0.0800,0.1374,0.2525,0.4738)\boldsymbol{Y}_t,\end{aligned}$$

a weighted average with weights that look much like the exponential smoothing weights

$$(1/32,1/16,1/8,1/4,1/2) = (0.03125,0.06250,0.12500,0.25000,0.50000).$$

A useful summary of the amount of smoothing is given by the degrees of freedom (df) of the smoother matrix,

$$\text{df} = \text{trace}(S) = \text{trace}\left(I_m - \Delta^T A^{-1}\Delta\right).$$

This value varies continuously between 1 and m. One df represents maximum smoothing; it corresponds to the simple moving average, or fitting of a common mean over the m-year window. No smoothing is represented by m df; this corresponds to the direct estimates, or fitting of separate means for each year. Other values of df correspond to different amounts of smoothing between the maximum and minimum values. A few examples are given in the following table:

df	α	
1	0	moving average: $S=(1/5)\mathbf{1}\mathbf{1}^T$
2	0.4206232	
3	1.545009	
4	5.380712	
5	∞	direct estimates: $S=I_5$

Note that the $\alpha = 0.4206232$ case considered above corresponds to two degrees of freedom and roughly corresponds to exponential smoothing with parameter 1/2.

C–3 RESULTS

C–3.1 General Results

Assume that changes in the signal are uncorrelated with the noise,

$$E(\Delta\theta_t)\boldsymbol{e}_t^T = 0_{(m-1)\times m} \tag{C.1}$$

and that

$$E(\Delta\theta_t)(\Delta\theta_t)^T = \sigma_e{}^2\psi M, \tag{C.2}$$

where the $(m-1)\times(m-1)$ matrix M does not depend on t. (That is, the covariance matrix for the differenced signal is time-invariant.) The matrix M depends on the model for the signal; several examples of M are given below under different models. The scalar ψ is interpreted as a signal-to-noise ratio (SNR).

The risk for estimation of the linear function $\boldsymbol{z}^T\theta$ under squared error loss is then $\boldsymbol{z}^T\Omega\boldsymbol{z}$, where

$$\begin{aligned}\Omega &= E\left(\theta-(I-\Delta^T A^{-1}\Delta)Y\right)\left(\theta-(I-\Delta^T A^{-1}\Delta)Y\right)^T \\ &= \sigma_e{}^2\left\{I-2\Delta^T A^{-1}\Delta+\Delta^T A^{-1}\left(\psi M+\Delta\Delta^T\right)A^{-1}\Delta\right\}. \end{aligned} \tag{C.3}$$

We can effectively take $\sigma_e{}^2 = 1$ and interpret all risks in units of noise variance. Note that (C.3) then depends on the smoothing parameter α through

A and on the true model through ψM. Also, observe that the term depending on ψM is the squared bias under model (C.2), and all other terms are attributable to variance from the noise.

Two special cases of this risk computation are worth considering, before moving on to consideration of the general case. First, for the direct estimates, $S = I$ and the risk is

$$\mathbf{z}^T \Omega \mathbf{z} = \sigma_e^{\,2} \mathbf{z}^T \mathbf{z} = \begin{cases} \sigma_e^{\,2}, & \text{level, midpoint} \\ 2\sigma_e^{\,2}, & \text{change} \\ \sigma_e^{\,2}/m, & \text{temporal average.} \end{cases} \tag{C.4}$$

These are useful benchmark values in looking at risk surfaces as functions of α.

The second special case arises in estimating the temporal average. The risk is given in the following result.

Result 1 *For any choice of the smoothing parameter α and for any model satisfying the conditions (C.1) and (C.2) above, the risk for estimating the temporal average $m^{-1}\mathbf{1}^T\theta_t$ is $\sigma_e^{\,2}m^{-1}$.*

This result is immediate from the fact that $\Delta \mathbf{1} = \mathbf{0}$. The result implies that if one is interested in estimating the temporal average only, then any strategy in this class is equally good, and the result does not depend on parameterization of the true model. The temporal average is unusual in this regard. In general, the risk surface depends nontrivially on the strategy and on the model. We need to remove the dependence on the model and then choose an optimal strategy.

One linear function of interest that is not included in the discussion above is "final" estimation of k-year change, computed as the difference of level estimates, as discussed in Section C–1.2. For $k = 1$, 1-year change is computed as follows: $\hat{\theta}_{t-1}$ is estimated on the basis of $\boldsymbol{Y}_{t-1}$, $\hat{\theta}_t$ is estimated on the basis of $\boldsymbol{Y}_t$, and the 1-year change is estimated as $\hat{\theta}_t - \hat{\theta}_{t-1}$. The prediction error is therefore

$$\begin{aligned}
&(0,\ldots,0,1)\{(\theta_t - \theta_{t-1}) - (S\boldsymbol{Y}_t - S\boldsymbol{Y}_{t-1})\} \\
&= (0,\ldots,0,1)\left\{\Delta_{m+1}\begin{bmatrix}\theta_{t-m} \\ \theta_t\end{bmatrix} - S\Delta_{m+1}\begin{bmatrix}Y_{t-m} \\ \boldsymbol{Y}_t\end{bmatrix}\right\} \\
&= (0,\ldots,0,1)\left\{-\Delta_{m+1}\begin{bmatrix}\epsilon_{t-m} \\ \boldsymbol{e}_t\end{bmatrix}\right. \\
&\qquad \left. + \Delta_m^T A^{-1}\Delta_m\left(\Delta_{m+1}\begin{bmatrix}\theta_{t-m} \\ \theta_t\end{bmatrix} + \Delta_{m+1}\begin{bmatrix}\epsilon_{t-m} \\ \boldsymbol{e}_t\end{bmatrix}\right)\right\}.
\end{aligned}$$

Extension to k-year change for $k > 1$ is straightforward.

C–3.2 Results for the I(1) Strategy Under the I(1) Model

The I(1) strategy was derived under an I(1) model, but the use of this strategy does not require that the I(1) model holds. Suppose for the moment that the I(1) model does hold. That is,

$$(1-B)\theta_t = \eta_t, \quad \{\eta_t\} \sim \mathrm{WN}(0, \sigma_\eta^2),$$

where "WN" signifies that $\{\eta_t\}$ are "white noise" or uncorrelated; B is the backshift operator ($B^k X_t = X_{t-k}$ for $k = 0, \pm 1, \pm 2 \ldots$); and the SNR ψ and model matrix M from (C.2) are given by

$$\psi = \frac{\sigma_\eta^2}{{\sigma_e}^2}, M = I.$$

Under this formulation, consider the risk surfaces in Figure C.1, which are functions of the strategy through df (or equivalently, through α) and are functions of the model (through the SNR $= \psi$).

Consider the upper left contour plot in Figure C.1, corresponding to the risk for estimation of current level. Note that this contour plot is in units of ${\sigma_e}^2$. The rightmost edge of this plot corresponds to the direct estimator, at 5 df. The risk for the direct estimator is identically 1, as given in equation (C.4). The lower left corner corresponds to the simple MA (1 df) with SNR $= 0$ (constant mean function). In this case, the risk is ${\sigma_e}^2/5$, or 0.2.

Similarly, the upper right contour plot corresponds to risk for estimation of the middle year values, or midpoint. Once again, the right edge is identically 1, and the lower left corner is 0.2. The bottom two plots correspond to estimates of 1-year change. The lower left plot is the "current" estimate of change computed from only the 5-year window, while the lower right plot is the "final" estimate of change computed from consecutive 5-year windows. This "final" estimate is the difference between current level estimates and presumably is the estimate that would be published. In both cases, the right edge is identically 2.

To choose an optimal strategy, it is necessary to remove the dependence on the model, which in this context means removing dependence on the SNR, ψ, since M is parameter-free. Two standard approaches are to compute the supremum risk over all models, or the average risk over all models.

The supremum risk corresponds to the worst-case scenario. The strategy that minimizes the maximum risk is the *minimax strategy*. For each strategy, find the model that maximizes the risk; that is, find the maximum on the risk surface along a vertical slice at a particular df. Then choose the strategy that minimizes this maximum risk curve. Clearly the minimax strategy is a very conservative approach.

For df $= 5$ in Figure C.1, the contour is identically 1, and so the maximum risk at df $= 5$ is 1. For any df < 5, the risk increases without bound as

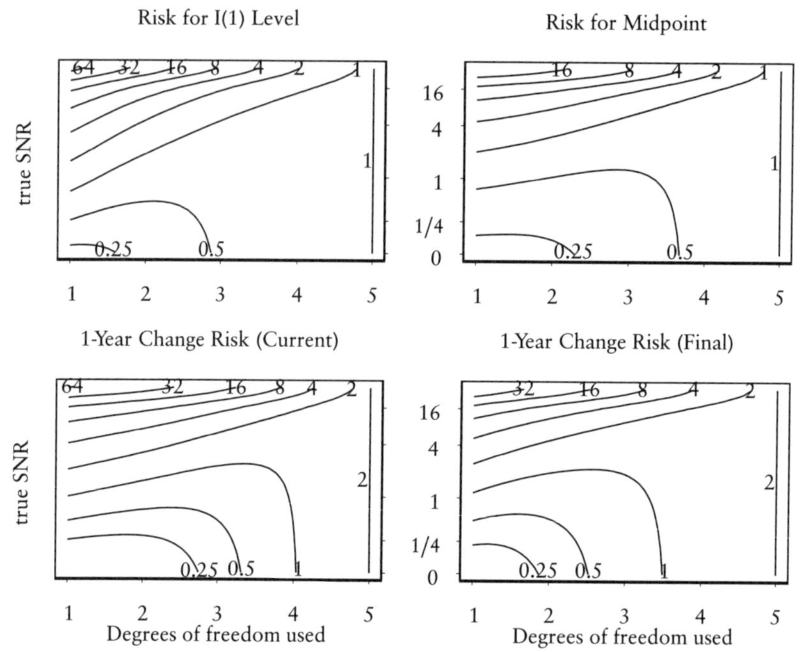

FIGURE C.1 Risk surfaces under I(1) model and I(1) strategy for estimation of current level, level at midpoint of 5-year time window, and 1-year change.

NOTE: "Current" estimate of change uses only 5-year time window, while "Final" uses consecutive 5-year time windows. Horizontal axis is degrees of freedom used in the smoother, from 1 = simple moving average to 5 = direct estimates. Vertical axis is true signal-to-noise ratio (SNR). Contours of risk surface are in units of σ_e^2.

the SNR increases. The supremum is therefore 1 for df = 5, and infinity for df < 5, so the minimax strategy is df = 5, the direct estimators.

This result can be generalized. Consider a model in which the matrix M does not depend on SNR ψ or on any other unknown parameter and in which all elements of M are finite. Then for any $\alpha < \infty$, maximizing the risk with respect to ψ is equivalent to maximizing

$$Q(\psi) = \psi \mathbf{z}^T \Delta^T A^{-1} M A^{-1} \Delta \mathbf{z} \geq 0$$

with respect to ψ. If $Q(\psi) \neq 0$, then $Q(\psi)$ is unbounded in ψ for any $\alpha < \infty$, but finite for $\alpha = \infty$. In this case, the direct estimates (df $= m$) are minimax.

The second approach to removing the dependence of the risk on the model is to consider the average or Bayes risk. Assume that the SNR has a

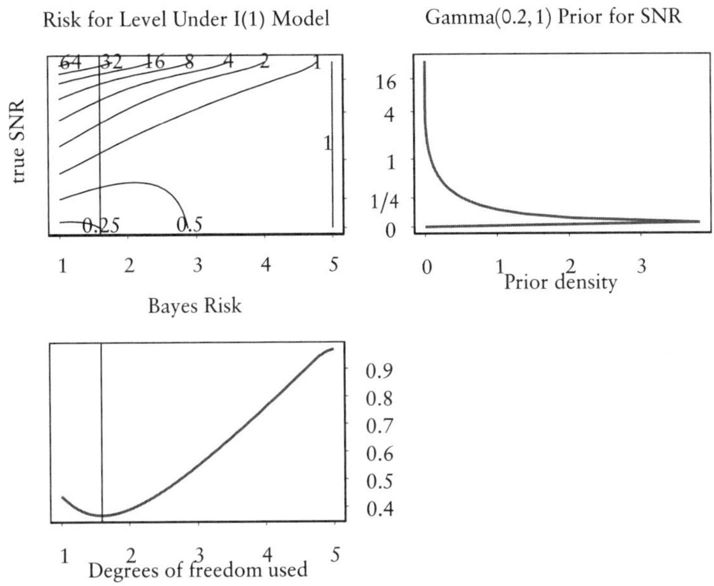

FIGURE C.2 Computation of Bayes risk for estimation of current level.

NOTE: Horizontal axis is degrees of freedom used in the smoother, from 1 = simple moving average to 5 = direct estimates. Vertical axis is true signal-to-noise ratio (SNR). Bottom left plot is Bayes risk, obtained by averaging risk surface in upper left with respect to prior density on SNR shown in upper right (Gamma(0.2,1) with prior mean 0.2). Bayes strategy under this prior is indicated by vertical line at 1.5394 df ($\alpha = 0.2$). Risk surface under I(1) model and I(1) strategy for estimation of current level is shown in upper left.

prior distribution, $\pi(\psi)$. For each strategy, find the average risk, computed with respect to the prior distribution. Then choose the strategy that minimizes this average risk curve. This is the *Bayes strategy*. The procedure is illustrated in Figure C.2. The upper right plot of the figure shows the prior density for the SNR (rotated 90 degrees since the prior density applies to the vertical axis of the upper left plot). The prior in this case, which is Gamma(0.2, 1) with mean $\psi_0 = 0.2$, assigns most mass to SNRs less than 0.5. Integrating the risk surface in the upper left along the vertical slice at df = 5 is the same as integrating the constant 1 against the prior, and so the average risk at df = 5 is 1, as shown in the lower left plot. For df slightly less than 5, the integral puts most mass on risks less than 1, and almost no mass on risks greater than 1, so the average risk is less than 1. For df = 1, the risks greater than one contribute nontrivially to the integral. At about 1.5934 df ($\alpha = 0.2$), the average risk attains its minimum, and so this is the Bayes strategy for estimation of current level under the given prior.

In fact, it is easy to find the Bayes strategy analytically under the I(1) model and to show that this same strategy is Bayes for estimation of any linear function. This is the content of the following result.

Result 2 *Assume that the I(1) model holds and that the SNR $\psi = \sigma_\eta^2 \sigma_e^{-2}$ has prior $\pi(\psi)$ with prior mean ψ_0. Then the Bayes strategy for estimation of any linear function $\mathbf{z}^T\theta$ using $\mathbf{Y}_t$ is obtained with $\alpha = \psi_0$:*

$$\mathbf{z}^T\left(I - \Delta^T\left(\psi_0 I + \Delta\Delta^T\right)^{-1}\Delta\right)\mathbf{Y}_t;$$

that is, this strategy minimizes expected risk among all rules of the form $\mathbf{z}^T\left(I - \Delta^T\left(\alpha I + \Delta\Delta^T\right)^{-1}\Delta\right)\mathbf{Y}_t$. The Bayes risk for this strategy is

$$\mathbf{z}^T\left(I - \Delta^T\left(\psi_0 I + \Delta\Delta^T\right)^{-1}\Delta\right)\mathbf{z}.$$

A consequence of Result 1 is that all strategies have the same average risk for estimating the temporal average, and so all strategies are equally successful. Ignoring this special case, an immediate consequence of Result 2 is that the simple MA is not Bayes for the I(1) model unless the prior is degenerate; that is, $\psi = 0$ with probability 1. (In other words, with an SNR = 0, the mean process is not changing in time, and therefore a simple average is optimal.)

Figure C.3 was constructed with the same prior shown in Figure C.2, but for a variety of linear functions. Figure C.3 illustrates the fact that all of these linear functions have minimum average risk at the same strategy, corresponding to $\alpha = \psi_0$.

C–3.3 Results for the I(1) Strategy Under Non-I(1) Models

I now turn to the robustness question of what happens if the proposed I(1) strategy is used with a non-I(1) model. For numerical illustration, it is convenient to consider models under which M is parameter-free, so that the risk depends on the model only through the single SNR parameter ψ. Some examples follow.

The I(2) Model

First, consider the I(2) model,

$$(1 - B)^2\theta_t = \eta_t,$$

where $\{\eta_t\} \sim \mathrm{WN}(0, \sigma_\eta^2)$ and $(1 - B)\theta_0$ is a constant. Then

$$M = \frac{\sigma_\eta^2}{\sigma_e^{\,2}} TT^T,$$

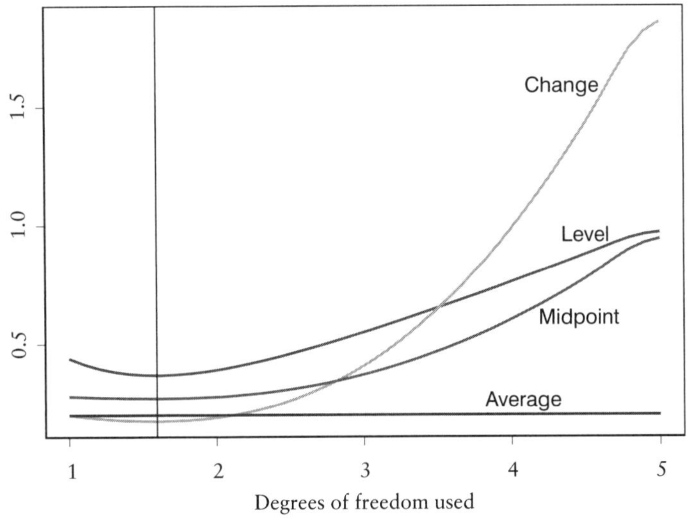

FIGURE C.3 Bayes risk of I(1) strategy for various linear functions under I(1) model with Gamma(0.2,1) prior on SNR.

NOTE: For each linear function, a Bayes strategy under this prior is to smooth with 1.5394 df ($\alpha = 0.2$), indicated by the vertical line.

with T being the lower triangular matrix of one's,

$$T = \begin{bmatrix} 1 & & 0 \\ \vdots & \ddots & \\ 1 & \cdots & 1 \end{bmatrix}.$$

The risk surfaces under this I(2) model are shown in Figure C.4. Note that the bottom edges of each of the plots in the figure agree with the corresponding bottom edges in Figure C.1, because the I(1) and I(2) models are identical when the SNR is zero.

Dependence of the risk on the model could be removed in the same way as in the I(1) case. The minimax strategy is again to use the direct estimators, and the Bayes strategy could be derived given a suitable prior on the SNR. Unlike in the I(1) case, the Bayes strategy will depend on which linear function is of interest. Since more than one linear function is usually of interest, some compromise strategy would need to be selected.

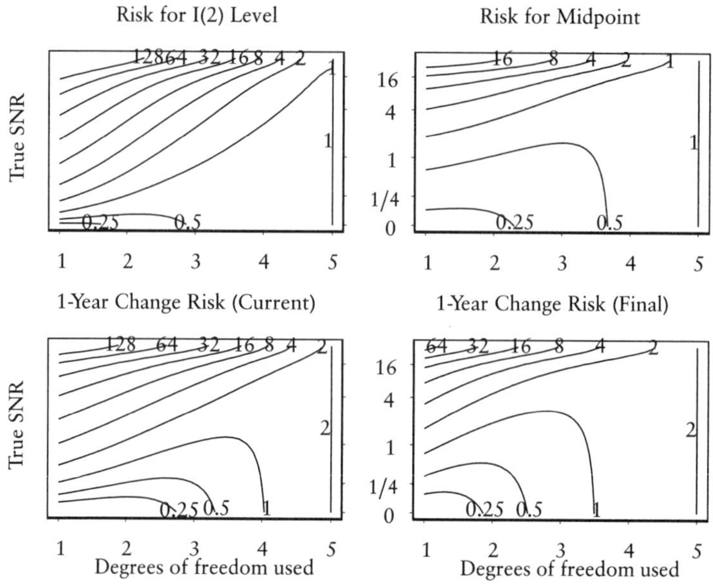

FIGURE C.4 Risk surfaces under I(12) model and I(1) strategy for estimation of current level, level at midpoint of 5-year time window, and 1-year change.

NOTE: "Current" estimate of change uses only 5-year time window, while "Final" uses consecutive 5-year time windows. Horizontal axis is degrees of freedom used in the smoother, from 1 = simple moving average to 5 = direct estimates. Vertical axis is true signal-to-noise ratio (SNR). Contours of risk surface are in units of σ_e^2.

The Linear Model With No Population Error

These same comments apply to all of the following models, derived for a linear model with no population-level error (admittedly unrealistic):

$$\theta_t = X_t \beta.$$

In particular, for a population that is perfectly linear, $\theta_t = \beta_0 + \beta t$,

$$X_t = \begin{bmatrix} 1 & t-m+1 \\ 1 & t-m+2 \\ \vdots & \vdots \\ 1 & t \end{bmatrix},$$

and

$$\psi = \frac{\beta^2}{{\sigma_e}^2}, M = \mathbf{1}\mathbf{1}^T.$$

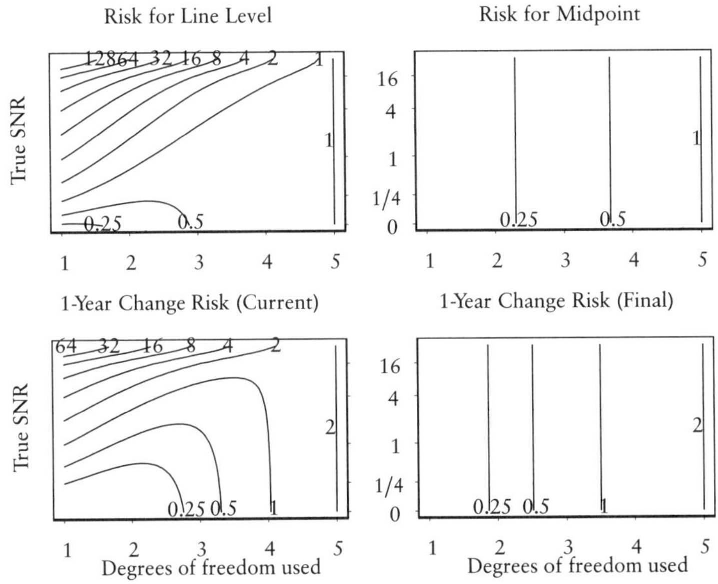

FIGURE C.5 Risk surfaces under Line model and I(1) strategy for estimation of current level, level at midpoint of 5-year time window, and 1-year change.

NOTE: "Current" estimate of change uses only 5-year time window, while "Final" uses consecutive 5-year time windows. Horizontal axis is degrees of freedom used in the smoother, from 1 = simple moving average to 5 = direct estimates. Vertical axis is true signal-to-noise ratio (SNR). Contours of risk surface are in units of σ_e^2.

The risk surface is given in Figure C.5. Note the vertical contours for risk in estimation of the midpoint, since the midpoint is unaffected by the slope of the line. Because of these vertical contours, the optimal strategy for estimation of the midpoint level is the simple MA. For a population that is constant until a 1-year level shift of size δ (in year $t-m+2$ or later, since a shift in year $t-m+1$ cannot be detected in a time window that only goes back to year $t-m+1$),

$$X_t = \begin{bmatrix} 1 & 0 \\ \vdots & \vdots \\ 1 & 0 \\ 1 & 1 \\ \vdots & \vdots \\ 1 & 1 \end{bmatrix},$$

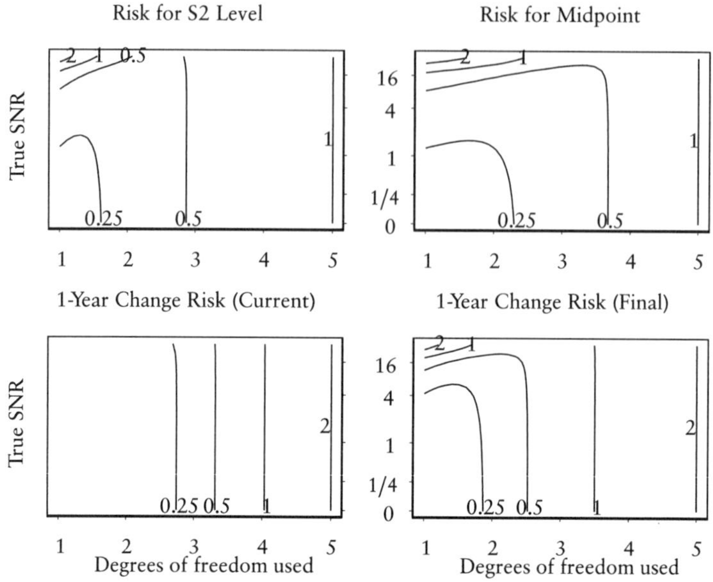

FIGURE C.6 Risk surfaces under model with level shift in year 2 (S2) and I(1) strategy for estimation of current level, level at midpoint of 5-year time window, and 1-year change.

NOTE: "Current" estimate of change uses only 5-year time window, while "Final" uses consecutive 5-year time windows. Horizontal axis is degrees of freedom used in the smoother, from 1=simple moving average to 5=direct estimates. Vertical axis is true signal-to-noise ratio (SNR). Contours of risk surface are in units of σ_e^2.

and

$$\psi = \frac{\delta^2}{{\sigma_e}^2}, M = \text{diag}\{0,\ldots,0,1,0,\ldots,0\},$$

with 1 in the $(k - 1)$th diagonal position. Figures C.6–C.9 show the risk surfaces associated with these level shift models. The later the shift, the more difficult the estimation of current level or 1-year change. Note also the symmetry in the risk surfaces for the midpoint between S2 and S5 and S3 and S4; that is, level shifts two years before and two years after the midpoint are equally difficult, and level shifts one year before or one year after the midpoint are equally difficult.

As David Binder pointed out in discussion of this paper, all of the models considered here can be considered as special cases of the local linear trend model (e.g., Harvey, 1989, p. 45). Thus, with a small number of parameters and a joint prior on those parameters, Bayes strategies could be derived.

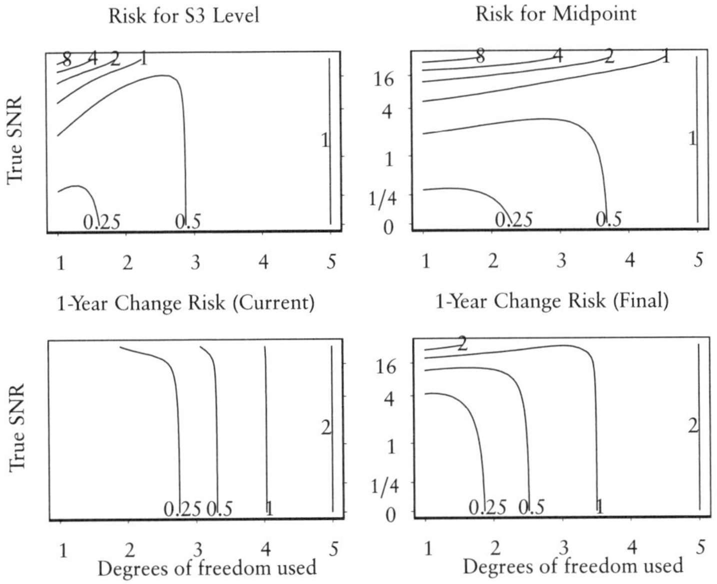

FIGURE C.7 Risk surfaces under model with level shift in year 3 (S3) and I(1) strategy for estimation of current level, level at midpoint of 5-year time window, and 1-year change.

NOTE: "Current" estimate of change uses only 5-year time window, while "Final" uses consecutive 5-year time windows. Horizontal axis is degrees of freedom used in the smoother, from 1=simple moving average to 5=direct estimates. Vertical axis is true signal-to-noise ratio (SNR). Contours of risk surface are in units of σ_e^2.

This would be an excellent model for further theoretical and empirical investigation. I did not consider it here and instead restricted attention to single-parameter models for purposes of illustration.

Prior Determination and Empirical Results for the I(1) Strategy

The previous sections have shown that it is possible to derive optimal strategies, given a model and a prior distribution for the model parameters. In practice, the ACS will produce multiyear estimates for many characteristics in governmental units at many different levels (e.g., states, counties, places, townships, school districts). These governmental units vary widely in size. As noted in the introduction, I focus here on uniform strategies. The results on *optimal* strategies in the previous sections can be used to choose sensible *uniform* strategies, which reflect a compromise among strategies.

To use the optimal strategy results, it is necessary to identify models for

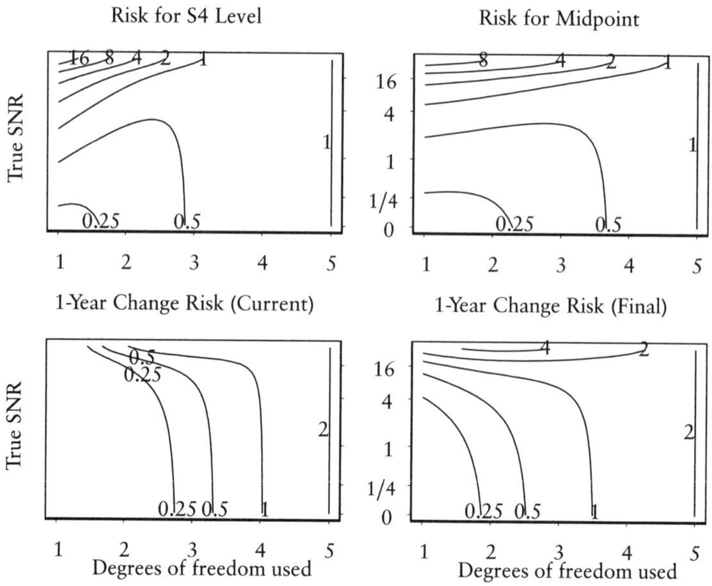

FIGURE C.8 Risk surfaces under model with level shift in year 4 (S4) and I(1) strategy for estimation of current level, level at midpoint of 5-year time window, and 1-year change.

NOTE: "Current" estimate of change uses only 5-year time window, while "Final" uses consecutive 5-year time windows. Horizontal axis is degrees of freedom used in the smoother, from 1=simple moving average to 5=direct estimates. Vertical axis is true signal-to-noise ratio (SNR). Contours of risk surface are in units of σ_e^2.

ACS characteristics, to determine numerical values for the associated model parameters, and to use the empirical distribution of model parameters as the prior distribution in identifying an optimal strategy. The optimal strategy under this empirical "prior" is then one uniform strategy that compromises among the optimal strategies for the various characteristics.

There is considerable information on the various characteristics studied by the ACS, from sources such as the ACS test sites during 1996–1999, the C2SS in 2000, the ACS test surveys in 2001 through 2004, and the ACS in 2005. In addition, there are other ongoing government surveys. It thus seems possible in principle to identify reasonable classes of models and reasonable numerical values for the associated SNRs. That is, for a given model class, determine a prior from historical data for which the model class is appropriate. Given the prior, compute the Bayes strategy for that model class. Finally, choose a compromise strategy from among the computed Bayes strategies.

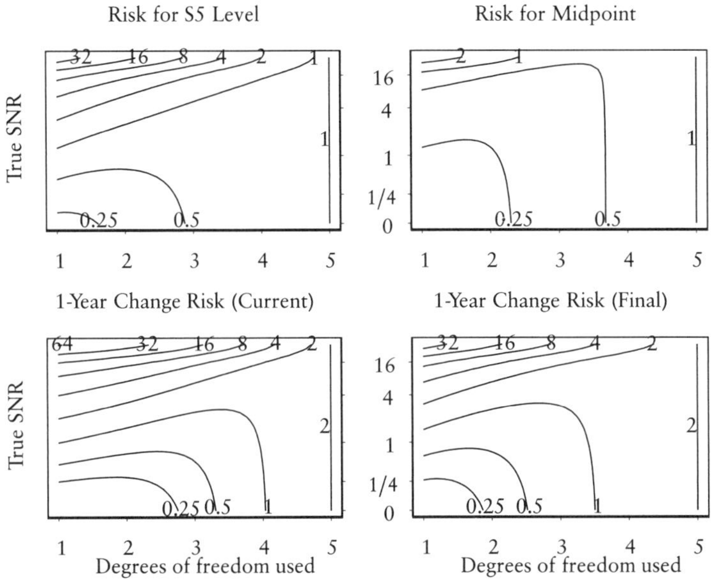

FIGURE C.9 Risk surfaces under model with level shift in year 5 (S5) and I(1) strategy for estimation of current level, level at midpoint of 5-year time window, and 1-year change.

NOTE: "Current" estimate of change uses only 5-year time window, while "Final" uses consecutive 5-year time windows. Horizontal axis is degrees of freedom used in the smoother, from 1=simple moving average to 5=direct estimates. Vertical axis is true signal-to-noise ratio (SNR). Contours of risk surface are in units of σ_e^2.

Alternatively, construct a prior as a mixture across model classes. Determine the frequency with which each model class is represented among the ACS characteristics of interest and then determine prior distributions for the model parameters in each model class. The final prior is then the mixture of these component priors, weighted by model class frequency.

As a simple numerical example (purely for illustration; not intended to be a realistic modeling exercise), consider the four years of demographic, social, economic, and housing characteristics from the ACS in Multnomah County, Oregon. Fitting the line+error model to each such series, we obtain the estimated SNR values:

$$\hat{\psi} = \hat{\beta}^2 / \hat{\sigma}_\epsilon^2.$$

Some of these values are estimated as infinity because the line fits perfectly. These infinities are trimmed from the set of estimated SNRs, and the stem-and-leaf plot of the remaining estimates is given in Figure C.10.

```
N = 225   Median = 0.2008032 Quartiles = 0.04122658, 0.548866

Decimal point is 1 place to the left of the colon

    0 : 000000000000000000011111111111111111122222223333333334444444555556666677778888889
    1 : 0000000011122222233556677788999
    2 : 0001111123344445666789
    3 : 0011223334455557789
    4 : 012345566789
    5 : 123456899
    6 : 0024478
    7 : 1226899
    8 : 0578
    9 : 0
   10 : 2346
   11 : 26
   12 : 13
   13 : 069
   14 : 2
   15 : 6

High:  1.699346  2.000000  2.027793  2.226016  2.321182  2.380952
2.434674  2.565172  2.659353 High:  2.804506  3.336079  3.555556
3.742215  3.830619  4.089362  4.339465  5.979310  5.980068 High:
9.161348 13.35308
```

FIGURE C.10 Stem-and-leaf plot of estimated signal-to-noise ratios in the line model fitted to four years of ACS data on demographic, social, economic, and housing characteristics in Multnomah County, Oregon.

With this empirical prior for SNR, the Bayes risk for various linear functions can be computed. These risks are shown in Figure C.11. The Bayes strategy for estimation of current level uses approximately 2.8 df. Since the model is a line, the Bayes strategy for estimation of the midpoint level uses 1 df. Any strategy is Bayes for estimation of the temporal average. Finally, the Bayes strategy for 1-year change uses approximately 2 df. Depending on the relative importance of estimation of the various linear functions, some compromise df could be chosen. In this numerical illustration, 2 df might be a sensible compromise: it is close to optimal for most of the linear functions considered and does not give up too much efficiency for current level.

This empirical example illustrates the choice of a data-driven compromise strategy, in which the compromise is across characteristics. The same sort of approach might be used to choose a single, compromise strategy across multiple areas in which the SNRs vary spatially.

C–4 SUMMARY

There are many possible sources of estimation strategies for a survey repeated over time, like the ACS. I have focused on the I(1) strategy in this

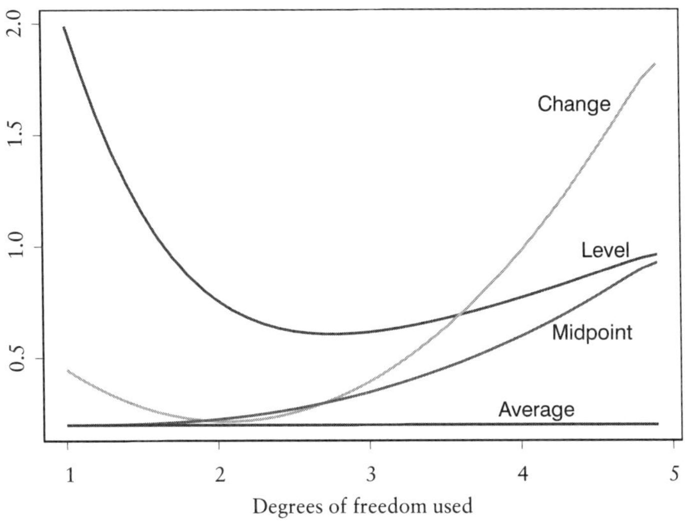

FIGURE C.11 Bayes risk for various linear functions using empirical prior fitted from Multnomah County ACS data.

NOTE: See Figure C.10 for derivation of the empirical prior using the "line" model as the true model.

case, evaluating it using simple decision-theoretic tools for various population characteristics, under various models, across a range of unknown model parameters.

The proposed MA strategy does poorly in this evaluation. It is not minimax (although this extremely conservative criterion is not very useful in practice). More importantly, it is generally not Bayes under any reasonable prior on the SNR. For example, under the I(1) model (and ruling out the temporal average for which all strategies are equally effective), MAs are Bayes only if the true SNR is zero, or equivalently if the true values are constant over time.

The question for this research was to determine if there are viable alternatives to the proposed MA strategy. The I(1) strategy meets the criteria set out at the beginning of this paper. It is simple and consistent. Its weights are unequal but fixed, so that large-scale implementation is no harder than MA, and comparability across domains is ensured. Its linear form means that tables add up. Guidance for users would seem to be no worse for a weighted MA than for an unweighted MA.

The I(1) strategy can be made robust. This paper has indicated methods by which compromise df can be chosen empirically for reasonable efficiency across a range of characteristics and population parameters. Finally, the I(1) strategy is defensible. It has a motivating statistical model but does not require correctness of that model. Choice of a particular strategy can build on extensive knowledge of related populations. If novel estimation problems are encountered, appropriate estimation techniques can be developed theoretically by going back to the motivating model, and then those techniques could be evaluated with decision-theoretic criteria when the motivating model does not hold.

Finally, it is important to note that although this paper has focused on the class of α-smoothers derived from an I(1) strategy, any other strategies could be evaluated with similar decision-theoretic criteria.

REFERENCES

Binder, D.A., and Hidiroglou, M.A. (1988), Sampling in time, in *Handbook of statistics*, Vol. 6, eds. P.R. Krishnaiah and C.R. Rao, pp. 187–211. Amsterdam: North-Holland.

Brockwell, P.J., and Davis, R.A. (1991). *Time series: Theory and methods,* 2nd ed. New York: Springer-Verlag.

Durbin, J., and Koopman, S.J. (2001). *Time series analysis by state space methods.* Oxford, England: Oxford University Press.

Gijbels, I., Pope, A., and Wand, M.P. (1999). Understanding exponential smoothing via kernel regression. *Journal of the Royal Statistical Society, Series B*, 61, 39–50.

Harvey, A.C. (1989). *Forecasting, structural time series models and the Kalman filter.* Cambridge, England: Cambridge University Press.

National Research Council. (2001). *The American Community Survey: Summary of a Workshop.* Committee on National Statistics. Washington, DC: National Academy Press.

Scott, A.J., and Smith, T.M.F. (1974). Analysis of repeated surveys using time series methods. *Journal of the American Statistical Association*, 69, 674–678.

APPENDIX D

Biographical Sketches of Panel Members and Staff

Graham Kalton (*Chair*) is chairman of the board, senior statistician, and senior vice president of Westat. He is also a research professor in the Joint Program in Survey Methodology at the University of Maryland. Previously he was a research scientist in the Survey Research Center and a professor of biostatistics and statistics at the University of Michigan, professor of social statistics at the University of Southampton, and a reader in social statistics at the London School of Economics. His research interests are in survey sampling and general survey methodology. He is a past member of the Committee on National Statistics (CNSTAT); chaired its Panel on Estimates of Poverty for Small Geographic Areas, the Panel to Evaluate the Survey of Income and Program Participation, and the Panel to Study the NSF Scientific and Technical Personnel Data System; and has served as a member of two other panels. A past president of the International Association of Survey Statisticians, he has been chair of the American Statistical Association's section on survey research methods and chair of the Royal Statistical Society's social statistics section and served on the Council of the Royal Statistical Society. He has a B.Sc. in economics and an M.Sc. in statistics from the University of London and a Ph.D. in survey methodology from the University of Southampton.

Barbara A. Bailar is an independent consultant on survey methodology who retired from the position of senior vice president for survey research at the National Opinion Research Center (NORC). Prior to joining NORC in 1995, Dr. Bailar was the executive director of the American Statistical

Association in Alexandria, Virginia. Most of her career was spent at the U.S. Census Bureau, where she was the Associate Director for Statistical Standards and Methodology. She has published numerous articles in such journals as the *Journal of the American Statistical Association*, *Demography*, and *Survey Research Methods*. She is a past president of the American Statistical Association and the International Association of Survey Statisticians, as well as a past vice president of the International Statistical Institute. She is a fellow of the American Statistical Association and the American Association for the Advancement of Science. She received a Ph.D. in statistics from American University in Washington, D.C.

Paul P. Biemer is distinguished fellow, statistics, at RTI International and associate director for survey research and development for the Odum Institute at the University of North Carolina at Chapel Hill. He has taught at the University of Maryland (Joint Program in Survey Methodology), the University of Michigan (Summer Institute), and George Washington University (Statistics Department). He was formerly head of the Department of Experimental Statistics and director of the Statistics Center at New Mexico State University; at the Census Bureau, he was assistant director for statistical research. His research has examined the relationships between survey design and survey error, statistical methods for assessing survey errors, particularly measurement errors and methods for the analysis of survey data. He co-developed computer audio recorded interviewing (CARI) and pioneered the field of latent class analysis for survey evaluation. At UNC, he established the Certificate Program in Survey Methodology, which he directs. He is the author of *Introduction to Survey Quality* and several edited volumes, including *Measurement Errors in Surveys*. He has a Ph.D. in statistics from Texas A&M University.

Constance F. Citro is the director of the Committee on National Statistics, a position she has held since May 2004. She began her career with CNSTAT in 1984 as study director for the panel that produced *The Bicentennial Census: New Directions for Methodology in 1990*. Previously she held positions as vice president of Mathematica Policy Research, Inc., and Data Use and Access Laboratories, Inc. She was an American Statistical Association/National Science Foundation/Census research fellow and is a fellow of the American Statistical Association and a member of the International Statistical Institute. For CNSTAT, she directed evaluations of the 2000 census, the Survey of Income and Program Participation, microsimulation models for social welfare programs, and the NSF science and engineering personnel data system, in addition to studies on institutional review boards and social science research, estimates of poverty for small geographic areas, data and methods for retirement income modeling, and alternative poverty measures.

She has a B.A. in political science from the University of Rochester and M.A. and Ph.D. degrees in political science from Yale University.

Michael L. Cohen is a senior program officer for CNSTAT, currently serving as director of the Panel on Census Coverage Measurement and the Panel on the Design of the 2010 Census Program of Evaluations and Experiments. He also served as co-study director for the Panel on Research on Future Census Methods and staff to the Panel to Review the 2000 Census. He previously assisted the Panel on Estimates of Poverty for Small Geographic Areas and directed the Panel on Statistical Methods for Testing and Evaluating Defense Systems. Formerly, he was a mathematical statistician at the Energy Information Administration, an assistant professor in the School of Public Affairs at the University of Maryland, and a visiting lecturer in statistics at Princeton University. His general area of research is the use of statistics in public policy, with a particular interest in the census undercount, model validation, and robust estimation. A fellow of the American Statistical Association and member of the International Statistical Institute, he has a B.S. in mathematics from the University of Michigan and M.S. and Ph.D. degrees in statistics from Stanford University.

Daniel L. Cork is a senior program officer for CNSTAT, currently serving as study director of the Panel to Review the Programs of the Bureau of Justice Statistics and senior program officer for the Panel on the Feasibility, Accuracy, and Technical Capability of a National Ballistics Database. He previously served as study director of the Panel on Residence Rules in the Decennial Census, co-study director of the Panel on Research on Future Census Methods, and program officer for the Panel to Review the 2000 Census. His research interests include quantitative criminology, particularly space-time dynamics in homicide; Bayesian statistics; and statistics in sports. He holds a B.S. in statistics from George Washington University and an M.S. in statistics and a joint Ph.D. in statistics and public policy from Carnegie Mellon University.

Nancy Dunton is associate research professor at the University of Kansas School of Nursing and associate research professor of health policy and management at the University of Kansas School of Medicine. She joined the University of Kansas faculty in 2001, having previously researched a wide variety of topics as principal social scientist at Midwest Research Institutes. Her research in health and social services has included evaluations of the "KIDS COUNT" program in New York and examination of barriers to self-sufficiency among welfare recipients. She is a member of the Kansas City Metro Outlook Technical Advisory Panel and the Mid American Regional Council, and is actively involved in various professional organizations. For CNSTAT, she served on the Panel on Estimates of Poverty for Small Geographic Areas. She has M.S. and Ph.D. degrees in sociology from the University of Wisconsin–Madison.

Martin R. Frankel is professor of statistics and computer information systems at Zicklin School of Business, Baruch College, City University of New York. Since 1996, he has also served as senior statistical scientist at Abt Associates. From 1974 to 1996, he was senior statistical scientist for NORC. He has published extensively on probability sampling and analysis of sample data. He has been involved in the development of sampling procedures for use in primary and secondary education as well as health care research. He is a fellow of the American Statistical Association and has chaired the association's section on survey research methods. He is a member of the International Statistical Institute and has served as president of the Market Research Council. From 1975 to 1981, he served as member and, ultimately, chair of the American Statistical Association's advisory committee to the Census Bureau regarding the 1980 census. He served on the CNSTAT Panel on Occupational Safety and Health Statistics. He has an M.A. in mathematical statistics and a Ph.D. in mathematical sociology from the University of Michigan.

D. Tim Holt is professor emeritus of social statistics at the University of Southampton, United Kingdom. In 1995, he left his academic position at Southampton to serve as director of the Office for National Statistics in London, head of the Government Statistical Service, and registrar general for England and Wales. He held the position of director of national statistics for the United Kingdom until 2000, when he returned to the University of Southampton. His research interests include methodology for official statistics as well as inference from clustered and aggregated data, small-area estimation, and ecological regression. He is the recipient of many honors in statistics and was made a Companion of the Bath on the Queen's New Years Honors List for 2000. He received a B.Sc. degree in mathematics and a Ph.D. in statistics, both from the University of Exeter.

Sharon Lohr is Thompson Industries dean's distinguished professor of statistics at Arizona State University. An active researcher in survey methodology, she is the author of *Sampling: Design and Analysis* (1999). She has served as a member of the Census Advisory Committee of Professional Associations and Statistics Canada's Advisory Committee on Statistical Methods. A fellow of the American Statistical Association, she has served as chair of its section on survey research methods and was the first recipient of the Washington Statistical Society's Gertrude M. Cox award for "making significant contributions to statistical practice." She has a Ph.D. in statistics from the University of Wisconsin–Madison.

Charles L. Purvis is principal transportation planner and analyst at the Metropolitan Transportation Commission in Oakland, California. An active user of census data for transportation planning, he currently chairs the Transportation Research Board's (TRB) section on travel analysis methods and a panel on statistical and methodological standards for metropolitan travel surveys; he was also a member of the TRB planning committee for a conference on census data for transportation planning held in May 2005. Previously, he chaired the TRB Committee on Urban Transportation Data and Information Systems. He has an M.A. in city and regional planning from Rutgers University.

Joseph J. Salvo is director of the Population Division at the New York City Department of City Planning, where he was previously deputy director and senior demographer. His background includes a year at the U.S. Census Bureau in 1981-1982. He has broad experience in immigration, the application of small-area data for policies and programs, and the use of census data. A past president of the Association of Public Data Users, he has experience with the Census Bureau's Master Address File and TIGER geographic database, as well as the American Community Survey. A member of the CNSTAT Panel on Research on Future Census Methods, he chaired the Local Update of Census Addresses (LUCA) working group jointly sponsored by that panel and the Panel to Review the 2000 Census. He is an adjunct associate professor in the Urban Affairs and Planning Department at Hunter College of the City University of New York. He is a recipient of the Sloan Public Service Award from the Fund for the City of New York, as well as a fellow of the American Statistical Association. He has M.A. and Ph.D. degrees in sociology from Fordham University.

Hal S. Stern is professor and founding chair of statistics at the University of California, Irvine. Prior to joining the Irvine faculty in 2002, he was Laurence H. Baker chair in biological statistics at Iowa State University and also held academic appointment at Harvard University. An expert in Bayesian modeling and techniques, he is coauthor of *Bayesian Data Analysis*, a leading text in the field. For CNSTAT, he earlier served on the Panel on Operational Test Design and Evaluation of the Interim Armored Vehicle (Stryker). A fellow of the American Statistical Association, he has served as editor of the association's magazine *Chance* and chair of the association's section on Bayesian statistical science and the section on statistics in sports. He has M.S. and Ph.D. degrees in statistics from Stanford University.

Meyer Zitter is an independent demographic consultant. Formerly, he was chief of the Census Bureau's Population Division and also served as assistant director for international programs. He is a fellow of the American Statistical Association and a member of the International Statistical Institute and the International Union for the Scientific Study of Population. He has a B.B.A. degree from City College of New York.

Index to Executive Summary and Chapters 1-7

C

D

E

F

G

L

O

P

Q

R

COMMITTEE ON NATIONAL STATISTICS

The Committee on National Statistics (CNSTAT) was established in 1972 at the National Academies to improve the statistical methods and information on which public policy decisions are based. The committee carries out studies, workshops, and other activities to foster better measures and fuller understanding of the economy, the environment, public health, crime, education, immigration, poverty, welfare, and other public policy issues. It also evaluates ongoing statistical programs and tracks the statistical policy and coordinating activities of the federal government, serving a unique role at the intersection of statistics and public policy. The committee's work is supported by a consortium of federal agencies through a National Science Foundation grant.